T0396424

Calculus

Formulations and
Solutions with Python

Series in Computational Methods

Series Editor: Gui-Rong Liu *(University of Cincinnati, USA)*

Published

Vol. 3   *Mechanics of Materials: Formulations and Solutions with Python*
by G R Liu

Vol. 2   *Calculus: Formulations and Solutions with Python*
by G R Liu

Vol. 1   *Numbers and Functions: Theory, Formulation, and Python Codes*
by G R Liu

Series in Computational Methods
Volume 2

Calculus

Formulations and Solutions with Python

G. R. Liu

University of Cincinnati, USA

 World Scientific

NEW JERSEY · LONDON · SINGAPORE · BEIJING · SHANGHAI · HONG KONG · TAIPEI · CHENNAI

Published by

World Scientific Publishing Co. Pte. Ltd.

5 Toh Tuck Link, Singapore 596224

USA office: 27 Warren Street, Suite 401-402, Hackensack, NJ 07601

UK office: 57 Shelton Street, Covent Garden, London WC2H 9HE

Library of Congress Control Number: 2025001826

British Library Cataloguing-in-Publication Data
A catalogue record for this book is available from the British Library.

Series in Computational Methods — Vol. 2
CALCULUS
Formulations and Solutions with Python

ISBN 978-981-98-0100-8 (hardcover)
ISBN 978-981-98-0101-5 (ebook for institutions)
ISBN 978-981-98-0102-2 (ebook for individuals)

For any available supplementary material, please visit
https://www.worldscientific.com/worldscibooks/10.1142/14046#t=suppl

Desk Editors: Nambirajan Karuppiah/Steven Patt

Typeset by Stallion Press
Email: enquiries@stallionpress.com

About the Author

Gui-Rong Liu received his Ph.D. from Tohoku University, Japan, in 1991. He was a Postdoctoral Fellow at Northwestern University, USA, from 1991 to 1993. He was a Professor at the National University of Singapore until 2010. He is currently a Professor at the Department of Aerospace Engineering and Engineering Mechanics, University of Cincinnati, USA. He was the Founder of the Association for Computational Mechanics (Singapore) (SACM) and served as the President of SACM until 2010. He served as the President of the Asia-Pacific Association for Computational Mechanics (APACM) (2010–2013) and an Executive Council Member of the International Association for Computational Mechanics (IACM) (2005–2010; 2020–2026). He authored a large number of journal papers and books, including two bestsellers, *Mesh Free Method: Moving Beyond the Finite Element Method* and *Smoothed Particle Hydrodynamics: A Meshfree Particle Methods*. He is the Editor-in-Chief of the *International Journal of Computational Methods* and served as an Associate Editor for *IPSE* and *MANO*. He is the recipient of numerous awards, including the Singapore Defence Technology Prize, NUS Outstanding University Researcher Award, NUS Best Teacher Award, APACM Computational Mechanics Award, JSME Computational Mechanics Award, ASME Ted Belytschko Applied Mechanics Award, Zienkiewicz Medal from APACM, AJCM Computational Mechanics Award, Humboldt Research Award, SACM Medal from the Association of Computational Mechanics (Singapore), and the Master Engineering Educator Award (UC, CEAS). He has been listed as one among the world's top 1% most influential scientists (Highly Cited Researchers) by Thomson Reuters for a number of years.

Contents

Chapter 1

Introduction

1.1 Computational methods

This book is part of a series that delves into the exciting world of computational methods. Our goal is to equip you with a solid foundation in this field, combining theoretical principles with practical applications.

The series encompasses general methods and techniques used throughout science, technology, engineering, and math (STEM) education, as well as research in various scientific and engineering disciplines. Think of it as your one-stop encyclopedia for computational methods, covering the theory, formulation, and even the code you'll need to get started. We've designed the series to be accessible. Readers who have completed elementary and middle school (roughly 9 years of education) can begin with the foundational volumes on mathematics. Once you progress through the mechanics volumes, you'll be equipped to tackle research and design projects that leverage computational methods.

Taking advantage of the rapid advancements in computer technology, we'll heavily integrate code examples throughout the series. This allows you to see concepts, theories, and formulations come to life through practical examples with clear visualizations. Initially, we'll primarily use Python for the fundamental volumes, potentially introducing other languages for more advanced topics. By effectively utilizing code, you can dedicate more time to understanding the core principles. Let the computer handle the complex calculations, tedious derivations for formulas, time-consuming operations, and vast number-crunching, allowing you to focus on the bigger picture.

This approach empowers you to explore the fascinating world of computational methods with a strong theoretical grounding and practical skills.

1.2 Why start and contribute to this book series

The Editor-in-Chief of this book series, Dr. G.R. Liu, has been working in areas related to computational methods for over 40 years. He developed his first FEM code for non-linear problems in 1980 and has published more than 600 journal papers and 12 monographs in this area since then. After all these years of studying, using, and developing computational methods, he started to think about a means to help other interested individuals learn computational methods in a more effective, systematic, and smooth way. He has concluded that developing this book series is the best way to achieve this objective.

1.3 Calculus: Differentiation and integration

Volume 1 of this book series covers the two most important building blocks in computational methods: *Numbers and Functions: Theory, Application, and Python Codes*. This book covers two of functions: Differentiation and Integration of functions. Both are mutually inverse and are essential in studying and understanding various types of functions used for problems in science and engineering. Differentiation explores the local behavior of a function: how fast it changes, is it at a stationary point, at a minimum or maximum point, how it curves, and so on. Integration, on the other hand, examines the global behavior of a function: what is the length of a curve, distance of a spacecraft traveled, the signed area under the function, the volume under a surface, the hypervolume of a higher-dimensional function, inertial of an object, strain energy accumulated in a structure, and so on.

Theories, concepts, and formulation of both differentiation and integration will be presented in detail. Differentiability and integrability of functions will be examined. Properties and features of derivatives, partial derivatives, integrals will be studied. Techniques for computing the derivatives, partial derivatives, definite integrals, and indefinite integrals will be introduced for various types of functions. Applications of differentiation and integration will be presented.

Python code is provided and used throughout the book so that readers can practice and interact with the codes during their studies, making it easier to comprehend the theory, concepts, formulations, techniques, and the outcome of the computational operations for differentiation and integration.

Proper connection between theory and real-world engineering problems will be made through a large number of examples.

The materials of this book can be digested through different forms of learning processes, including classroom teaching, online courses, and also self-study. Since Python codes are provided, readers can easily see how the theory is formulated and how the solutions are obtained in terms of formulas, numerical numbers, and graphs. Readers may also deepen their understanding by playing with codes and even further develop their own codes for solving other related problems.

The book is written in Jupyter notebook format, so that description of theory, formulation, coding, and real-time interaction with codes can all be done in a unified document. This provides an environment for easy reading, exercise, practicing, and further exploration.

This book is written as a reference to Refs. [1–4], which were the textbooks used when the author was a university student, while [5] is a reference book for self study and research work. Some example problems given in these textbooks are used in this book. Both NumPy and SymPy are used in the development of code for the demonstration examples. Wikipedia pages have been a useful source of reference. "Discussions" with ChatGPT, Gemini, and Bing have also helped greatly in writing some of the codes and in the preparation of this volume. Experiences, insights, new viewpoints and understandings, new theories, and new formulations, algorithms and utility codes developed in the research life of the author have also been included in the book. Therefore, although calculus is a classic topic, it is presented in this book afresh in many ways with novel discussions and modern tools, aiming to establish a solid foundation for effective computational methods for the analysis and design of advanced systems in science and engineering.

Numerous examples are presented in this book. However, the application of derivatives of functions is vast, and the examples and case studies given are far from exhaustive.

This book in general deals with real functions unless specified otherwise.

1.4 Who may read this book

This book is written for beginners interested in learning computational methods for solving problems in science, engineering, and nature. The target audience includes high-school students, university students, graduate students, researchers, and professionals in any STEM discipline. Engineers and practitioners may also find the book useful for establishing a strong foundation in core computational methods and concepts.

For beginners, it is recommended to first read or skim through Volume 1 of this book series: *Numbers and Functions: Theory, Application, and Python Codes*.

This book is written for human readers. It cannot be used for training and testing any automatic systems without the permission from the author.

1.5 Codes used in this book

Readers who purchased the book may contact the author directly at liugr100@gmail.com to request a free softcopy of the book in Jupyter notebook format with codes (which may be updated) for academic use after registration.

Conditions of use: The following conditions apply to the use of the book and codes developed by the author in hardcopy, softcopy, and any other media and format:

- **User responsibility:** You are solely responsible for any risks associated with using any part of the codes and techniques.
- **Code purpose:** The codes are primarily written to demonstrate concepts and may not be optimized for efficiency or robustness. Many of the codes have not been thoroughly tested by a third party.
- **Limited use:** The book and codes are for your personal use only. You cannot distribute them further without the author's permission.
- **No user support:** No user support is provided for the codes.
- **Citation:** Proper citation and acknowledgment are required when using the book, codes, ideas, and techniques.

These codes often rely on various external packages/modules. Their behavior can be affected by the versions of Python and these packages/modules. Version mismatch could be the reason if the code doesn't run as expected. If you encounter issues running a code, check the versions of the packages/modules used. You can check the versions within a code cell of the Jupyter Notebook. For example, to check the current Python environment version, you can use the following:

```
1  !python -V                    #! is used to execute an external command
```

Python 3.9.16

```
1  !jupyter notebook --version
```

6.1.5

Troubleshooting version issues: If you encounter a version mismatch error, you can either adapt the code to the specific version or install the required version on your system. It's often helpful to search the web using the error message to find solutions or leads. Large language models can also be quite helpful in this regard. This is the author's preferred approach when encountering code execution issues.

Learning Python: This book won't cover Python basics, as there are many resources available online. Interested readers can refer to Chapters 2 and 3 in Ref. [6] for a concise description of using Python for scientific computations. Since Python is relatively easy to learn, you can leverage the codes and examples in this book to get a head start on learning Python while studying the technical subjects.

1.6 Use of external modules or dependences

Module imports in Jupyter notebooks: Frequently used modules in a chapter are imported in the first code cell of the chapter, which you may execute at the beginning. If a name error (such as `NameError: name 'sp' is not defined`) is encountered, you can jump back to the beginning of the chapter and execute the cell:

```
1  #Often used external modules are in commonImports placed in folder grbin
2  #Place cursor in this cell, and press Ctrl+Enter to import dependences.
3
4  import sys                              # for accessing the computer system
5  sys.path.append('../grbin/')           # Add in the path to your system
6
7  from commonImports import *            # Import dependences from '../grbin/'
8  import grcodes as gr                    # Import the module of the author
9  importlib.reload(gr)                    # For in case, when grcodes is modified
10
11 from continuum_mechanics import vector
12 from continuum_mechanics.solids import sym_grad, strain_stress
13 init_printing(use_unicode=True)         # For Latex-like quality printing
14
15 # Digits in print-outs
16 np.set_printoptions(precision=4,suppress=True,
17                     formatter={'float_kind': '{:.4e}'.format})
```

The above cell includes a common import file, `commonImports.py` (given in following). It contains all the frequently used modules in this book to avoid repeatedly importing lengthy external modules and reduce redundancy:

```
 1  # commonImports.py:
 2  from __future__ import print_function
 3  import numpy as np                    # for numerical computation
 4  import sympy as sp                     # sympy module for computation
 5  import numpy.linalg as lg              # numpy linear algebra module
 6  import scipy.linalg as sg              # scipy linear algebra module
 7  import scipy.integrate as si
 8  from scipy.stats import ortho_group
 9  import importlib
10  import itertools
11  import inspect
12  import csv
13  import pandas as pd
14  import random
15  from IPython.display import display, Math
16
17  import autograd.numpy as anp                      # Thinly-wrapped numpy
18  from autograd import grad
19
20  from grcodes import drawArrow, plotfig, printM, printx # frequently used
21
22  import math as ma
23  from sympy import sin, cos, exp, symbols, lambdify, init_printing, latex
24  from sympy import pi, Matrix, sqrt, oo, integrate, diff, Derivative
25  from sympy import MatrixSymbol, simplify, nsimplify, Function
26  from sympy import factor, expand, nsimplify, Matrix, ordered, hessian
27  from sympy.plotting import plot as splt
28  init_printing(use_unicode=True) # for latex-like qualityprinting formate
29
30  from matplotlib.ticker import MultipleLocator
31  import matplotlib.pyplot as plt        # for plotting figures
32  import matplotlib as mpl
```

Module reloading: While importing the same module multiple times generally doesn't cause issues, Python utilizes a cache and ignores subsequent imports if the module is already loaded. However, if you modify an imported module and want those changes to reflect in your code, you need to reload the module. This can be achieved using the `importlib.reload()` function:

```
 1  # grcodes was imported as gr, reload it when grcodes is modified
 2  importlib.reload(gr)        # This is also included in the first code cell
```

To view the codes in the imported module, one may just use the following by uncommenting it (It may produce a long output):

```
 1  import inspect
 2  #source_code = inspect.getsource(gr)     # to view everything in gr module
 3  source_code = inspect.getsource(gr.cheby_T) # to view any function in gr
 4  print(source_code)
```

```
def cheby_T(n, x):
    '''Generate the first kind Chebyshev polynomials of degree (n-1) '''
    if   n == 0: return sp.S.One
    elif n == 1: return x
    else:        return (2*x*cheby_T(n-1,x)-cheby_T(n-2,x)).expand()
```

Editing codes:

- You can use any text editor to view and modify the codes.
- Alternatively, copy the code into a cell within a Jupyter Notebook and edit it directly.

Using functions from imported modules: To use a function from an imported module, you typically combine the module name with the function name using a dot (.). For example, if you import a module named **gr** containing a function called **printx()**, you would use it like as follows:

```
1 x = 8
2 gr.printx('x')       # when gr. is used, printx() is from grcodes module
```

```
x = 8
```

1.7 Use of help()

Exploring modules and objects: To access detailed information about a module or an object within a module after importing it, you can use the help() function. This is a helpful way to explore functionalities and attributes available in different modules. For example:

```
1 help(gr.volume_H8)
```

```
Help on function volume_H8 in module grcodes:

volume_H8(nodeX, nodeY, nodeZ)
    Compute the volume of a general hexahedron, using its eight nodal
    coordinates (H8), with node-0 removed.
    It uses the vetor-matrix-vector (vMv) dot-product formula.
```

Code comments: The code cells throughout the book incorporate comments to provide additional explanations. These comments, denoted by a hash symbol "#" at the beginning, are placed on the right side of the cell to minimize disruption to the code itself while offering quick reference and guidance when needed.

References

[1]   Y.C. Fan, *Advanced Mathematics*, 1979.

[2]   L.X. Yang and B.Y. Bi, *Analytic Mathematics Practices*, Boris Demidovich, 2005.

[3]   Stephen Timoshenko and J.N. Goodier, *Theory of Elasticity*, 1970. Available online (http://books.google.com/books?id=yFISAAAAIAAJ&dq=theory+of+elasticity&ei=ICiKSsr3G4jwkQSbxMyPCg).

[4]   Z.L. Xu, *Elasticity*, Vol. 1&2, People's Publisher, China, 1979.

[5]   Gilbert Strang, *Calculus*, 1991. Available online (http://ocw.mit.edu/OcwWeb/resources/RES-18-001Spring-2005/ResourceHome/index.htm).

[6]   G.R. Liu, *Machine Learning with Python: Theory and Applications*, World Scientific, New York, 2023.

Chapter 2

Derivatives of One-Dimensional Functions

```
1  # Import necessary dependences.
2  import sys
3  sys.path.append('../grbin/')
4  from commonImports import *              # import dependences
5  import grcodes as gr                     # import own modules
6  importlib.reload(gr)                 # when grcodes is modified
7  np.set_printoptions(precision=4,suppress=True,  # Digits in print-outs
8                      formatter={'float_kind': '{:.4e}'.format})
```

In Volume 1, *Numbers and Functions* [1] of this book series, we covered various aspects of functions in detail, including types, definitions, domains, codomains, limits, continuity, basis functions, and methods for function approximation. We established that functions represent numbers and can undergo arithmetic and transcendental operations, resulting in new functions.

Now, we delve into another critical concept: operations on functions with respect to their independent variables. One such operation is differentiation, which will be the focus of this chapter. In this book, the domain and codomain of functions are both real: $\mathbb{X} \in \mathbb{R}$ and $\mathbb{Y} \in \mathbb{R}$, unless otherwise specified. This implies that all the parameters or constants used in the functions are also real unless specified otherwise.

This chapter is written with reference to textbooks in Refs. [1–4]. Many of the proofs can be found in the first three texts. Wikipedia pages on derivative (https://en.wikipedia.org/wiki/Derivative) and product rule (https://en.wikipedia.org/wiki/Product_rule) serve as valuable additional references. Both NumPy and SymPy are used in the development of code for the demonstration examples. "Discussions" with ChatGPT, Gemini, and Bing have also greatly helped in coding and preparation of this chapter.

2.1 Definition

2.1.1 *The derivative: Rate of change*

Since functions can exhibit varying behaviors in response to changes in their independent variable, one way to quantify this behavior is through the concept of its derivative. The derivative, for a continuous function $f(x)$ with independent variable x, measures how rapidly the function's value changes with respect to a change in x.

Mathematically, the derivative of $f(x)$ is defined as

$$\frac{df(x)}{dx} \equiv f'(x) = \lim_{\Delta x \to 0} \frac{\Delta f}{\Delta x} = \lim_{\Delta x \to 0} \frac{f(x + \Delta x) - f(x)}{\Delta x} \tag{2.1}$$

where

- Δx is a real number, representing a small change in x around the point x.
- Δf is the corresponding change in the function f at the point x due to the change Δx.

The concept of limits, used here, is explained in detail in Ref. [1].

2.1.2 *Indeterminate form and local change*

Since $f(x)$ is assumed continuous, as Δx approaches zero, Δf (the change in the function) must also approach zero. The ratio $\frac{\Delta f}{\Delta x}$ in Eq. (2.1) becomes an indeterminate form $\frac{0}{0}$ when $\Delta x \to 0$. Note that another possible indeterminate form could be $\frac{\infty}{\infty}$.

Therefore, for Eq. (2.1) to be meaningful, we need to manipulate the expression $\frac{f(x+\Delta x)-f(x)}{\Delta x}$ in a way that cancels out Δx in the denominator (if possible) and then take the limit as $\Delta x \to 0$. The derivative defined in Eq. (2.1) is essentially the rate of change (Δf divided by Δx) at the limit of Δx approaching zero. This implies that the derivative is a local property of a function. It reflects the function's behavior at a specific point at x and its immediate surroundings.

It's crucial to note that Δx in Eq. (2.1) can be positive or negative. The derivative, if it exists, must have a finite limit and remain the same regardless of the sign of Δx.

Key point

The **locality** of the derivative implies that it can be determined solely by analyzing its local behavior, specifically the rate of change at the limit of $\Delta x \to 0$.

This locality behavior of derivative contrasts with integration (covered in later chapters), which examines the function's global behavior.

2.2 Example: Derivative and slope

Let us look at an example.

2.2.1 *A car traveling at a constant speed*

Consider a car cruising at a constant speed v [m/s]. The distance that the car travels during time t should be

$$f(t) = vt \tag{2.2}$$

where $f(t)$ is a function of the independent variable t, which is the distance that the car traveled. Following the definition in Eq. (2.1), we find that

$$\frac{df(t)}{dt} = \lim_{\Delta t \to 0} \frac{\Delta f}{\Delta t} = \lim_{\Delta t \to 0} \frac{f(t + \Delta t) - f(t)}{\Delta t}$$

$$= \lim_{\Delta t \to 0} \frac{v[t + \Delta t] - vt}{\Delta t} = \lim_{\Delta t \to 0} \frac{vt + v\Delta t - vt}{\Delta t}$$

$$= \lim_{\Delta t \to 0} \frac{v\Delta t}{\Delta t} = \lim_{\Delta t \to 0} v = v \tag{2.3}$$

Note that the limit is carried along throughout the derivation process and executed only at the last step. In this simple case, the Δts in red are canceled out entirely in the final expression before the limit is taken. Let's call this the red-cancellation. This red-cancellation is made possible by the prior cancellation of the two blue terms: $vt - vt$. Let's call this the blue-cancellation.

The above manipulation has successfully avoided the indeterminate form $\frac{0}{0}$ and made it determinate. It involves only the standard arithmetic operations resulting in the blue-cancellation and then the red-cancellation. We obtain a finite value v for the derivative.

The result given in Eq. (2.3) means that the derivative of the distance function is the speed of the car. Assume that the speed of the car is $v = 72$ km/hour, which is $v = 2$ m/s, we can then draw the speed and the distance traveled by the car as functions of time t in Fig. 2.1.

In this case, due to the constant speed, the function $v(t)$ is a simple horizontal straight line, as shown in Fig. 2.1(a). The distance function $f(t)$ is linear in t, and hence it is a straight inclined line, as shown in Fig. 2.1(b).

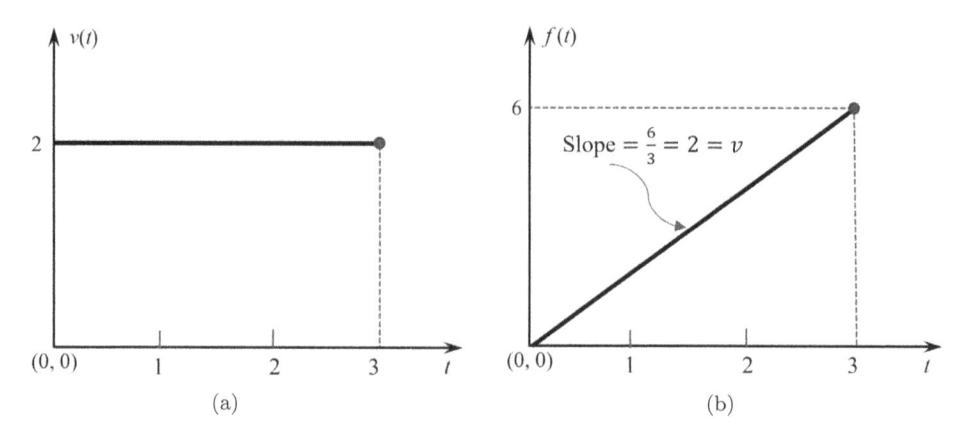

Figure 2.1. Distance traveled, $f(t)$, by a car cruising at a constant speed, $v(t)$. Both are functions of time t.

The distance traveled by the car in three seconds should be $f = vt = 2 \times 3 = 6$ [m]. The slope of the straight inclined line is calculated as $\frac{\text{distance traveled}}{\text{time duration}} = \frac{6}{3} = 2\,\text{m/s}$, which is the speed of the car. This implies that the derivative of the distance function is the **slope** of the function. It is also the **tangent** of the line.

The following is clear:

The derivative of a function is essentially the (local) slope of the function.

2.2.2 Significance of the limit process

In the previous example, the speed v is constant, and hence the slope of the $f(t)$ is also constant, as shown in Fig. 2.1(b). Therefore, Δt in Eq. (2.3) (in the last step) can be any value. Thus, the limit process does not really play a role in obtaining the slope for such linear functions.

If the car is accelerating at a constant rate starting from time $t = 0$, where the speed $v = 0$ and distance $f = 0$, the acceleration is denoted as a. Assumeing that $a = 4\,\text{m/s}^2$, the distance that the car traveled from the start to time t is

$$f(t) = \frac{1}{2}at^2 = \frac{1}{2} \cdot 4t^2 = 2t^2 \tag{2.4}$$

where $f(t)$ is the new distance function for the accelerating car up to time t. We can draw the speed and the distance as functions of the independent variable t in Fig. 2.2.

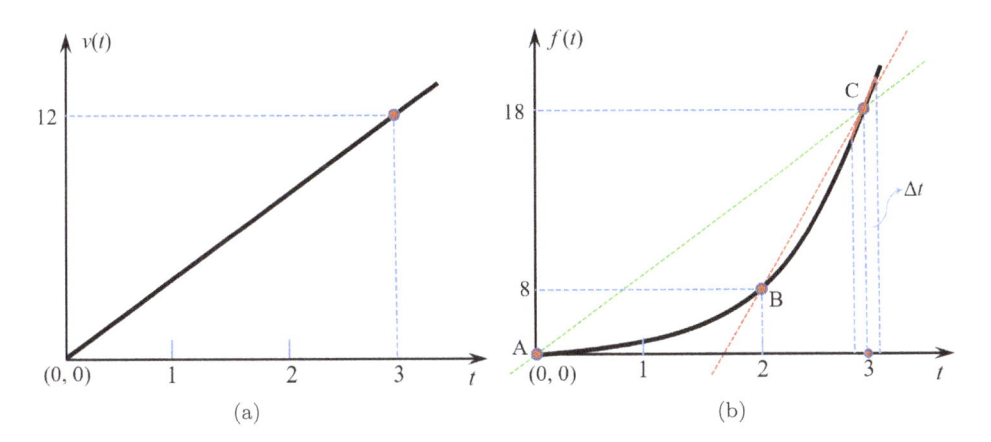

(a) (b)

Figure 2.2. Distance function, $f(t)$, for an accelerating car with a linear speed $v(t)$.

Now, following the definition in Eq. (2.1), we can find the **local slope** for $f(t)$ at point C (at $t = 3$):

$$\frac{df(t)}{dt} = \lim_{\Delta t \to 0} \frac{\Delta f}{\Delta t} = \lim_{\Delta t \to 0} \frac{f(t + \Delta t) - f(t)}{\Delta t}$$

$$= \lim_{\Delta t \to 0} \frac{2(t + \Delta t)^2 - 2t^2}{\Delta t} = \lim_{\Delta t \to 0} \frac{2t^2 + 4t\Delta t + 2\Delta t^2 - 2t^2}{\Delta t}$$

$$= \lim_{\Delta t \to 0} \frac{4t\Delta t + 2\Delta t\Delta t}{\Delta t} = \lim_{\Delta t \to 0} \frac{4t + 2\Delta t}{1}$$

$$= \underbrace{4t}_{\text{function of } t} = \underbrace{12}_{t=3} \ [\text{m/s}] \tag{2.5}$$

Here, we see again two key steps:

1. **Blue-cancellation:** The function $f(t)$ in the numerator ($2t^2$ in blue) is first canceled out.
2. **Red-cancellation:** Δt in both the numerator and the denominator (in red) is canceled out. The limit finally takes effect, forcing the remaining Δt in the numerator to zero, leading to $4t$.

These two key steps avoided the indeterminate form $\frac{0}{0}$ and made it determinate.

In this more general case, the limit process plays an important role in obtaining the local slope. In other words, we cannot use a finite value for Δt arbitrarily. A different Δt will give a different slope value that is not the local slope at point C. For example, if we use $\Delta t = -1$, we get $10 \, \text{m/s}$, which is the slope of the straight line BC (red dashed line in Fig. 2.2(b)). Using $\Delta t = -3$ gives $6\,\mathrm{m/s}$, which is the slope of the straight line AC (green dashed line in Fig. 2.2(b)). Both values are underestimates. If we choose a positive finite Δt, the value will be an overestimate in this example. Readers may give it a try.

Can we directly take $\Delta t = 0$ at the start? No, we cannot. Since it would give $\frac{\Delta f}{\Delta t} = \frac{0}{0}$ at the first step in Eq. (2.5), which is indeterminate, as we discussed earlier. In other words, its value cannot be determined. We thus note the following:

> The limit operation is, in general, a must for obtaining the derivative of a function at a point. After $f(t + \Delta t)$ is expanded explicitly in terms of Δt, $f(t)$ in the numerator is neutralized and Δt in the denominator is canceled out, and the value of $\frac{f(t+\Delta t)-f(t)}{\Delta t}$ can be determined by taking the limit as $\Delta t \to 0$.

It is clear now that the definition in Eq. (2.1) is carefully thought out and constructed. This was largely due to the original work of Isaac Newton, among others. Readers are encouraged to take time to carefully examine Eqs. (2.3) and (2.5) until they make sense. Once this is done, the concept of the derivative is well established and the rest of this chapter is much easier to comprehend.

Can we use a very small Δt to avoid the zero division? Yes, we can. In this case, however, we let the numerical algorithm work out the cancellations, and hence the result will be an approximation depending on the size of Δt used. It is a frequently used numerical method to estimate the derivative of a function.

In general, **the derivative of a function becomes a new function** that changes with the independent variable, as shown in Eq. (2.5).

Equation (2.1) is the basic formula for deriving a derivative of a function at x. The derivative of a function $f(x)$ can often be written as $f'(x)$. Equation (2.1) can also be written as

$$df(x) = f'(x)dx \tag{2.6}$$

Since $f(x)$ is often written as y, one may see expressions like $\frac{dy}{dx} = y'$, $dy = f'(x)dx$, $dy = \frac{df}{dx}dx$ in the literature. All these carry the same meaning. They differ only in notation.

In summary, the derivative $f'(x)$ is the **local slope** at x of the curve defined by $f(x)$. It is also called the **tangent** or **gradient** of $f(x)$ at x.

It represents how fast the function is changing locally with respect to the change in the independent variable. The operation used to obtain the derivative of a function is also called **differentiation**.

Let us derive the derivatives of some basic functions and show further how Eq. (2.1) is used for a given general function.

2.3 Derivative of elementary functions

All elementary functions discussed in Ref. [1] have their derivatives provided, and there are formulas for all of them. We shall discuss a few more functions to further demonstrate how to use the definition to obtain the derivative of a function and reveal some simple and widely used rules and formulas. We will also show how to use SymPy to compute the derivative formulas symbolically for any elementary function when needed.

2.3.1 *Derivative of linear functions*

We start with the simplest functions that are linear in the independent variable x.

First, if the function is a constant, by the definition of the derivative, its derivative must be zero because the function does not change with the independent variable. Its slope, tangent, and gradient are all zero. We are done.

Consider now a non-constant general linear function defined as

$$f(x) = kx - b \tag{2.7}$$

It is continuous in $(-\infty, \infty)$. The definition of derivative gives

$$\frac{df(x)}{dx} = \lim_{\Delta x \to 0} \frac{\left[k[x + \Delta x] - b\right] - (kx - b)}{\Delta x}$$

$$= \lim_{\Delta x \to 0} \frac{k\Delta x}{\Delta x} = \lim_{\Delta x \to 0} k = k \tag{2.8}$$

This means that the derivative of a linear function is a constant that is the slope of the straight line. It is a new function that no longer depends on x.

We note that the constant b in a function has no effect on the derivative of the function. It is the same value everywhere in the function's domain. It contributes nothing to the slope of the function. In the substitution in Eq. (2.1), the constant will be canceled.

The following is clear:

A constant can be simply ignored when computing the derivative of the function.

On the other hand, when finding an antiderivative (or integral) of a function (to be studied in Chapters 4 and 5), one adds an arbitrary constant for the same reason.

2.3.2 *Example: A spring loaded by a force*

Consider a linear elastic string fixed at point A and subjected to a force F, as shown in Fig. 2.3, which is familiar to many readers. The spring has a spring constant K that is known as the **stiffness** of the spring. The force F at point B can be expressed as a function of displacement d:

$$F(d) = Kd \tag{2.9}$$

This time, the independent variable is displacement d. Following the definition in Eq. (2.1), we find that

$$\frac{\mathrm{d}F(d)}{\mathrm{d}d} = \lim_{\Delta d \to 0} \frac{\Delta F}{\Delta d} = \lim_{\Delta d \to 0} \frac{F(d + \Delta d) - F(d)}{\Delta d}$$

$$= \lim_{\Delta d \to 0} \frac{K[d + \Delta d] - Kd}{\Delta d} = \lim_{\Delta d \to 0} \frac{Kd + K\Delta d - Kd}{\Delta d}$$

$$= \lim_{\Delta d \to 0} \frac{K\Delta d}{\Delta d} = \lim_{\Delta d \to 0} K = K \tag{2.10}$$

The result given in Eq. (2.10) means that the derivative of the force function is the stiffness of the spring.

These examples given above show that the derivative of a function has a physical meaning, which depends on the type of the actual problem. The derivative definition in Eq. (2.1) is universal and applicable to any function that is differentiable. It applies to problems in various disciplines.

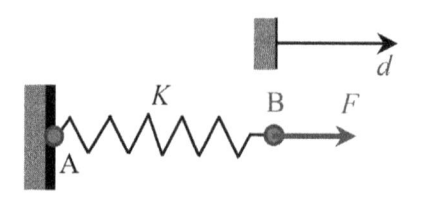

Figure 2.3.   A stretched spring by a force.

2.3.3 *Example: Derivative of functions using SymPy*

Let us use SymPy to compute directly the derivative of a linear function without using the definition.

```
1  # Import sympy tools:
2  from sympy import symbols, diff, sin
3  x, K, b = symbols('x, K, b ')                    # define symbolic variables
4
5  f_linear = K*x - b                               # define a linear function
6  df_linear = f_linear.diff(x)    # compute the derivative using sp.diff()
7  print(f" Derivative of linear functions = {df_linear}")   # print it out
8
9  df_linear = diff(f_linear, x)                         # An alternative
10 print(f" An alternative method = {df_linear}")
```

```
Derivative of linear functions = K
An alternative method = K
```

We see again that the derivative of the linear function is a constant function K, which is the slope of the straight line. It no longer depends on the independent variable x.

The `sp.diff()` function in SymPy is a useful tool to obtain the derivative of a given function. There are two methods to use it, and it will be used frequently in this book.

2.3.4 *Derivative of monomials*

Consider a **monomial** function $f(x) = x^n$, where n is an integer. It is examined in great detail in Ref. [1]. The formula for the derivative of a monomial function is

$$f'(x) = \frac{dx^n}{dx} = nx^{n-1} \tag{2.11}$$

This can be proven as follows. Using Eq. (2.1), we obtain

$$\frac{df(x)}{dx} = \lim_{\Delta x \to 0} \frac{(x + \Delta x)^n - x^n}{\Delta x} \tag{2.12}$$

Using the binomial theorem (https://en.wikipedia.org/wiki/Binomial _theorem), we have the following expansion:

$$(x + \Delta x)^n = x^n + nx^{n-1}\Delta x + \binom{n}{2}x^{n-2}(\Delta x)^2 + \binom{n}{3}x^{n-3}(\Delta x)^3 + \cdots \tag{2.13}$$

where $\binom{n}{k} = \frac{n!}{k!\,(n-k)!}$. Substituting (2.13) into (2.12), x^n is canceled out (the blue-cancellation), which gives

$$\frac{dx^n}{dx} = \lim_{\Delta x \to 0} \frac{nx^{n-1}\Delta x + \binom{n}{2}x^{n-2}(\Delta x)^2 + \binom{n}{3}x^{n-3}(\Delta x)^3 + \cdots}{\Delta x}$$

$$= nx^{n-1} + \lim_{\Delta x \to 0}\left[\binom{n}{2}x^{n-2}(\Delta x) + \binom{n}{3}x^{n-3}(\Delta x)^2 + \cdots\right]$$

$$= nx^{n-1} \tag{2.14}$$

The first term in the second line of the foregoing equation is the result of the red-cancellation. The limit of the second term simply vanishes.

In the above proof, we used the binomial theorem, which can be proven using defined arithmetic operations of scalars and an induction approach. Interested readers are referred to the following site: http://amsi.org.au/ESA_Senior_Years/SeniorTopic1/1c/1c_2content_6.html. We will derive this formula when discussing Taylor's expansion. For now, we omit the proof.

Note also that Eq. (2.11) can be derived using the product rule of differentiation, which will be discussed later.

Equation (2.11) is known as the **power rule**. Its more general form is

$$\frac{d}{dx}f^n = nf^{n-1}\frac{df}{dx} \tag{2.15}$$

where f is a differentiable function. We will prove this in Section 2.5.

2.3.5 *Derivative of trigonometric functions*

Since there are a large number of trigonometric functions, as discussed in Ref. [1], we will not derive the derivative formulas for all of them. As a typical example, we prove that *the derivative of the sine function is the cosine function*, using the derivative definition.

Given the sine function, its derivative is

$$\frac{d}{dx}\sin(x) = \lim_{\Delta x \to 0}\frac{\sin(x + \Delta x) - \sin(x)}{\Delta x}$$

$$= \lim_{\Delta x \to 0}\frac{\sin(x)\cos(\Delta x) + \cos(x)\sin(\Delta x) - \sin(x)}{\Delta x}$$

$$= \sin(x)\underbrace{\lim_{\Delta x \to 0}\frac{\cos(\Delta x) - 1}{\Delta x}}_{0} + \cos(x)\underbrace{\lim_{\Delta x \to 0}\frac{\sin(\Delta x)}{\Delta x}}_{1}$$

$$= \cos(x) \tag{2.16}$$

Here, we used the angle addition formula for the sine function, for which the proof is given in Ref. [1]. We used the fact that $\lim_{\Delta x \to 0} \frac{\sin(\Delta x)}{\Delta x} = 1$. This can be easily proven using the Taylor expansion (see Section 2.12). A solely geometry-based proof can be found in the following online article: https://medium.com/however-mathematics/a-beautiful-proof-why-the-limit-of-sin-x-x-as-x-approaches-0-is-1-c9709e72fda. We also used $\lim_{\Delta x \to 0} \frac{\cos(\Delta x)-1}{\Delta x} = 0$. This can also be easily proven using the Taylor expansion; a proof can be found in Section 2.3 of Ref. [4]. Here, let us use Python to "prove" both limits:

```
1  x = symbols('x')                                # define symbolic variables
2  print(sp.limit( sp.sin(x)/x,    x, 0, dir='+-'))    # limits, both sides
3  print(sp.limit((sp.cos(x)-1)/x, x, 0, dir='+-'))
```

```
1
0
```

```
1  # The derivative of the sine function is the cosine function:
2  f = sp.sin(x)                                    # define the function
3  dfdx = f.diff(x)                                 # use sp.diff()
4  display(Math(f" \\frac{{d}}{{dx}}\\big({latex(f)}\\big)={latex(dfdx)}"))
```

$$\frac{d}{dx}(\sin(x)) = \cos(x)$$

We see that the derivative of the sine is the cosine. Using a similar procedure, it is easy to prove that *the derivative of the cosine is the negated sine*:

```
1  f = sp.cos(x)                                    # define the function
2  dfdx = f.diff(x)                                 # use sp.diff()
3  display(Math(f" \\frac{{d}}{{dx}}\\big({latex(f)}\\big)={latex(dfdx)}"))
```

$$\frac{d}{dx}(\cos(x)) = -\sin(x)$$

Therefore, both the sine and cosine functions are **differentiable** infinitely many times.

Derivatives of other functions can be derived in a similar manner. Using Eq. (2.1) is basic but often very tedious for more complicated functions. The most difficult part is to properly expand $f(x + \Delta x)$ so that the blue-cancellation can take place. Fortunately, many techniques have been developed in the past to overcome difficulties, and derivatives of many commonly used differentiable functions have already been found, with tables available in many texts and online. Therefore, there is no need to rederive using Eq. (2.1).

Most conveniently, we can obtain derivatives of known functions via symbolic computation using SymPy, as we have done for a few cases. Let us see more SymPy examples.

2.3.6 *Examples: Derivative of exponential functions*

The following codes compute the derivatives of functions with exponential functions using SymPy:

```
1  x, k = symbols('x, k')                    # define symbolic variables
2  f = sp.exp(k*x)*x                              # define the function
3  dfdx = sp.diff(f, x)                        # diff with respect to x
4  display(Math(f" \\frac{{d}}{{dx}}({sp.latex(f)}) = {sp.latex(dfdx)}"))
```

$$\frac{d}{dx}(xe^{kx}) = kxe^{kx} + e^{kx}$$

```
1  f = sp.exp(k*x**2)
2  dfdx = sp.diff(f,x)
3  display(Math(f" \\frac{{d}}{{dx}}({sp.latex(f)}) = {sp.latex(dfdx)}"))
```

$$\frac{d}{dx}(e^{kx^2}) = 2kxe^{kx^2}$$

Using SymPy, the derivative of a "strange" exponential function $f(x) = x^x$ studied in Ref. [1] can be found as follows:

```
1  f = x**x
2  dfdx = sp.diff(f, x)
3  display(Math(f" \\frac{{d}}{{dx}}({sp.latex(f)}) = {sp.latex(dfdx)}"))
```

$$\frac{d}{dx}(x^x) = x^x(\log(x) + 1)$$

2.3.7 *Examples: Derivative of trigonometric functions*

The derivatives of some basic trigonometric functions and their inverse functions are computed using SymPy:

```
1  α = symbols('α')                          # define symbolic variables
2
3  print(f"tan'(α) = ",sp.tan(α).diff())               # diff w.r.t α
4  print(f"cot'(α) = ",sp.cot(α).diff())
5  print(f"atan'(α)=",sp.atan(α).diff())
6  print(f"asin'(α)=",sp.asin(α).diff())
7  print(f"csc'(α) = ",sp.csc(α).diff())
8  print(f"acos'(α)=",sp.acos(α).diff())
9  print(f"sec'(α) = ",sp.sec(α).diff())
```

```
tan'(α)  = tan(α)**2 + 1
cot'(α)  = -cot(α)**2 - 1
atan'(α) = 1/(α**2 + 1)
asin'(α) = 1/sqrt(1 - α**2)
csc'(α)  = -cot(α)*csc(α)
acos'(α) = -1/sqrt(1 - α**2)
sec'(α)  = tan(α)*sec(α)
```

Since the derivative of the sine function is cosine, the cosine function is thus the slope of the sine function. Also, the derivative of the cosine function is negated sine, implying that the sine function is the negated slope of the cosine function. This is a useful property of these two trigonometric functions and is used frequently in function analysis.

The following are examples of functions that combine polynomials with trigonometric functions:

```
1  f = x*sin(x**2) + 1            # define a more complicated function
2  dfdx = f.diff(x)
3  display(Math(f" \\frac{{d}}{{dx}}({sp.latex(f)}) = {sp.latex(dfdx)}"))
```

$$\frac{d}{dx}(x\sin(x^2) + 1) = 2x^2\cos(x^2) + \sin(x^2)$$

To obtain the numerical values of the derivative function at locations, we use sp.subs():

```
1  [dfdx.subs({x:xi}).evalf(6) for xi in [-1.8, -0.8, 0.,0.8, 1.8, 2.8]]
```

$$[-6.5469,\ 1.62388,\ 0,\ 1.62388,\ -6.5469,\ 1.21913]$$

2.3.8 *Examples: Derivative of frequently used functions*

The following code snippets produce derivatives of some frequently encountered functions:

```
1  x = symbols('x', real = True)
2  f = 1/sp.sqrt(1+x)
3  display(Math(f" \\frac{{d}}{{dx}}\\big({sp.latex(f)}\\big)=\
4                       {sp.latex(f.diff())}"))
5  f = sp.sin(x)/x
6  display(Math(f" \\frac{{d}}{{dx}}\\big({sp.latex(f)}\\big)=\
7                       {sp.latex(f.diff())}"))
8  f = sp.sin(1/x)
9  display(Math(f" \\frac{{d}}{{dx}}\\big({sp.latex(f)}\\big)=\
10                       {sp.latex(f.diff())}"))
```

$$\frac{d}{dx}\left(\frac{1}{\sqrt{x+1}}\right) = -\frac{1}{2(x+1)^{\frac{3}{2}}}$$

$$\frac{d}{dx}\left(\frac{\sin(x)}{x}\right) = \frac{\cos(x)}{x} - \frac{\sin(x)}{x^2}$$

$$\frac{d}{dx}\left(\sin\left(\frac{1}{x}\right)\right) = -\frac{\cos\left(\frac{1}{x}\right)}{x^2}$$

One can also conveniently generate a table of derivatives of functions:

```
1  # A list of functions often used in machine learning:
2  functions=[sp.log(1+exp(x)),sp.tanh(x),x/(1+sp.Abs(x)),1/(1+exp(-x))]
3
4  # get the table of the derivative functions of these functions:
5  [f.diff(x).simplify() for f in functions]
```

$$\left[\frac{e^x}{e^x+1}, \frac{1}{\cosh^2(x)}, \begin{cases} 1 & \text{for } x=0 \\ \frac{|x|}{x^2|x|+2x^2+|x|} & \text{otherwise} \end{cases}, \frac{1}{4\cosh^2\left(\frac{x}{2}\right)}\right]$$

As seen, with Python, one can obtain the derivatives of functions with ease. The conventional practice of looking up tables in handbooks is no longer necessary. Furthermore, subsequent operations involving these functions and derivatives become significantly easier.

2.3.9 *Locality of derivatives*

Figure 2.4 shows a nicely made movie (https://en.wikipedia.org/wiki/Derivative#/media/File:Tangent_function_animation.gif) that demonstrates the change of derivative of a curve defined by a function as the x location changes. The locality of derivatives is vividly revealed.

2.4 Differentiability of functions

Not all functions are differentiable, and many functions lack differentiability at certain points within their domain. A function is **differentiable** at a point if formula (2.1) yields a unique finite number, regardless of whether Δx is positive or negative. Since Eq. (2.1) involves a limit, the differentiability of a function is closely related to its continuity.

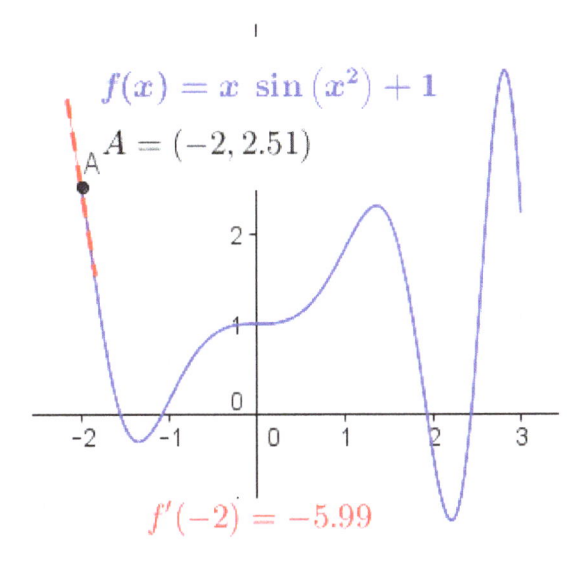

$$f(x) = x \, \sin\left(x^2\right) + 1$$

$$A = (-2, 2.51)$$

$$f'(-2) = -5.99$$

Figure 2.4. Demonstration movie for derivatives at different locations on the curve defined by a function.
Source: Movie from en.wikipedia Wikimedia Commons under the CC BYSA 3.0 license by User Lfahlberg.

2.4.1 *On continuity of functions*

Continuity of functions is discussed in detail in Ref. [1]. The key points are summarized as follows:

A function $f(x)$ is said to be continuous at a point $x = a$ if and only if the following three conditions are met:

1. a is in the domain of $f(x)$.
2. The limit of $f(x)$ as x approaches a from any direction exists.
3. The limit equals $f(a)$.

In other words, a function is continuous at a point if its value at that point equals the limit of the function at that point.

If a function is continuous at every point in a domain, then we say that it is a continuous function in that domain.

2.4.2 *Basic conditions for functions to be differentiable*

1. A function is differentiable at a point if it is defined at that point.
2. It must be continuous at that point. This follows directly from the definition in Eq. (2.1) and the continuity conditions for a function. If the

function is not continuous, the limit in Eq. (2.1) cannot be established. Therefore, a function is not differentiable at a point where it 'jumps'. Continuity is necessary but not sufficient for a function to be differentiable.

2.4.3 *Examples of not differentiable functions*

1. The power function $f(x) = \frac{1}{x}$ is not defined and not differentiable at $x = 0$. It is differentiable in $(-\infty, 0)$ and $(0, \infty)$.
2. The absolute value function $f(x) = |x|$ is defined and continuous at $x = 0$, but not differentiable at $x = 0$. This is because the limits in the definition in Eq. (2.1) from two sides of $x = 0$ are not equal. Let us call this type of point a kink point. The absolute value function is differentiable in $(-\infty, 0)$ and $(0, \infty)$, excluding the kink point.
3. The (leaky) ReLU function widely used in machine learning (ML) is

$$f(x) = \begin{cases} k_n x & x < 0 \\ k_p x & x \geq 0 \end{cases} \tag{2.17}$$

where k_n and k_p are given constants, but $k_n \neq k_p$. It is defined and continuous at $x = 0$, but not differentiable at $x = 0$ for the same reason as the absolute value function. It also exhibits a kink point. The ReLU function is differentiable in $(-\infty, 0)$ and $(0, \infty)$ after excluding the kink point. When used, a sub-gradient approach is employed [5].
4. Step functions

$$f(x) = \begin{cases} a & x < 0 \\ b & x \geq 0 \end{cases} \tag{2.18}$$

where a and b are given real constants, but $a \neq b$. It is defined and discontinuous at $x = 0$. It is not differentiable at $x = 0$ due to being a jumping point. The widely used Heaviside step function given in Eq. (4.37) belongs to this category.

2.4.4 *Code for examining differentiability of a function*

We write the following simple code to examine the differentiability of a simple function at $x = a$:

```
1  # Extra modules needed for this example:
2  from sympy import S
3  from sympy.calculus.util import continuous_domain
4  x, a = sp.symbols('x, a', real=True)
5  # at x=a the differentiability is examined
```

```python
 1  def diff_check(f, x, a):
 2      '''Check the differentiability of a give function f (sympy) at x=a.
 3         On the way, it checks the continuity of f at x=a.
 4         a: must be in the continuous domain of the function f.
 5      '''
 6      h = sp.symbols('h', real=True)  # h stands for Δx in the definision
 7      f_left = sp.limit(f.subs(x,a+h), h, 0, dir='-')
 8      f_right= sp.limit(f.subs(x,a+h), h, 0, dir='+')
 9      print(f"f_left={f_left}, f_right={f_right}")
10      print(f"Continuous at {a}? {f_left==f_right}")
11
12      df_left = sp.limit((f.subs(x,a+h)-f.subs(x,a))/h, h, 0, dir='-')
13      df_right= sp.limit((f.subs(x,a+h)-f.subs(x,a))/h, h, 0, dir='+')
14
15      print(f"df_left={df_left}\ndf_right={df_right}")
16      print(f"Differentiable at {a}? "
17      f"{df_left==df_right and df_left.is_finite and df_right.is_finite}")
```

We examine first the sine function that is known continuous and differentiable over $\mathbb{R}$:

```python
1  # The sine function:
2  f = sp.sin(x)
3  continuous_domain(f, x, S.Reals)        # Check the continuous_domain
```

$\mathbb{R}$

```python
1  a = 0
2  diff_check(f, x, a)                     # Check the differentiability at x = a
```

```
f_left=-oo, f_right=oo
Continuous at 0? False
df_left=Limit((zoo + 1/h)/h, h, 0, dir='-')
df_right=Limit((zoo + 1/h)/h, h, 0)
Differentiable at 0? False
```

Let us check the absolute function, which is black curve plotted in Fig. 2.5. It is continuous, but not differentiable at the kink point:

```python
1  f = sp.Abs(x)-1
2  continuous_domain(f, x, S.Reals)        # Check the continuous_domain
```

$\mathbb{R}$

```python
1  a = 0
2  diff_check(f, x, a)                     # Check the differentiability at x = a
```

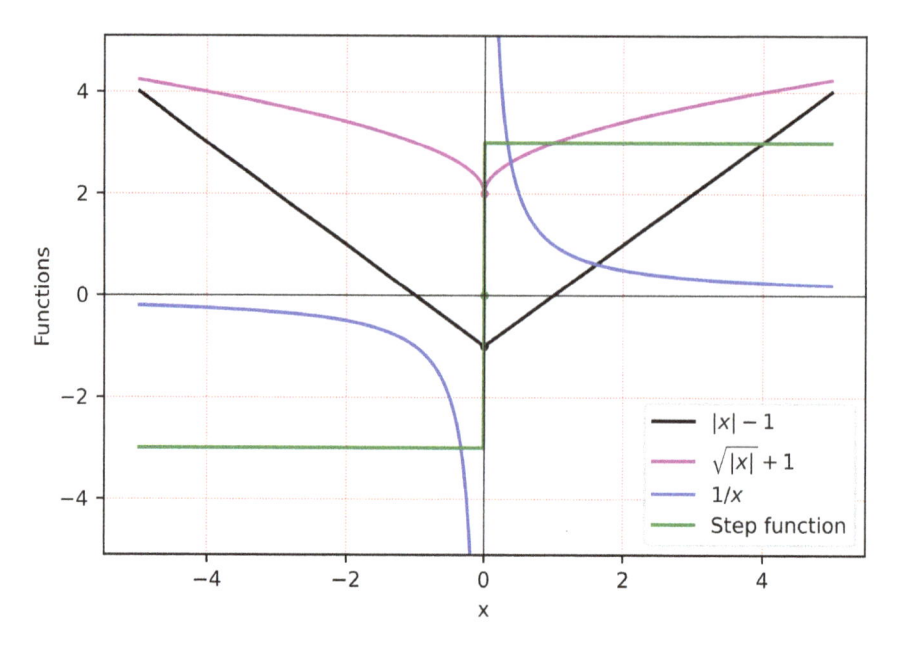

Figure 2.5. Examples of functions not differentiable at $x = 0$, which is a kink point or a jumping point or a singular point.

```
f_left=-oo, f_right=oo
Continuous at 0? False
df_left=Limit((zoo + 1/h)/h, h, 0, dir='-')
df_right=Limit((zoo + 1/h)/h, h, 0)
Differentiable at 0? False
```

Let us check a square-root absolute function, which is magenta curve plotted in Fig. 2.5. It is also continuous, but not differentiable at the kink point:

```
1  f = sp.sqrt(sp.Abs(x))+1
2  a = 0
3  diff_check(f, x, a)                    # Check the differentiability at x = a
```

```
f_left=1, f_right=1
Continuous at 0? True
df_left=-oo
df_right=oo
Differentiable at 0? False
```

The rational function, denoted by the blue curve plotted in Fig. 2.5, is not continuous and not defined at 0:

```
1  x, a = sp.symbols('x, a', real=True)
2  f = 1/x
3  continuous_domain(f, x, S.Reals)          # Check the continuous_domain
```

$(-\infty, 0) \cup (0, \infty)$

We see that the function is not defined at $x = 0$. It is a singular point. Thus, it is not differentiable, and hence there is no need to check. If we check it anyway using diff_check(), we obtain the following:

```
1  a = 0                              # function is not defined at a = 0
2  diff_check(f, x, a)               # Check the differentiability at x = a
```

```
f_left=-oo, f_right=oo
Continuous at 0? False
df_left=Limit((zoo + 1/h)/h, h, 0, dir='-')
df_right=Limit((zoo + 1/h)/h, h, 0)
Differentiable at 0? False
```

When using SymPy for computations involving functions that are not defined at certain points, such as $f(x) = \frac{1}{x}$, SymPy handles these cases by defaulting to complex numbers because the complex set is closed. This is why you see complex infinity results in the output when $x = 0$.

If you evaluate the function for any non-zero $x = a$, where $f(x)$ is defined, SymPy will return a definite result, indicating that the function is differentiable. This illustrates how SymPy manages computations in cases where functions are undefined at specific points:

```
1  x, a = sp.symbols('x, a', real=True)
2  a = 1                          # when a=1, the function is differentiable
3  diff_check(f, x, a)
```

```
f_left=1, f_right=1
Continuous at 1? True
df_left=-1
df_right=-1
Differentiable at 1? True
```

2.4.5 *Plots of functions with non-differentiable point*

The following code plots some often encountered functions with a discontinuous point for easy viewing:

```
 1  plt.figure(); plt.ioff()
 2  plt.rcParams.update({'font.size': 9})
 3  fig, ax = plt.subplots(1,1)#,figsize=(4,2.5))
 4
 5  xL, xR = -5., 5.                              # domain for the functions
 6  n_x = 1000                                    # number of points on x axis
 7  X = np.linspace(xL, xR, n_x)                        # x coordinates
 8
 9  # Plot the curves for the functions:
10  ax.set_xlabel('x')
11  ax.set_ylabel('Functions')
12  ax.grid(color='r', linestyle=':', linewidth=0.5)
13  ax.plot(X, np.abs(X)-1, c="k", lw=1.5, label="$|x|-1$")
14  ax.plot(X, np.sqrt(np.abs(X))+2.,c="m", lw=1.5, label=r"$\sqrt{|x|}+1$")
15
16  power_f = 1/X
17  mask = np.abs(X)<=.1                          # Define a masked in plot
18  power_f[mask] = np.nan                   # to avoid the singular point
19  ax.plot(X, power_f, c="b", lw=1.5, label=r"$1/x$")
20
21  def step_f(x):
22      return np.piecewise(x, [x < 0, x >= 0], [-3, 3])
23  ax.plot(X, step_f(X), c="g", lw=1.5, label="Step function")
24
25  ax.scatter(0,  0., c="g", s=8)
26  ax.scatter(0, -1., c="k", s=8)
27  ax.scatter(0,  2., c="m", s=8)
28
29  ax.axvline(x=0, c="k", lw=0.6); ax.axhline(y=0, c="k", lw=0.6)
30  ax.legend(loc='lower right') #, bbox_to_anchor=(.2, .8))
31  ax.set_ylim(-5.1, 5.1)
32
33  plt.savefig('imagesDI/Undiff.png', dpi=500)
34  #plt.show()
```

Figure 2.5 shows four example functions that are not differentiable at $x = 0$.

If a function is differentiable in a domain, then it must be continuous in that domain, a principle known as *differentiability implies continuity*.

However, it is not always true that a continuous function is differentiable, as shown in the previous examples. Another well-known case is the specially constructed Weierstrass function (https://en.wikipedia.org/wiki/Weierstrass_function). It is continuous everywhere but differentiable nowhere due to its construction involving kink points.

2.4.6 *On techniques for handling non-differentiable points*

In computational methods, we frequently encounter functions that are not differentiable at points, including singular points, jump discontinuities, and kink points. However, we often need to obtain estimates of the derivatives at these points. There are several techniques to achieve this objective:

1. **Piecewise differentiation:** This technique excludes non-differentiable points by splitting the domain into a number of sub-domains, each of which is differentiable. This technique is simple but needs to be used with care. If the function has only kink points, piecewise differentiation can work well for many problems and is used in the finite element method (FEM) [6] based on the so-called weak formulations.
2. **Gradient smoothing:** For functions with kink points, some kind of smoothing operation is needed for stable solutions. The simplest way is to use the so-called **sub-derivative** or **sub-gradient**, which averages the derivatives on both sides of a kink point. This is widely used in ML models when the ReLU activation function is used [5]. A well-formulated method is the **gradient smoothing** technique, which creates a smoothing domain that straddles the kink point. By introducing a smoothing function, the derivative (or gradient) at the kink point (or in the smoothing domain) can be obtained. This is used in both strong formulations [7,8] and weak formulations [9, 10].
3. **Generalized gradient smoothing:** For functions with kink points and certain types of jump discontinuities, the generalized gradient smoothing technique is used [11, 12]. It is used in meshfree methods that employ radial basis functions [13, 14] based on the so-called weakened weak formulation.

2.5 Rules for derivatives of functions

2.5.1 *Linear combination rule*

According to the definition in Eq. (2.1), the derivative of a linear combination of differentiable functions is the linear combination of their derivatives:

$$\frac{d}{dx}(af(x) + bg(x)) = a\frac{df(x)}{dx} + b\frac{dg(x)}{dx} \qquad (2.19)$$

where a and b are arbitrary constants (independent of x). The proof is straightforward and relies solely on using the definition in Eq. (2.1).

When a and b are both 1, we have the sum rule: The derivative of a sum of differentiable functions is the sum of their derivatives.

```
1  # Linear combination rule:
2  x, a, b = symbols('x, a, b', real=True)       # define symbolic variables
3  f = Function('f')(x)
4  g = Function('g')(x)
5  display(Math(f" \\frac{{d}}{{dx}}({sp.latex(a*f+b*g)}) =\
6                  {sp.latex((a*f+b*g).diff(x))}"))
```

$$\frac{d}{dx}(af(x) + bg(x)) = a\frac{d}{dx}f(x) + b\frac{d}{dx}g(x)$$

2.5.2　*Product rule*

The derivative of a product of two differentiable functions is given as follows:

$$\frac{d}{dx}(f(x)g(x)) = g(x)\frac{df(x)}{dx} + f(x)\frac{dg(x)}{dx} \tag{2.20}$$

The proof uses the definition in Eq. (2.1):

$$\frac{d}{dx}(f(x)g(x))$$

$$= \lim_{\Delta x \to 0} \frac{\Delta(f(x)g(x))}{\Delta x} = \lim_{\Delta x \to 0} \frac{f(x+\Delta x)g(x+\Delta x) - f(x)g(x)}{\Delta x}$$

$$= \lim_{\Delta x \to 0} \frac{\begin{array}{c} f(x+\Delta x)g(x+\Delta x) \\ -f(x)g(x+\Delta x) + f(x)g(x+\Delta x) - f(x)g(x) \end{array}}{\Delta x}$$

$$= \lim_{\Delta x \to 0} g(x+\Delta x)\frac{f(x+\Delta x) - f(x)}{\Delta x} + \lim_{\Delta x \to 0} f(x)\frac{g(x+\Delta x) - g(x)}{\Delta x}$$

$$= g(x)\frac{df(x)}{dx} + f(x)\frac{dg(x)}{dx} \tag{2.21}$$

In the process, we add and subtract $f(x)g(x + \Delta x)$, which is a simple trick that leads to two differentials.

One can confirm this using SymPy:

```
1  # Product rule:
2  display(Math(f" \\frac{{d}}{{dx}}\\big({latex(f*g)}\\big) = \
3                    {latex(diff(f*g))}"))
```

$$\frac{d}{dx}(f(x)g(x)) = f(x)\frac{d}{dx}g(x) + g(x)\frac{d}{dx}f(x)$$

Note that Eq. (2.20) can be expressed in the following more concise form:

$$d(f\,g) = f\,dg + g\,df \tag{2.22}$$

The reason was given by Leibniz:

$$\begin{aligned} d(f\,g) &= (f + df)(g + dg) - (f\,g) \\ &= f\,dg + g\,df + df\,dg \end{aligned} \tag{2.23}$$

The last term $df\,dg$ is higher order and hence is negligible.

2.5.3 *Power rule*

With the product rule, we can effortlessly derive the power rule. First, consider $g(x) = f(x)$. Equation (2.21) gives

$$\frac{d}{dx}\left([f(x)]^2\right) = 2f(x)\frac{df(x)}{dx} \tag{2.24}$$

Next, consider $f(x) = [f(x)]^2$ and $g(x) = f(x)$. Equation (2.21) gives

$$\frac{d}{dx}\left([f(x)]^3\right) = \frac{d}{dx}\left([f(x)]^2 f(x)\right) = 3[f(x)]^2\frac{df(x)}{dx} \tag{2.25}$$

Finally, a simple induction gives the **power rule**:

$$\frac{d}{dx}\left([f(x)]^n\right) = n[f(x)]^{n-1}\frac{df(x)}{dx} \tag{2.26}$$

This power rule is given earlier in Eq. (2.15). The following SymPy code confirms this rule:

```
1  # Power rule for d[f(x)^n]/dx:
2  n = symbols('n', integer=True)
3  dfn = (f**n).diff(x).simplify()
4  display(Math(f" \\frac{{d}}{{dx}}\\big({latex(f**n)}\\big) = \
5                          {latex(dfn)}"))
```

$$\frac{d}{dx}\left(f^n(x)\right) = nf^{n-1}(x)\frac{d}{dx}f(x)$$

2.5.4 Reciprocal rule

By setting $n = -1$, the power rule leads to the reciprocal rule:

$$\frac{d}{dx}\left(\frac{1}{f(x)}\right) = \frac{d}{dx}\left([f(x)]^{-1}\right) = (-1)[f(x)]^{-2}\frac{df(x)}{dx}$$

$$= \frac{-1}{[f(x)]^2}\frac{df(x)}{dx} \qquad (2.27)$$

This can also be obtained using SymPy:

```
1  # Reciprocal rule:
2  drf = (1/f).diff(x).simplify()
3  display(Math(f" \\frac{{d}}{{dx}}\\big({latex(1/f)}\\big) = \
4                          {latex(drf)}"))
```

$$\frac{d}{dx}\left(\frac{1}{f(x)}\right) = -\frac{\frac{d}{dx}f(x)}{f^2(x)}$$

2.5.5 Quotient rule

Using the product rule and the reciprocal rule, one can easily obtain the quotient rule:

$$\frac{d}{dx}\left(\frac{f(x)}{g(x)}\right) = \frac{g(x)\frac{df(x)}{dx} - f(x)\frac{dg(x)}{dx}}{[g(x)]^2} \qquad (2.28)$$

Let us confirm this using SymPy:

```
1  #  Reciprocal rule:
2  df_g = (f/g).diff(x).simplify()
3  display(Math(f" \\frac{{d}}{{dx}}\\big({latex(f/g)}\\big) = \
4                          {latex(df_g)}"))
```

$$\frac{d}{dx}\left(\frac{f(x)}{g(x)}\right) = \frac{-f(x)\frac{d}{dx}g(x) + g(x)\frac{d}{dx}f(x)}{g^2(x)}$$

2.5.6 *Chain rule of differentiation*

Consider a nested composite function

$$y = f\big(g(x)\big) \tag{2.29}$$

where $f(g)$ and $g(x)$ are all differentiable. The derivative of the composite function is

$$\frac{dy}{dx} = \frac{d}{dx}f\big(g(x)\big) = \frac{df}{dg}\frac{dg}{dx} \tag{2.30}$$

This is the widely used **chain rule of differentiation**.

Assuming that $f(g) = \sin(g)$ and $g(x) = ax^2 + bx + c$, in which a, b and c are constants, we have

$$f\big(g(x)\big) = \sin(\overbrace{ax^2 + bx + c}^{f(g)})_{g(x)} \tag{2.31}$$

The derivative becomes

$$\frac{d}{dx}f\big(g(x)\big) = \frac{df}{dg}\frac{dg}{dx}$$

$$= \cos(g)(2ax + b) = (2ax + b)\cos(ax^2 + bx + c) \tag{2.32}$$

This can be obtained using SymPy:

The chain rule of differentiation is used in all machine learning models based on neural networks, which use backpropagation [5].

2.5.7 *SymPy example for chain rule of differentiation*

```
1  g, x, a, b, c = sp.symbols('g, x, a, b, c')          # define variables
2  g = sp.Function('g')(x)                               # define functions
3  f = sp.Function('f')(g)
4  display(Math(f" \\frac{{d}}{{dx}}\\big({latex(f)}\\big) = \
5                          {latex(f.diff(x))}"))
```

$$\frac{d}{dx}\left(f(g(x))\right) = \frac{d}{dg(x)}f(g(x))\frac{d}{dx}g(x)$$

```
1  g = a*x**2 + b*x + c
2  f = sp.sin(g)                           # Define the nested function
3  display(Math(f" \\frac{{d}}{{dx}}\\big({latex(f)}\\big) = \
4                          {latex(f.diff(x))}"))
```

$$\frac{d}{dx}\left(\sin(ax^2 + bx + c)\right) = (2ax + b)\cos(ax^2 + bx + c)$$

2.5.8 *Proof of the chain rule of differentiation*

The proof uses the definition in Eq. (2.1):

$$\frac{d}{dx}f(g(x)) = \lim_{\Delta x \to 0} \frac{f(g(x + \Delta x)) - f(g(x))}{\Delta x}$$

$$= \lim_{\Delta x \to 0} \left[\frac{f(g(x + \Delta x)) - f(g(x))}{g(x + \Delta x) - g(x)} \frac{g(x + \Delta x) - g(x)}{\Delta x}\right]$$

$$= \frac{df}{dg}\frac{dg}{dx} \tag{2.33}$$

In the process, we multiplied and divided $g(x + \Delta x) - g(x)$. This trick leads to two differentials.

The proof shows that there is no limit on the number of functions nested. For example, if

$$y = f\left(g(h(x))\right) \tag{2.34}$$

where $f(g)$, $g(h)$, and $h(x)$ are all differentiable. The derivative of the composite function becomes

$$\frac{dy}{dx} = \frac{d}{dx}f\left(g(h(x))\right) = \frac{df}{dg}\frac{dg}{dh}\frac{dh}{dx} \tag{2.35}$$

2.5.9 *Summary of derivative rules*

Linear combination rule:	$(af + bg)' = af' + bg'$
Product rule:	$(fg)' = gf' + fg'$
Power rule:	$(f^n)' = nf^{n-1}f'$
Reciprocal rule:	$\left(\frac{1}{f}\right)' = -\frac{f'}{f^2}$
Quotient rule:	$\left(\frac{f}{g}\right)' = \frac{gf'-fg'}{g^2}$
Chain rule of differentiation:	$f(g(x)) = f'g'$

$$(2.36)$$

Note that f' stands for derivative of $\frac{df(x)}{dx}$, in which x is the immediate independent variable of f. The same applies to g.

2.6 Useful properties and theorems

There are some useful properties and theorems for theoretical study and for applications in computation, including the closure property, symmetry property, the extreme value theorem (EVT), Rolle's theorem, the mean value theorem (MVT), and L'Hôpital's rule.

2.6.1 *The closure property*

2.6.1.1 *Elementary functions, closed under differentiation*

For real elementary functions [1], we note the following:

> All differentiable elementary functions are closed under the differential operation.

This closure property means the following:

1. The derivative of a differentiable elementary function is still an elementary function.
2. We can always find the derivative of an elementary function, but the domain and codomain may change.

This is because of the following:

1. The operations used in Eq. (2.1) are essentially basic arithmetic operations (addition, subtraction, multiplication, defined division) applied **locally** around a given point of an elementary function. The result of

these operations will still be an elementary function, although the form of the function will change.

2. These rules of differentiation involve essentially only these basic arithmetic operations.

3. The change of the domain is due to the possible creation of singularity points or the potential behavioral change at the non-differentiable points caused by differentiation.

This closure property is important for many applications. When one defines a differentiable elementary function in SymPy, `sp.diff()` can always produce its derivative. When we use autograd (Section 2.7) for a complicated loss function that is a composition of chained elementary functions [5], a derivative function object will be generated.

Note that this closure property does not apply to integral operations or antiderivatives (to be discussed in Chapters 4 and 5) because integration is of **global** nature.

2.6.1.2 *Examples: Domain change by differentiation*

The following are two examples where the domain of the derivative of a function is different from that of the function. We need to import `continuous_domain` to check the continuity domains of these functions:

```
 1  from sympy import S
 2  from sympy.calculus.util import continuous_domain
 3
 4  # case I:
 5  f = sqrt(x)
 6  display(Math(f" \\text{{The continuous domain for f(x)}}={latex(f)}:\
 7                  {latex(continuous_domain(f, x, S.Reals))}"))
 8
 9  display(Math(f" \\text{{The continuous domain for f'(x)}}=\
10     {latex(diff(f))}:{latex(continuous_domain(diff(f), x, S.Reals))}"))
11
12  # case II:
13  f = sp.log(x)
14  display(Math(f" \\text{{The continuous domain for f(x)}}={latex(f)}:\
15                  {latex(continuous_domain(f, x, S.Reals))}"))
16  display(Math(f" \\text{{The continuous domain for f'(x)}}=\
17     {latex(diff(f))}:{latex(continuous_domain(diff(f), x, S.Reals))}"))
18
19  # case IIb:
20  f = sp.log(sp.Abs(x))
21  display(Math(f" \\text{{The continuous domain for f(x)}}={latex(f)}:\
22                  {latex(continuous_domain(f, x, S.Reals))}"))
```

The continuous domain for $f(x) = \sqrt{x} : [0, \infty)$

The continuous domain for $f'(x) = \dfrac{1}{2\sqrt{x}} : (0, \infty)$

The continuous domain for $f(x) = \log(x) : (0, \infty)$

The continuous domain for $f'(x) = \dfrac{1}{x} : (-\infty, 0) \cup (0, \infty)$

The continuous domain for $f(x) = \log(|x|) : (-\infty, 0) \cup (0, \infty)$

In case I, the derivative of the function becomes singular at $x = 0$, and hence its domain must exclude 0. In case II, the domain of the derivative of the function is extended to $(-\infty, 0)$, which is the same as that of $f(x) = \ln(|x|)$. The kink point is the root cause of this domain change.

2.6.2 *Symmetry property, odd to even*

An **even function** $f_s(x)$ is symmetric with respect to the vertical line $x = 0$. It is also called a **symmetric function**. It satisfies the following condition:

$$f_s(-x) = f_s(x) \tag{2.37}$$

An **odd function** $f_{as}(x)$ is antisymmetric with respect to the vertical line $x = 0$. It is also called an **antisymmetric function**. It satisfies the following condition:

$$f_{as}(-x) = -f_{as}(x) \tag{2.38}$$

We note the following:

The derivative of an odd function is an even function.

$$f'_{as}(-x) = f'_{as}(x) \tag{2.39}$$

This can be easily proven using the chain rule of differentiation to Eq. (2.38). The left side of Eq. (2.38) gives

$$\left(f_{as}(-x)\right)' = f'_{as}(-x)(-x)' = -f'_{as}(-x) \tag{2.40}$$

The right side of Eq. (2.38) gives

$$\left(-f_{as}(x)\right)' = -f'_{as}(x) \tag{2.41}$$

We thus have

$$-f'_{as}(-x) = -f'_{as}(x) \quad \text{or} \quad f'_{as}(-x) = f'_{as}(x) \tag{2.42}$$

which is Eq. (2.39). This completes the proof.

A typical example is $\sin'(x) = \cos(x)$, where $\sin(x)$ is an odd function, but its derivative $\cos(x)$ is an even function.

2.6.3 *Symmetry property, even to odd*

On the other hand, for an **even function**, we have

$$f_s'(-x) = -f_s'(x) \tag{2.43}$$

This means that $f_s'(x)$ is an odd function. We note the following:

The derivative of an odd function is an even function.

This can be proven in exactly the same manner. A typical example is $\cos'(x) = -\sin(x)$, where $\cos(x)$ is an even function, but its derivative $-\sin(x)$ is an odd function.

It is important to note the following:

1. The same symmetry property holds if a function is symmetric (or antisymmetric) with respect to a vertical line $x = a$.
2. The same symmetry property holds locally if a function is symmetric (or antisymmetric) locally with respect to a vertical line $x = a$. This is useful when applying L'Hôpital's rule (to be discussed at the end of this section).

2.6.4 *Extreme value theorem*

To study these theorems in a logical path, we first introduce the EVT, which is a fundamental result in calculus. It holds for all functions that are continuous in a closed interval. It states as follows:

If $f(x)$ is continuous on a closed interval $[a, b]$, then $f(x)$ attains its maximum and minimum values on $[a, b]$. This implies that there exist points c and d in $[a, b]$, such that $f(c) \leq f(x) \leq f(d)$ for all $x \in [a, b]$.

There are a number of ways to prove this theorem rigorously. The essential idea is to use the continuity and boundedness of the function, the concepts of supremum, infimum, and the closure properties of compact intervals. One can then show the existence of the maximum and minimum values and also that these values can be found on the interval (including the boundary points). Interested readers may visit the Wikipedia page on EVT (https://en.wikipedia.org/wiki/Extreme_value_theorem). Here, we use a simple intuitive pictorial proof:

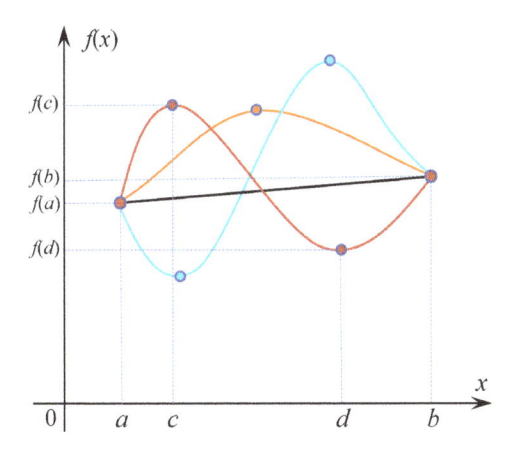

Figure 2.6. Any continuous function in the closed interval always attains its maximum and minimum values in the closed interval.

Consider a closed interval $[a, b]$ for the independent variable x, as shown in Fig. 2.6. A function $f(x)$ is defined on $[a, b]$. There are an infinite number of such functions. Since any such function must be continuous on $[a, b]$, it is thus bounded. We draw some of these functions in Fig. 2.6.

One can simply imagine a pencil drawing such a continuous curve that ends at $x = a$ and $x = b$. Regardless of how it is drawn, one can always find a minimum and a maximum value on the curve, as long as the curve is continuous on $[a, b]$. Note that these values can be right at the boundary points $x = a$ or $x = b$, as shown by the black straight line and also the orange line on which the minimum is at $x = a$. Since the closed interval includes $x = a$ or $x = b$, the theorem holds always.

We are now ready to present Rolle's theorem.

2.6.5 *Rolle's theorem*

The theorem states the following:

> If function $f(x)$ is continuous on the closed interval $[a, b]$, differentiable on the open interval (a, b), and $f(a) = f(b)$, then there exists at least one point $c \in (a, b)$, such that $f'(c) = 0$.

This implies that there is at least one stationary point. The picture presentation of Rolle's theorem is given in Fig. 2.7.

Proof. Since $f(x)$ is continuous on the closed interval $[a, b]$, it has a maximum and a minimum on $[a, b]$ based on the EVT. If both the maximum and minimum are at $x = a$ or $x = b$, which must be the special case of the

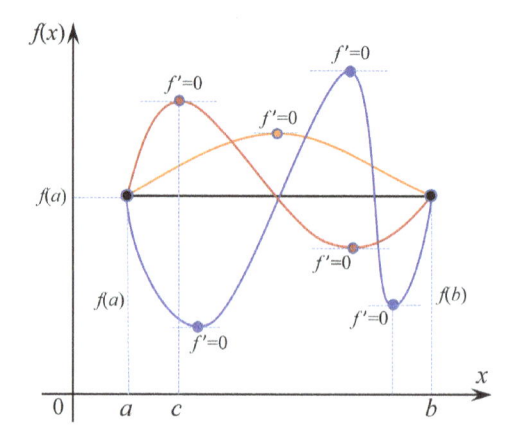

Figure 2.7. Any continuous function defined in the closed interval and differentiable in the open interval has at least one stationary point in the open interval.

black line, this is because of the $f(a) = f(b)$ condition and the fact that $f(x)$ is differentiable. In such a case, any point in (a, b) is a stationary point. If either the maximum or minimum is in the open interval (a, b), we denote it as $x_{\max}$ or $x_{\min}$. Since the function is differentiable everywhere in (a, b), we must have $f'(x_{\max}) = 0$ or $f'(x_{\min}) = 0$ as a stationary point. This completes the proof.

Rolle's theorem leads to the widely used MVT.

2.6.6 *The mean value theorem*

The MVT states the following:

> If function $f(x)$ is continuous on a closed interval $[a, b]$ and differentiable on the open interval (a, b), there exists at least one point $c \in (a, b)$, such that

$$\frac{f(b) - f(a)}{b - a} = f'(c) \tag{2.44}$$

This gives an important relation between the function values at two points and its derivative at a point between these two points. The picture presentation of the MVT is given in Fig. 2.8.

Proof. Comparing Fig. 2.8 with Fig. 2.7, the MVT is a generalized Rolle's theorem. Therefore, our proof can make use of the proven Rolle's theorem. We first define a new function by substituting the linear function

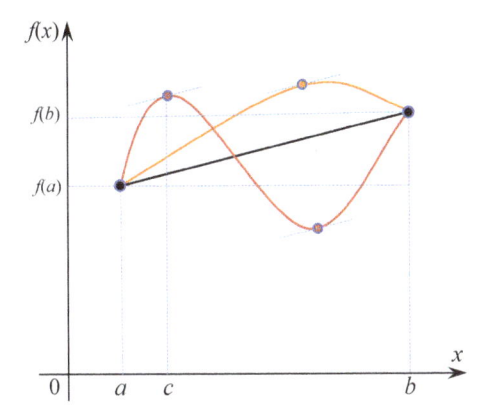

Figure 2.8. Any continuous function defined on a closed interval and differentiable on the open interval has at least one point in the open interval where the derivative equals the average slope of the function over the interval.

(the black line in Fig. 2.8) from $f(x)$:

$$g(x) = f(x) - \left(f(a) + \frac{f(b) - f(a)}{b - a}(x - a) \right) \tag{2.45}$$

This new function $g(x)$ satisfies

$$g(a) = f(a) - f(a) = 0$$

$$g(b) = f(b) - \left(f(a) + \frac{f(b) - f(a)}{b - a}(b - a) \right) = 0 \tag{2.46}$$

Due to the continuity and differentiability of $f(x)$, $g(x)$ is also continuous on $[a, b]$ and differentiable on (a, b). We can now apply Rolle's theorem to $g(x)$: There must exist at least one $c \in (a, b)$, such that $g'(c) = 0$. The derivative of $g(x)$ is

$$g'(x) = f'(x) - \frac{f(b) - f(a)}{b - a} \tag{2.47}$$

At $x = c$,

$$g'(c) = f'(c) - \frac{f(b) - f(a)}{b - a} = 0 \tag{2.48}$$

which is Eq. (2.44). This completes the proof.

2.6.7 *A concise exact presentation of a function*

Utilizing the MVT, we express a continuous and differentiable function as follows by letting $b = x$:

$$f(x) = f(a) + f'(c)(x - a) \tag{2.49}$$

where $c \in (a, x)$. This representation of a function is valuable in function analysis. In numerical computations, the challenge lies in determining such a c for complex functions. Although this expression may resemble a linear form, it is not strictly linear because c generally depends on x unless $f(x)$ is itself a linear function.

2.6.8 *Generalized mean value theorem*

The generalized MVT states the following:

> If two functions $f(x)$ and $g(x)$ are both continuous on a closed interval $[a, b]$ and differentiable on the open interval (a, b), there exists at least one point $c \in (a, b)$, such that

$$\frac{f(b) - f(a)}{g(b) - g(a)} = \frac{f'(c)}{g'(c)} \tag{2.50}$$

This theorem extends the MVT to consider the ratio of function values and their derivatives at corresponding points. When $g(x) = x$, Eq. (2.50) simplifies to Eq. (2.44).

Proof. We construct a new function

$$F(x) = \big(f(b) - f(a)\big)g(x) - \big(g(b) - g(a)\big)f(x) \tag{2.51}$$

This new $F(x)$ satisfies the conditions for Rolle's theorem. Since

$$F(a) = \big(f(b) - f(a)\big)g(a) - \big(g(b) - g(a)\big)f(a)$$
$$= f(b)g(a) - g(b)f(a)$$

$$F(b) = \big(f(b) - f(a)\big)g(b) - \big(g(b) - g(a)\big)f(b)$$
$$= -f(a)g(b) + g(a)f(b) \tag{2.52}$$

we have $F(a) = F(b)$. Also, $F(x)$ is continuous on $[a, b]$ and differentiable on (a, b) because of the same property of $f(x)$ and $g(x)$. Rolle's theorem gives the following: There must exist at least one $c \in (a, b)$, such that $F'(x) = 0$, which gives Eq. (2.50).

2.6.9 *L'Hôpital's rule*

Using the generalized MVT, we can derive L'Hôpital's rule as follows:

If two differentiable functions $f(x)$ and $g(x)$ both approach zero as $x \to a$, then

$$\lim_{x \to a} \frac{f(x)}{g(x)} = \lim_{x \to a} \frac{f'(x)}{g'(x)} \tag{2.53}$$

if the limit on the right-hand side exists.

Proof. Since $f(x)$ and $g(x)$ both approach zero as $x \to a$, substituting $f(a)$ and $g(a)$ with zero in Eq. (2.50) gives

$$\frac{f(b)}{g(b)} = \frac{f'(c)}{g'(c)} \tag{2.54}$$

Now, when taking the limit of $b \to a$, we shall have $c \to a$ because $c \in (a, b)$. This leads to Eq. (2.53).

L'Hôpital's rule is useful when evaluating $\lim_{x \to a} \frac{f(x)}{g(x)}$ encounters difficulty, typically when both $f(x)$ and $g(x)$ approach zero as $x \to a$. This rule leverages the derivatives to simplify the limit. The mechanism is given as follows:

1. Utilizing the symmetry property of functions mentioned earlier, if $g(x)$ approaches zero as $x \to a$, it behaves locally like an antisymmetric function at $x = a$. Consequently, $g'(x)$ behaves symmetrically and is likely non-zero at $x = a$. This transforms the limit $\lim_{x \to a} \frac{f'(x)}{g'(x)}$ from an indeterminate $\frac{0}{0}$ form to a definite value, resolving the problem in one step.
2. If both $f'(x)$ and $g'(x)$ approach zero as $x \to a$, $g'(x)$ behaves antisymmetrically locally. Here, two scenarios arise:
   - $f'(x)$ is non-zero at $x = a$, leading $\lim_{x \to a} \frac{f'(x)}{g'(x)}$ to infinity, resolving the limit.
   - $f'(x)$ also approaches zero at $x = a$, necessitating further examination with higher derivatives $\frac{f''(x)}{g''(x)}$, and so on.

L'Hôpital's rule applies strictly under the condition that $f(x)$ and $g(x)$ each approach zero as $x \to a$. If not, one may need to reformulate the limit to meet these criteria. For more details, refer to Ref. [4] and the Wikipedia page on L'Hôpital's rule.

2.6.9.1 *Example: Solution found via single step*

Let us look at an example encountered in Eq. (2.16) when we need to evaluate $\lim_{\Delta x \to 0} \frac{\sin(\Delta x)}{\Delta x}$. We can now use L'Hôpital's rule to get it done:

$$\lim_{\Delta x \to 0} \frac{\sin(\Delta x)}{\Delta x} = \lim_{\Delta x \to 0} \frac{\left(\sin(\Delta x)\right)'}{(\Delta x)'} = \lim_{\Delta x \to 0} \frac{\cos(\Delta x)}{1} = 1 \qquad (2.55)$$

This is the same result obtained before. In this simple case, Δx is an odd function, and hence $(\Delta x)'$ becomes an even function that is simply 1. We found the solution with single step.

In Eq. (2.16), we also need to evaluate $\lim_{\Delta x \to 0} \frac{\cos(\Delta x)-1}{\Delta x}$. Let us use L'Hôpital's rule to get it done:

$$\lim_{\Delta x \to 0} \frac{\cos(\Delta x) - 1}{\Delta x} = \lim_{\Delta x \to 0} \frac{(\cos(\Delta x) - 1)'}{(\Delta x)'} = \lim_{\Delta x \to 0} \frac{-\sin(\Delta x)}{1} = 0 \quad (2.56)$$

This is also the same result obtained before. Here, we obtain the same results effortlessly using L'Hôpital's rule.

Let us confirm all these using SymPy:

```
1  x = symbols('x')                        # define symbolic variables
2  f = sp.sin(x)
3  g = x
4  print('f_limt=', sp.limit(f, x, 0), ' g_limt=', sp.limit(g, x, 0))
5  print("(f'/g')_limt=", sp.limit(f.diff()/g.diff(), x, 0))
```

```
f_limt= 0  g_limt= 0
(f'/g')_limt= 1
```

```
1  f = sp.cos(x)-1
2  g = x
3  print('f_limt=', sp.limit(f, x, 0), ' g_limt=', sp.limit(g, x, 0))
4  print("(f'/g')_limt=", sp.limit(f.diff()/g.diff(), x, 0))
```

```
f_limt= 0  g_limt= 0
(f'/g')_limt= 0
```

These two typical examples shows the effectiveness of L'Hôpital's rule for this type of problems. Only single step is needed.

2.6.9.2 *Example: Solution found via two steps*

There are situations where we may need more than one step to obtain the solution. The following example requires two steps, implying that we need to compute up to the second derivatives. We present two cases: Case I gives an infinite solution, and Case II gives a finite solution, both found after two steps.

```
1  # Case I:
2  f = x**2
3  g = sp.sin(x) - x
4
5  print('f_limit= ', sp.limit(f, x, 0), ' g_limt=', sp.limit(g, x, 0))
6  print("f'_limit=", sp.limit(f.diff(), x, 0),           # 1st step: f'/g'
7         " g'_limit=", sp.limit(g.diff(), x, 0))
8  print('f"_limit=', sp.limit(f.diff(x,2), x, 0),        # 2nd step: f"/g"
9         ' g"_limit=', sp.limit(g.diff(x,2), x, 0))
10 print('(f"/g")_limit=', sp.limit(f.diff(x,2)/g.diff(x,2), x, 0))
```

```
f_limit=  0  g_limt= 0
f'_limit= 0  g'_limit= 0
f"_limit= 2  g"_limit= 0
(f"/g")_limit= -oo
```

The solution is found via two steps, and $\lim_{x \to a} \frac{f(x)}{g(x)} = \lim_{x \to a} \frac{f''(x)}{g''(x)} = -\infty$. This is because the rate at which $f(x) \to 0$ is of second order, requiring differentiation twice to obtain a finite value. The rate at which $g(x) \to 0$ is of third order, implying that it approaches zero one order faster than $f(x)$, resulting in the infinite limit.

```
1  # Case II, f and g swapped:
2  f = sp.sin(x) - x
3  g = x**2
4
5  print('f_limit= ', sp.limit(f, x, 0), ' g_limt=', sp.limit(g, x, 0))
6  print("f'_limit=", sp.limit(f.diff(), x, 0),           # 1st step: f'/g'
7         " g'_limit=", sp.limit(g.diff(), x, 0))
8  print('f"_limit=', sp.limit(f.diff(x,2), x, 0),        # 2nd step: f"/g"
9         ' g"_limit=', sp.limit(g.diff(x,2), x, 0))
10 print('(f"/g")_limit=', sp.limit(f.diff(x,2)/g.diff(x,2), x, 0))
```

```
f_limit=  0  g_limt= 0
f'_limit= 0  g'_limit= 0
f"_limit= 0  g"_limit= 2
(f"/g")_limit= 0
```

The solution is found via two steps, and in this case, $\lim_{x \to a} \frac{f(x)}{g(x)} = \lim_{x \to a} \frac{f''(x)}{g''(x)} = \frac{0}{2} = 0$. The reasons are the same as in the previous case.

In summary, L'Hôpital's rule is a powerful tool for finding the limit of the ratio of two functions. It is also useful for finding the derivative of a complicated function using Eq. (2.1), where an evaluation of the limit of the ratio of two expressions is required.

2.7 Autograd for differentiations

2.7.1 *Numerical but exact differentiation*

In computational methods, we frequently need to evaluate the derivatives of various types of functions. The formulas presented above are analytical and useful for analyzing elementary functions.

For complicated functions consisting of a large number of chained composite elementary functions, we require practical and efficient methods to compute derivatives quickly, following the rules of differentiation, including the chain rule. Such applications are typical in training ML models [5]. Methods called automatic differentiation have become extremely useful. In NumPy, these methods are implemented in modules such as **autograd** or similar to compute derivatives of chained composite functions. The results will be **numerical but exact** (to machine accuracy) as long as each of the sub-functions in the chain has an analytical formula for its derivative. For more detailed discussions, see Ref. [5] or documentation on various autograd modules for ML models. Autograd can be applied to any other applications, not just in ML. The following provides a simple example to demonstrate how autograd is typically performed.

Note that the term "gradient" of a function is more appropriate for functions in higher dimensions. It includes all the partial derivatives of a function with respect to each of the dimensional coordinates. In this chapter, which deals only with functions in one dimensional (1D) coordinates, the gradient and derivative carry exactly the same meaning. Thus, we use them interchangeably.

2.7.2 *Example: Use of autograd*

```
1   # Using numpy to compute the numerical value of gradient directly.
2
3   import autograd.numpy as anp                    # Thinly-wrapped numpy
4   from autograd import grad
5
6   def f_xsin(x):                      # Define a chained composite function.
7       f1 = (0.8*x - 5.0)                 # There is no limit on number of
8       f2 = anp.sin(f1)              # functions be chained in a composite form,
9       f  = x*f2                          # and hence can be very complicated.
10      return f
11
12  gf_xsin = grad(f_xsin)          # compute the gradient (derivative for 1D)
13  gf_xsin                              # print it out as an pyhon object
```

```
<function autograd.wrap_util.unary_to_nary.<locals>.nary_operator.<locals>.nar
y_f(*args, **kwargs)>
```

It is evident that the derivative of a function (object) is generated as a new function using NumPy's **grad()**. This function acts as a sophisticated tool for computing the derivative at a specified x.

To retrieve the value of this derived function at a specific point, such as $x = 5.0$, one simply invokes the derivative function with the desired argument:

```
1  gf_value = gf_xsin(0.5)
2  gr.printx('gf_value')
```

```
gf_value = 0.9488299928594425
```

Note that the value obtained is exactly the same as when using the analytic formula in SymPy. One can compute the derivative's values at multiple locations simply by using a for-loop:

```
1  for x in np.linspace(0., 10., 3):          # values at more locations
2      print(f'The value of df/dx at x={x}: {gf_xsin(x):.4f}')
```

```
The value of df/dx at x=0.0: 0.9589
The value of df/dx at x=5.0: 1.3197
The value of df/dx at x=10.0: -7.7788
```

We can also simply use a one-line list comprehension:

```
1  list_gf = [gf_xsin(x) for x in np.linspace(0., 10., 5)]
2  gr.printl(list_gf, 'df/dx values:')
```

```
df/dx values: [0.9589, -2.1211, 1.3197, 4.0833, -7.7788]
```

Note again that the results obtained using autograd are numerical but exact (to machine accuracy). This method will be used frequently in our later studies, including the following chapter which deals with a complex model in higher dimensions.

2.8 Implicit differentiation

In many applications, a function $y = f(x)$ is given implicitly in the form of an expression $F(x, y) = 0$. In these cases, we can perform implicit differentiation to obtain the required derivatives. The **basic idea** is to treat each term in $F(x, y)$ as a composite function and perform differentiation using the rules presented earlier.

2.8.1 *Example: Tangent and curvature of circles*

For example, the function for a circle is given as

$$F(x, y) = y^2 + x^2 - R^2 = 0 \qquad (2.57)$$

where R is a given constant (radius).

If our interest is in the slope at a point on the circle in Cartesian coordinates, we can perform implicit differentiation using the rules of differentiation. Let us take the first derivative with respect to each of the terms in the expression in Eq. (2.57):

$$2y\frac{dy}{dx} + 2x = 0 \qquad (2.58)$$

Simple manipulations give the tangent at any point on the circle:

$$\frac{dy}{dx} = -\frac{y}{x} \qquad (2.59)$$

Clearly, it is negative in the first and third quadrants, and positive in the second and fourth quadrants. All are correctly obtained automatically.

Furthermore, we may ask for the second derivative of the circle, which relates to the curvature. This can also be calculated implicitly again in exactly the same manner using Eq. (2.59):

$$\frac{d^2y}{dx^2} = -\frac{y\frac{dy}{dx} - x\frac{dy}{dx}}{y^2} = -\frac{y^2 + x^2}{y^3} \qquad (2.60)$$

At any point on the upper part of the circle, y is positive, and hence the curvature is negative (curving down). At any point on the lower part, y is negative, and hence the curvature is positive (curving up), as expected. Implicit differentiation has conveniently given us both the first and second derivatives of a circle.

Alternatively, for this problem, we have the option to first solve Eq. (2.57) for $y = \pm\sqrt{R^2 - x^2}$ and then calculate the derivatives as usual, but this method is a little more cumbersome.

Implicit differentiation allows for the direct computation of derivatives of a function that is implicitly given in an expression. There is no need to first solve the expression for the function and then differentiate it.

2.8.2 *Example: Implicit differentiation, unsolvable expression*

In more general cases, we may not be able to solve the equation $F(x, y) = 0$ explicitly for y. Let's consider an example examined in Ref. [4]:

$$F(x, y) = y^5 + xy - 3 = 0 \tag{2.61}$$

Clearly, we cannot solve this expression explicitly for y; it requires a rather complicated procedure, as shown in Ref. [1].

Let's obtain the derivatives without solving Eq. (2.61) for y using implicit differentiation. This time, we will use SymPy to demonstrate how this can be done for complicated expressions with ease:

```
1  x = symbols('x')              # define independent variable
2  y = Function('y')(x)          # define the function y=f(x)
3  F = y**5 + x*y -3             # define the implicit expression F(x,y)
4  dFdx = F.diff(x)              # derivative of F(x,y)
5  dFdx
```

$$x\frac{d}{dx}y(x) + 5y^4(x)\frac{d}{dx}y(x) + y(x)$$

We see $\frac{dy}{dx}$ in the output expression, and hence we can solve the expression for it. To do so, we shall first locate the term $\frac{dy}{dx}$ in expression. This is done using sp.args:

```
1  dFdx.args
```

$$\left(x\frac{d}{dx}y(x),\ 5y^4(x)\frac{d}{dx}y(x),\ y(x)\right)$$

SymPy organizes all terms into a tuple, with the derivative located in the first element of the tuple. Let's delve deeper into this:

```
1  dFdx.args[0].args
```

$$\left(x,\ \frac{d}{dx}y(x)\right)$$

The term $\frac{dy}{dx}$ is the second element of this tuple. Let us assign it to a new variable $dydx$ and then solve the expression $dFdx$ for $dydx$:

```
1  dydx = dFdx.args[0].args[1]     # get the term dy/dx in SymPy variables
2  dy_dx = sp.solve(dFdx, dydx)[0]           # find the formula for dy/dx
3  display(Math(f" \\frac{{dy}}{{dx}} = {latex(dy_dx)}"))
```

$$\frac{dy}{dx} = -\frac{y(x)}{x + 5y^4(x)}$$

We can now compute the value of $\frac{dy}{dx}$ by simple substitutions:

```
1  dy_dx_ = dy_dx.subs({x:2, y:1})   # find the values at for given x and y
2  dy_dx_                  # we use _ to indicate the variable is evaluated
```

$$-\frac{1}{7}$$

One can pursuit further for higher derivatives:

```
1  d2y_dx2 = dy_dx.diff(x)                    # comput the 2nd derivative
2  display(Math(f" \\frac{{d^2y}}{{dx^2}} = {latex(d2y_dx2)}"))
```

$$\frac{d^2y}{dx^2} = -\frac{\frac{d}{dx}y(x)}{x + 5y^4(x)} - \frac{\left(-20y^3(x)\frac{d}{dx}y(x) - 1\right)y(x)}{(x + 5y^4(x))^2}$$

```
1  (d2y_dx2.subs(dydx,dy_dx_)).subs({x:2, y:1})      # values at given x & y
```

$$-\frac{6}{343}$$

For complicated expressions, the above procedure is very useful. Most importantly, it avoids manual derivations by hand, which are prone to errors.

2.8.3 *Example: Implicit differentiation, complicated expression*

Let us look at a more complicated implicit expression:

$$F(x, y) = y^5 \sin(y) + x \cos(y) + x^2 = 0 \qquad (2.62)$$

Clearly, there is no hope that we can solve this expression explicitly for y. We obtain the derivatives via implicit differentiation using SymPy:

```
1  x = symbols('x')                       # define independe variable
2  y = Function('y')(x)                          # define function y
3  F = y**5*sp.sin(y) + x*sp.cos(y) + x**2     # define the expression
4  dFdx = F.diff(x)
5  dFdx
```

$$- x \sin(y(x)) \frac{d}{dx} y(x) + 2x + y^5(x) \cos(y(x)) \frac{d}{dx} y(x)$$

$$+ 5y^4(x) \sin(y(x)) \frac{d}{dx} y(x) + \cos(y(x))$$

Let us find this derivative term in the expression by quiring its arguments:

```
1  dFdx.args
```

$$\left(2x, \ y^5(x) \cos(y(x)) \frac{d}{dx} y(x), \ -x \sin(y(x)) \frac{d}{dx} y(x), \right.$$

$$\left. 5y^4(x) \sin(y(x)) \frac{d}{dx} y(x), \ \cos(y(x)) \right)$$

It is inside the second item of this tuple. We require further the following:

```
1  dFdx.args[1].args
```

$$\left(y^5(x), \ \frac{d}{dx} y(x), \ \cos(y(x)) \right)$$

```
1  # It is in the 2nd term of the tuple:
2  dFdx.args[1].args[1]                    # get the derivative term
```

$$\frac{d}{dx} y(x)$$

We find the derivative term and can now solve the expression dFdx for it:

```
1  dydx = dFdx.args[1].args[1]    # get the term dy/dx in SymPy variables
2  dy_dx = sp.solve(dFdx, dydx)[0]          # find the formula for dy/dx
3  display(Math(f" \\frac{{dy}}{{dx}} = {latex(dy_dx)}"))
```

$$\frac{dy}{dx} = \frac{2x + \cos(y(x))}{x \sin(y(x)) - y^5(x) \cos(y(x)) - 5y^4(x) \sin(y(x))}$$

We have successfully found it and can further compute its value at a point:

```
1  dy_dx.subs({x:2, y:1}).evalf(8)    # find the values at given x and y
```

$$-1.4814761$$

One can pursue further for higher derivatives. The code is as follows:

```
1  #d2y_dx2 = dy_dx.diff(x)              # This computes the 2nd derivative
2  #d2y_dx2                          # but the obtained formula is very long.
```

These two examples above show the power of implicit differentiation, which is useful for handling complicated functions, especially those defined implicitly. SymPy can effectively accomplish this task.

2.9 Derivatives of inverse functions

We discussed inverse functions in great detail in Ref. [1], including how to find the inverse function of a given function, possible multiple solutions, and changes of domain and codomain for the inverse function. Let us now study the derivative of inverse functions. First, we look at the simplest linear function written as

$$y = kx + b \qquad (2.63)$$

Its derivative is k. The corresponding inverse function can be found by simply solving Eq. (2.63) for x, which gives

$$x = \frac{y - b}{k} \qquad (2.64)$$

2.9.1 *Line of symmetry*

If we rewrite the inverse function obtained in Eq. (2.64) as $y = \frac{x-b}{k}$ and plot x versus y together with the original function $y = kx - b$, these two functions are symmetric with respect to the line $y = x$. This symmetry can be easily demonstrated using the following code:

```
1  plt.figure(); plt.ioff()
2  mpl.rcParams.update({'font.size': 8})
3  fig, ax = plt.subplots(1, 1)#, figsize=(6,3))
4  colors = ['r', 'b', 'g']
5
6  def f(x):
7      return lambda k,b,: k*x+b              # f(x) with parameters k and b
8
9  def f_inv(x):
10     return lambda k,b,: (x-b)/k                   # inverse of f(x)
11
12 x = np.arange(-5, 5, .1)                       # domain of x
13 k = np.array([2., 4.])                    # a few parameters k
14 b = 1
15
```

```
16  ax.plot(x, x, "k", linestyle='-.', label="Line of symmetry, y=x, k=1")
17
18  for i in range(len(k)):
19      y = f(x)(k[i], b)
20      ax.plot(x, y, c=colors[i], label="f, slope="+str(k[i]))
21      y_inv = f_inv(x)(k[i], b)
22      ax.plot(x, y_inv, c=colors[i], linestyle='--', \
23              label="f_inv, slope="+str(1/k[i]))
24
25  ax.set_xlabel('x'); ax.set_ylabel('f  and  f_inv ')
26  ax.grid(color='r', linestyle=':', linewidth=0.5)
27  ax.axis('equal'); ax.set_xlim(-5, 5); ax.set_ylim(-5, 5)
28  ax.legend()
29  plt.savefig('imagesDI/f_inv_kx-b.png', dpi=500)
30  #plt.show()
```

Figure 2.9 plots two linear functions with different slopes, along with their inverse functions, which are also linear. Each function and its inverse are symmetric with respect to the line $y = x$, which is a dash-dot line at a $45°$ angle to the x-axis.

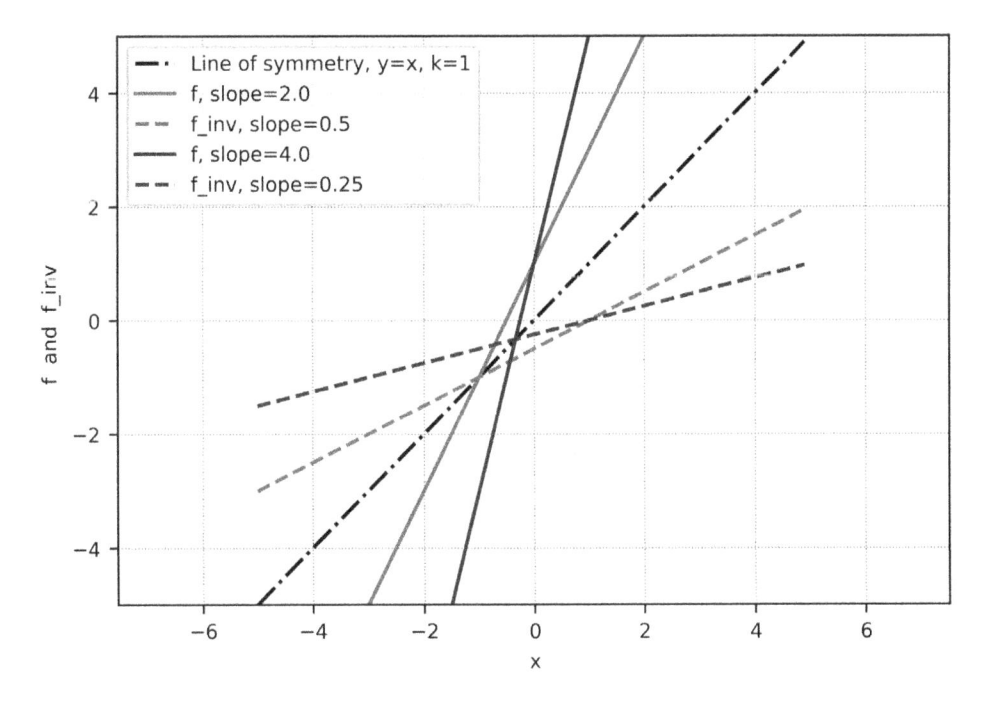

Figure 2.9. Linear functions and their corresponding inverse functions.

2.9.2 *Unity property*

The derivative of the inverse function can be obtained using Eq. (2.64):

$$\frac{dx}{dy} = \frac{1}{k} \qquad (2.65)$$

This simple example reveals the important unity property:

$$\underbrace{\frac{dy}{dx}}_{\text{derivative of the function}} \qquad \underbrace{\frac{dx}{dy}}_{\text{derivative of the inverse function}} \qquad = k\frac{1}{k} = 1 \qquad (2.66)$$

This means the following:

> The product of the derivative of a function and that of its inverse function equals one. These two derivatives are mutually inverse.

This implies that if we know the derivative of a function, we immediately know the derivative of its inverse function through inversion. This leads to an important principle for inverse problems: If the forward problem is well posed, the corresponding inverse problem may be ill posed. Interested readers can refer to Ref. [15] for studies on inverse problems in mechanics.

Let's now explore some examples using SymPy to obtain the inverse functions of more complicated functions and their derivatives. We'll start with a SymPy code example for a linear function.

2.9.3 *Example: Inverse function and derivative, linear*

```
1  x, y, k, b = symbols('x, y, k, b', real=True)        # define variables
2  fx = k*x + b                              # define the given function f(x)
3  dydx = fx.diff(x)                                   # derivative of f(x)
4  print(f"The derivative of f(x) = {dydx}")
5
6  expr = y-k*x-b                          # define expression involving y and x
7  fy = sp.solve(expr, x)    # solve for x to obtain inverse func, f^{-1}(y)
8
9  dxdy = fy[0].diff(y)        # compute the derivative of the inverse func.
10 print(f"The derivative of f^(-1)(x) = {dxdy}")
11
12 print(f"The unity relation = {dydx*dxdy}")
```

```
The derivative of f(x) = k
The derivative of f^(-1)(x) = 1/k
The unity relation = 1
```

The results from the code confirm our discussion.

2.9.4 *Example: Inverse function and derivative, quadratic*

Consider a quadratic function:

$$y = x^2 + bx + c \tag{2.67}$$

where b and c are real numbers. It has one stationary point where $\frac{dy}{dx} = 0$, given by $x = -\frac{b}{2}$, at which the function reaches an extremum (maximum or minimum). This implies that for every y value on the curve defined by Eq. (2.67), there are two corresponding x values, including repeated values. This indicates that there exist two branches of the function — one on each side of the stationary point. Its inverse function will also have two branches. These branches can be determined by solving Eq. (2.67) for x, as demonstrated in the following SymPy code snippet:

```
1  x, y, b, c = symbols('x, y, b, c', real=True)        # define variables
2  fx = x**2 + b*x + c                          # define the given function f(x)
3  dydx = fx.diff(x)                                  # derivative of f(x)
4  display(Math(f" \\frac{{d}}{{dx}}({latex(fx)}) = {latex(dydx)}"))
5
6  expr = (y-c) - x**2 -b*x              # define expression involving y and x
7  fy = sp.solve(expr, x)   # solve for x to obtain inverse func, f^{-1}(y)
8  display(Math(f" \\text{{{$f(y)$ = }}{latex(fy)}"))
```

$$\frac{d}{dx}(bx + c + x^2) = b + 2x$$

$$f(y) = \left[-\frac{b}{2} - \frac{\sqrt{b^2 - 4c + 4y}}{2}, \quad -\frac{b}{2} + \frac{\sqrt{b^2 - 4c + 4y}}{2} \right]$$

We found two branches of the inverse function. One corresponds to $y(x)$ on the left side and the other on the right side of the stationary point. Let's consider the first branch and compute its derivative:

```
1  dxdy = fy[0].diff(y)       # derivative of the 1st branch of inverse func.
2  display(Math(f" \\text{{{$(f^{{-1}})'(y)$ = }}{latex(dxdy)}"))
3  print(f"The unity relation = {(dydx*dxdy).subs(x,fy[0])}")
```

$$(f^{-1})'(y) = -\frac{1}{\sqrt{b^2 - 4c + 4y}}$$

```
The unity relation = 1
```

Let us do the same for the second branch:

```
1  dxdy = fy[1].diff(y)      # derivative of the 2nd branch of inverse func.
2  display(Math(f" \\text{{$(f^{{-1}})'(y)$ = }}{latex(dxdy)}"))
3  print(f"The unity relation = {(dydx*dxdy).subs(x,fy[1])}")
```

$$(f^{-1})'(y) = \frac{1}{\sqrt{b^2 - 4c + 4y}}$$

```
The unity relation = 1
```

It is observed that the derivatives of these pairs all satisfy the unity property. This property is fundamentally based on the symmetry about $y = x$.

In the following, we provide SymPy code to plot both branches of the quadratic function alongside its inverse function's branches:

```
1  plt.figure(); plt.ioff()
2  mpl.rcParams.update({'font.size': 8})
3  fig, ax = plt.subplots(1, 1)#, figsize=(6,3))
4  colors = ['r', 'b', 'g']
5
6  def f(x):
7      return lambda b,c: x**2 + b*x + c     # f(x) with parameters b and c
8
9  def f_inv_left(x):       # inverse funtion w.r.t the left branch of f(x)
10     return lambda b,c: -b/2-np.sqrt(b**2-4.*c+4.*x)/2    #
11
12 def f_inv_right(x):      # inverse funtion w.r.t the right branch of f(x)
13     return lambda b,c: -b/2+np.sqrt(b**2-4.*c+4.*x)/2
14
15 x = np.arange(-3., 3., .05)  # x-domain for plotting the symmetric line
16 xf = np.arange(-3., 1., .02)            # x-domain for plotting f(x)
17 xf_inv = np.arange(-1., 3., .02)      # x-domain for plotting f_inv(x)
18 b = 2.1; c = 0.1         # set the parameters, one may use other values
19
20 ax.plot(x, x, "k", linestyle='-.', label="Line of symmetry, y=x, k=1")
21
22 ax.plot(xf, f(xf)(b, c), c='r', label="f, b="+str(b)+", c="+str(c))
23
24 ax.plot(xf_inv, f_inv_left(xf_inv)(b, c), c='r', linestyle='--', \
25         label="f_inv_left, b="+str(b)+", c="+str(c))
26
27 ax.plot(xf_inv, f_inv_right(xf_inv)(b, c), c='r', linestyle='--', \
28         label="f_inv_right, b="+str(b)+", c="+str(c))
29
30 ax.set_xlabel('x'); ax.set_ylabel('f  and  f_inv ')
31 ax.grid(color='r', linestyle=':', linewidth=0.5)
32 ax.axis('equal'); ax.set_xlim(-3, 3); ax.set_ylim(-3, 3)
33 ax.legend(loc='lower right', bbox_to_anchor=(1, 0.2))
34 plt.savefig('imagesDI/quadraticInv.png', dpi=500)
35 #plt.show()
```

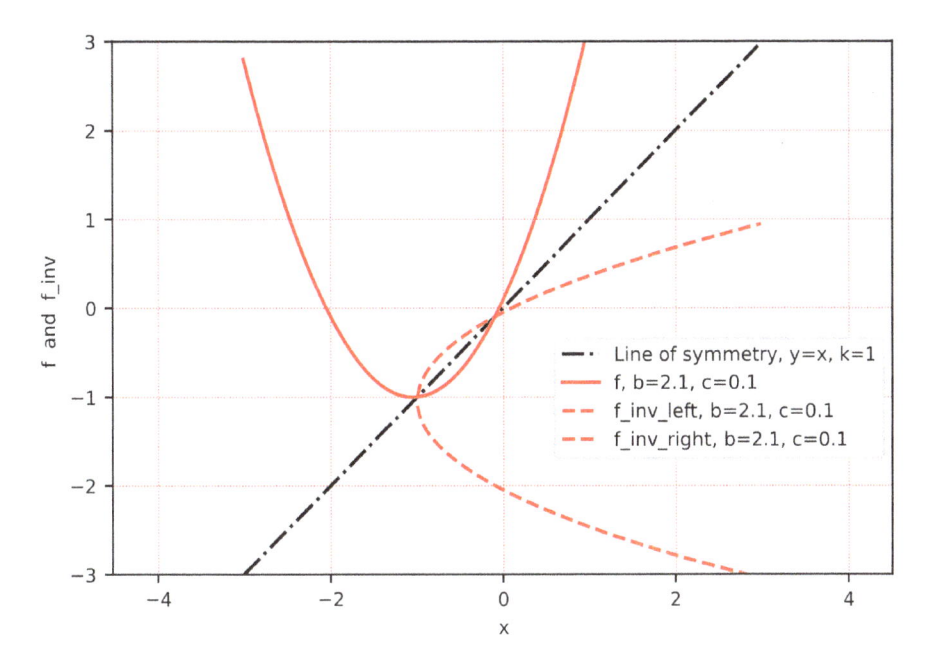

Figure 2.10. A quadratic function and its corresponding inverse function.

The quadratic function and its corresponding inverse function are plotted in Fig. 2.10. They are symmetric with respect to $y = x$.

2.9.5 *Example: Inverse function and derivative, trigonometric*

As discussed in Ref. [1], the arcsine function is the principal inverse of the sine function, and vice versa. Let us derive this relationship using SymPy and plot the curves of both mutually inverse functions. Due to the bounded nature of the sine function, our study focuses on the domain $(-1, 1)$:

```
1  x, y = symbols('x, y', real=True)          # define variables
2  fx = sp.sin(x)                             # define the given function f(x)
3  dydx = fx.diff(x)                           # derivative of f(x)
4  display(Math(f" \\frac{{d}}{{dx}}({latex(fx)}) = {latex(dydx)}"))
5
6  expr = y-sp.sin(x)               # define expression involving y and x
7  fy = sp.solve(expr, x)    # solve for x to obtain inverse func, f^{-1}(y)
8  display(Math(f" \\text{{{$(f^{{-1}})(y)$ = }}{latex(fy)}"))
```

$$\frac{d}{dx}(\sin(x)) = \cos(x)$$

$$(f^{-1})(y) = [\pi - \text{asin}\,(y),\ \text{asin}\,(y)]$$

We found two solutions. Due to the periodicity of the trigonometric functions. Let us choose the second one:

```
1  dxdy = fy[1].diff(y)        # compute the derivative of the inverse func.
2  display(Math(f" \\text{{$(f^{{-1}})'(y)$ = }}{latex(dxdy)}"))
3  print(f"The unity relation = {(dydx*dxdy).subs(x,fy[1])}")
```

$$(f^{-1})'(y) = \frac{1}{\sqrt{1-y^2}}$$

```
The unity relation = 1
```

We observe again the unity property for these derivatives. Let us plot these curves for easy examination:

```
1  plt.figure(); plt.ioff()
2  fig, ax = plt.subplots(1, 1)#, figsize=(6,3))
3
4  xf = np.arange(-3, 3., .02)                      # domain of x for f(x)
5  xf_inv = np.arange(-1, 1., .02)             # domain of x for f_inv(x)
6
7  ax.plot(xf, xf, "k", linestyle='-.', label="Line of symmetry, y=x, k=1")
8
9  ax.plot(xf, np.sin(xf), c='r', label="sin(x)")
10 ax.plot(xf_inv,np.arcsin(xf_inv),c='r',linestyle='--',label="arcsin(x)")
11
12 ax.set_xlabel('x'); ax.set_ylabel('sin(x)  and  arcsin(x)')
13 ax.grid(color='r', linestyle=':', linewidth=0.5)
14 ax.axis('equal'); ax.set_xlim(-1, 1); ax.set_ylim(-1.5, 1.5)
15 ax.legend()
16 plt.savefig('imagesDI/sineInv.png', dpi=500)
17 #plt.show()
```

The sine function and its corresponding inverse function are plotted in Fig. 2.11. They are symmetric with respect to $y = x$.

As discussed in Ref. [1], the logarithmic and exponential functions are mutual inverses. Their curves are plotted in Fig. 4.34 in Ref. [1]. Their derivatives also satisfy the unity property.

These three examples above demonstrate how straightforward it is to find an inverse function using SymPy once the concept is established.

2.10 Applications of derivatives of functions

2.10.1 *Linearization of a function locally*

2.10.1.1 *Formulation*

The derivative of a given function $f(x)$ can be used to approximate the function locally with a straight line. The slope of this straight line is the local

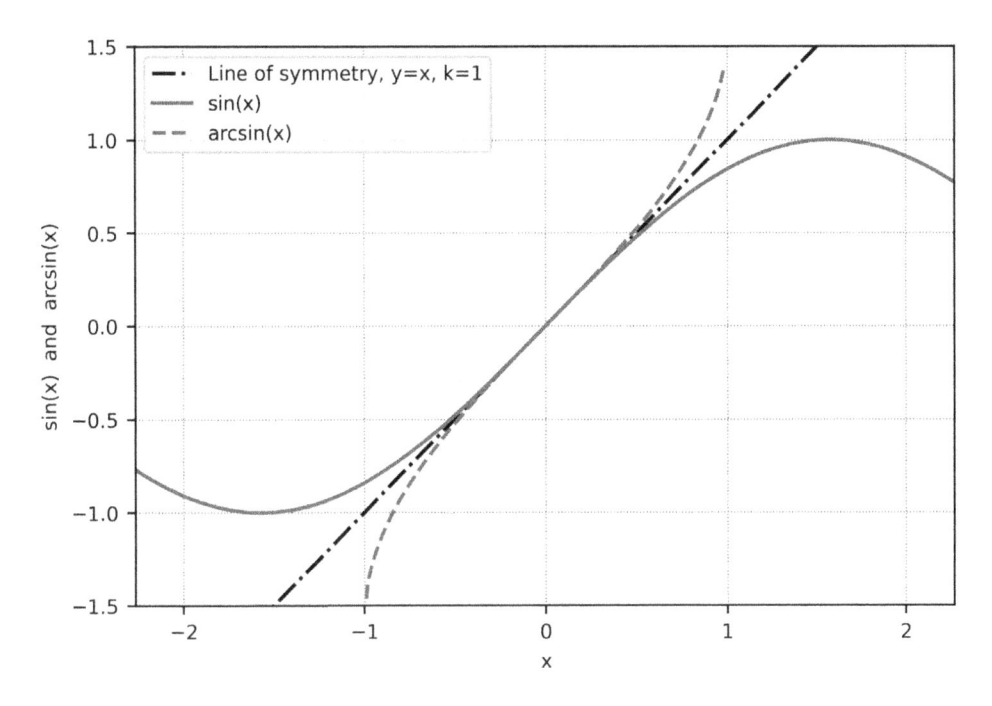

Figure 2.11. The sine function and its corresponding inverse function.

derivative of the function at a point. This process is known as **linearization of a function**. To approximate $f(x)$ around $x = a$, where the function is differentiable, the formula is

$$f_{\text{linearized}}(x) \approx f(a) + \left.\frac{df}{dx}\right|_{x-a} (x - a) \tag{2.68}$$

Moving $f(a)$ to the left side, this equation can be expressed simply as follows:

$$\Delta f \approx \frac{df}{dx} \Delta x \tag{2.69}$$

where Δx is a small change in around x at which $\frac{df}{dx}$ is computed. The change in the function Δf is estimated using Eq. (2.69).

Let us write the following Python code to approximate the sine function at $x = 0$. Readers may use it for other functions.

2.10.1.2 *Python code for linearization of functions*

The following code approximates a function in the vicinity of a point using a straight line. The slope of this line is obtained by computing the derivative of the function at that point. In this example, we utilize NumPy and autograd to compute the derivative numerically:

```
1  import autograd.numpy as anp                        # Thinly-wrapped numpy
2  from autograd import grad
3
4  plt.figure(); plt.ioff()
5  plt.rcParams.update({'font.size': 8})
6  fig, ax = plt.subplots(1,1) #,figsize=(6,4.5))
7  a = 0.0 # 0.5                      # at x = a, the function is linearized
8  xL, xR = -1.2+a, 1.2+a                  # domain for the approximation
9
10 n_x = 1000      # number of points for ploting the functions on x axis
11 X = np.linspace(xL, xR, n_x)          # sample x in the domain for plots
12
13 # Plot the original curve and linearized line:
14 ax.set_xlabel('x')
15 ax.grid(color='r', linestyle=':', linewidth=0.5)
16 f  = anp.sin                              # function to be linearized
17 gf = grad(f)                   # gradient (derivative) of the function
18 f_linearized = f(a)+gf(a)*(X-a)
19 ax.plot(X, f(X), c="k", lw=2., label="Original function $sin(x)$")
20 ax.plot(X, f_linearized, "r:", lw=2., label=f"Linearized around x={a}")
21
22 ax.axvline(x=0, c="k", lw=0.6); ax.axhline(y=0, c="k", lw=0.6)
23 ax.legend(loc='center left', bbox_to_anchor=(.2, .8))
24 #ax.set_ylim(-1.1, 1.1)
25
26 plt.savefig('imagesDI/sin_linearized.png', dpi=500)
27 #plt.show()
```

It is evident from the plots above that the sine function is well approximated near $x = 0$. The accuracy holds within the interval $[-0.3, 0.3]$, as illustrated in Fig. 2.12. However, the approximation deviates beyond this range. Therefore, a practical approach is to locally linearize the function at various points, which is known as **piecewise linearization**.

Since linearization is generally effective only in the vicinity of the point where the derivative is computed, achieving accurate approximation requires computing derivatives at multiple locations.

2.10.1.3 *Local linearization of functions at locations*

The following code linearizes function $f(x) = x\sin(5x) + 1$ locally at multiple locations:

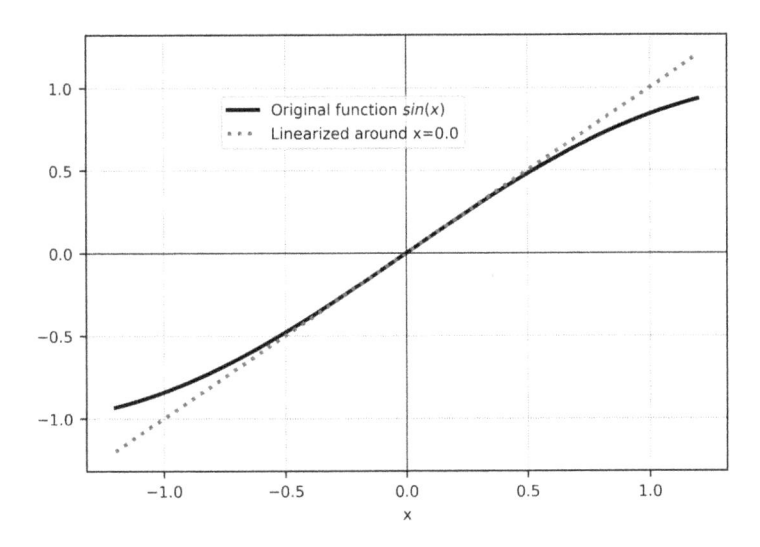

Figure 2.12. The sine function and its linearized function near $x = 0$.

```
1  import autograd.numpy as anp              # Thinly-wrapped numpy
2  from autograd import grad
3
4  plt.figure(); plt.ioff()
5  plt.rcParams.update({'font.size': 8})
6  fig, ax = plt.subplots(1,1) #,figsize=(6,4.5))
7  xL, xR = 0., 5.              # domain boundary for plotting the function
8
9  n_x = 500        # number of points for ploting the functions on x axis
10 X = np.linspace(xL, xR, n_x)         # sample x in the domain for plots
11
12 f  = lambda x: x*anp.sin(5*x) + 1          # define the function
13 gf = grad(f)                        # gradient (derivative) of the function
14
15 # Plot the function over the domain:
16 ax.set_xlabel('x')
17 ax.grid(color='r', linestyle=':', linewidth=0.5)
18 ax.plot(X, f(X), c="k", lw=1., label="Given function $x*sin(5*x)+1$")
19
20 na = 20; l = (xR-xL)/(3*na)
21 xa = np.linspace(xL+0.5, xR-0.5, na)             # sample x for plots
22
23 for a in xa:                           # plot these straight lines
24     X_local = np.linspace(a-l, a+l, 3)
25     f_local = f(a)+gf(a)*(X_local-a)  # compute derivative at locations
26     plt.plot(X_local, f_local, "b")
27
28 ax.scatter(xa, f(xa), c="r", s=8.)
29 ax.axvline(x=0, c="k", lw=0.6); ax.axhline(y=0, c="k", lw=0.6)
30 ax.legend(loc='upper left') #, bbox_to_anchor=(.2, .8))
31
32 plt.savefig('imagesDI/sin_local_df.png', dpi=500)
33 #plt.show()
```

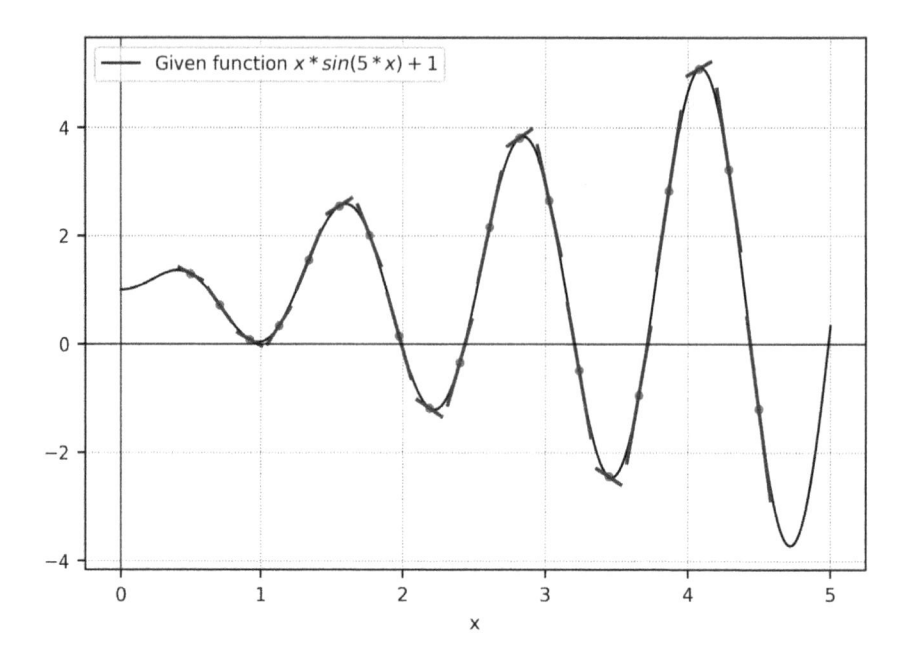

Figure 2.13. A function and its approximation at local points with linear functions.

The function is accurately approximated locally using linearization, as depicted in Fig. 2.13. To further enhance the accuracy of approximation, one can increase the number of points where linearization is applied, resulting in denser piecewise linear approximations. Since linear functions are simple and easy to manage, such piecewise linear approximations find utility in computational methods like the FEM [6] and the smoothed finite element method (S-FEM) [10].

On the other hand, Eq. (2.69) can also be written as

$$\frac{df}{dx} \approx \frac{\Delta f}{\Delta x} \tag{2.70}$$

This also provides the expression for approximating the first derivative of a function at x by taking a small Δx away from it and computing Δf.

2.10.2 *Quadratic approximation of functions at locations*

To further improve accuracy, higher-order approximations can be employed. When approximating $f(x)$ around $x = a$ using a second-order approximation, the formula is

$$f_{\text{approximated}}(x) \approx f(a) + \frac{df}{dx}\bigg|_{x=a}(x - a) + \frac{1}{2}\frac{d^2 f}{dx^2}\bigg|_{x=a}(x - a)^2 \tag{2.71}$$

The factor $\frac{1}{2}$ is necessary to ensure the equation holds after double differentiation. The following code demonstrates a quadratic approximation:

```python
1  plt.figure(); plt.ioff()
2  plt.rcParams.update({'font.size': 8})
3  fig, ax = plt.subplots(1,1) #,figsize=(6,4.5))
4  xL, xR = 0., 5.                        # domain for plotting the function
5
6  n_x = 500        # number of points for ploting the functions on x axis
7  X = np.linspace(xL, xR, n_x)           # sample x in the domain for plots
8
9  f   = lambda x: x*anp.sin(5*x) + 1             # define the function
10 gf = grad(f)                    # gradient (derivative) of the function
11 g2f = grad(gf)                      # 2nd derivative of the function
12
13 # Plot the function over the domain:
14 ax.set_xlabel('x')
15 ax.grid(color='r', linestyle=':', linewidth=0.5)
16 ax.plot(X, f(X), c="k", lw=1., label="Given function $x*sin(5*x)+1$")
17
18 na = 20; l = (xR-xL)/(3*na)
19 xa = np.linspace(xL+0.5, xR-0.5, na)               # sample x for plots
20
21 for a in xa:                            # plot these straight lines
22     X_local = np.linspace(a-l, a+l, 10)
23     f_local = f(a) + gf(a)*(X_local-a) + 0.5*g2f(a)*(X_local-a)**2
24     plt.plot(X_local, f_local, "b", lw=2.)
25
26 ax.scatter(xa, f(xa), c="r", s=8.)
27 ax.axvline(x=0, c="k", lw=0.6); ax.axhline(y=0, c="k", lw=0.6)
28 ax.legend(loc='upper left') #, bbox_to_anchor=(.2, .8))
29
30 plt.savefig('imagesDI/sin_local_quad.png', dpi=500)
31 #plt.show()
```

The quadratic approximation yields significantly improved accuracy, as illustrated in Fig. 2.14. Further enhancement can theoretically be achieved with higher-order approximations by modifying the aforementioned codes. However, caution is advised as higher-order approximations can lead to over-fitting and stability issues, as discussed in Ref. [1]. Therefore, we conclude our approximation study here.

2.10.3 *On approximation of complicated functions*

The previous examples highlight an important observation regarding improving the accuracy of approximating complex functions: There are two primary approaches. Firstly, dense local node placement, known as **h-convergence**, focuses on refining the discretization grid. Secondly, higher-order but sparser node placement, referred to as **p-convergence**, involves using fewer nodes

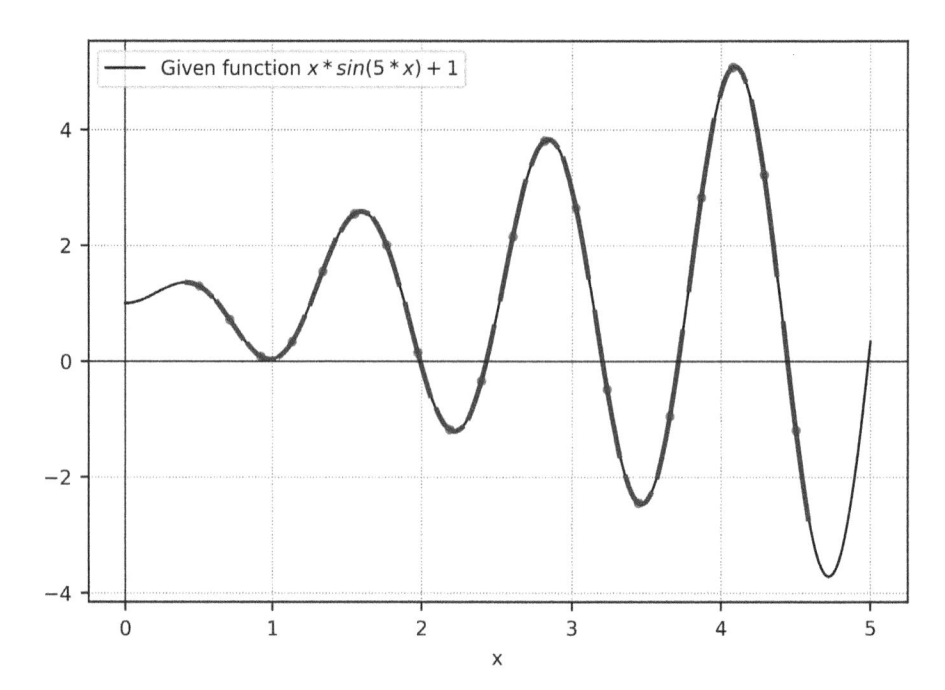

Figure 2.14.   A function and its approximation at local points with quadratic functions.

with higher-order approximations. Both approaches are integral to well-established FEM [6] and S-FEM [10] techniques.

In terms of convergence rate, p-convergence typically offers superior performance, while h-convergence excels in scalability and simplicity. From the author's perspective, the h-convergence approach is more commonly utilized. The prevailing trend favors h-convergence with dense discretization, employing simple basis functions in abundance, which often proves more robust and effective for addressing complex problems. This approach mirrors many FEM and S-FEM models as well as prevalent methodologies in ML [5], where the affine transformations tend to be predominantly linear.

2.10.4 *Newton iteration: Roots finding using the derivative*

One of the primary applications of the derivative of a function is to find roots x, such that

$$f(x) = 0 \qquad\qquad (2.72)$$

This is achieved using the widely adopted Newton iteration algorithm. The method applies to general functions provided

1. its derivative is available
2. the initial guess of the root is not hindered by a steep ascent or descent of the function.

An animation illustrating this process can be found here (https://en.wikipedia.org/wiki/Newton%27s_method#/), offering an excellent intuitive insight into the method.

The basic idea is schematically shown in Fig. 2.15. The function, shown in blue, has a root (marked by the red dot). At the nth iteration, the derivative of the function is computed at x_n, which represents the local slope of the tangent line (shown in red) at x_n. This tangent line, serving as the local linear approximation of the function, intersects the x-axis at x_{n+1} (black dot), which is closer to the target root.

Next, the derivative of the function is computed at x_{n+1} to find a new tangent line, resulting in another linear approximation. This new line intersects the x-axis at x_{n+2} (blue dot), which is even closer to the root. This process is repeated iteratively until the intersection point is sufficiently close to the root (red dot).

As we can see, the Newton iteration leverages the derivative of the function as a guide to quickly approach the root, leading to quadratic convergence. This makes it significantly faster than other methods that do not use derivatives, such as the Secant method.

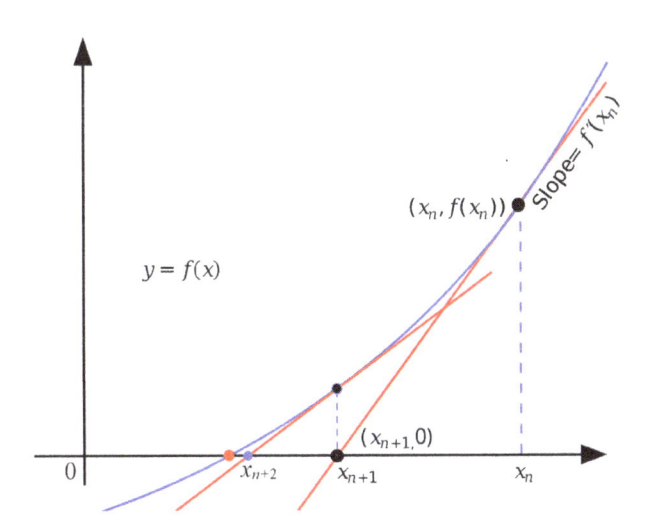

Figure 2.15. Schematic drawing of the Newton iteration to find a root of a function, via stages of linearization of the function.
Source: Image is modified based on the one in the public domain by Olegalexandrov (https://commons.wikimedia.org/).

The following section provides a Python code for a basic implementation of the Newton iteration for a given function and its derivative along with initial guesses.

2.10.4.1 *Python code for the Newton iteration*

```
 1  def newton_iter(f, df, guess_roots, limit=1000, atol=1e-6):
 2      ''' Finds the roots of a function via the Newton iteration. This is
 3          primitive code for proof of concepts.
 4          inputs: f--the given numpy function;
 5                  df--the given derivative of f;
 6                  guess_roots--possible locations of the roots;
 7                  limit: max number of interations;
 8                  atol:  tolerance to control convergence;
 9          return: the roots found.'''
10      i = 0;   roots = np.array([])
11      for r in guess_roots:
12          while lg.norm(f(r)) > atol and i < limit:
13              r = r - f(r)/df(r)
14              i += 1
15          roots = np.append(roots, r)
16
17      if lg.norm(f(r)) > atol:
18          print(f'Not yet converged:{atol}!!!')
19      else:
20          print(f'Converged at {i}th iteration to {atol}')
21
22      return roots
```

2.10.4.2 *Python code for finding multiple roots*

We know that a differentiable function implies continuity. A continuous function can have multiple roots, each separated by an ascent (or descent) due to its continuity. Therefore, an effective method is needed to isolate these roots. This can be achieved by providing quality initial guesses for the Newton iteration to find these possible roots one by one, which can be quite challenging. In this section, we propose an efficient algorithm using **dense sampling** and **sign detection**. The algorithm works as follows:

1. Based on the knowledge of the given function, compute the function values on sufficiently dense grids. This involves only function evaluations, which can be done quickly on modern computers, allowing for the use of very dense grids.
2. Compute the sign of the function values using **np.sign()**, a built-in NumPy function.

3. Identify locations where the sign changes using `np.where()`, another NumPy built-in tool. These locations become the initial guesses for the Newton iteration, allowing it to zoom in and accurately find the roots. The expensive Newton iteration is called only at locations near which there is a root.

The Python code for this algorithm is given as follows:

```
 1  def roots_finder(f, df, xL, xR, n_x, limit=1000, atol=1e-6):
 2      '''Find the roots of a given function f and its derivative df
 3         using the Newton's method, in a given domain [xL, xR].
 4      f:       the given numpy function;
 5      df:      the derivative of f (numpy):
 6      n_x:     number of points for sampling the domain (resolution)
 7      limit:   max number of interations in the Newton method.
 8      atol:    tolerance to control convergence.
 9      return:  f_value--the values of the function;
10               df_value--the values of the derivative of the funcion;
11               roots--roots of the function in the domain;
12               dfr--the derivative values at these roots.'''
13
14      X = np.linspace(xL, xR, n_x)        # discretization of the interval
15
16      # Step 1: compute the values of f, and df (optional)
17      f_value = np.array([])
18      # df_value = np.array([]) # optional
19      for x in X:
20          f_value = np.append(f_value , f(x) )
21          #df_value= np.append(df_value, df(x))               # optional
22
23      # Step 2: compute the sign of f
24      signs = np.sign(f_value)
25
26      # Step 3: find the indices where the sign changes, which
27      # gives the intial guesses of the roots.
28      indices = np.where(np.diff(signs))[0] + 1
29
30      # Step 4: use Newton iteration to find the roots near the guesses:
31      roots = newton_iter(f, df, X[indices], limit=1000, atol=1e-6)
32
33      df_roots = np.array([])  # compute df values at the roots #optional
34      for x in roots:
35          df_roots = np.append(df_roots, df(x))
36
37      return f_value, roots, df_roots, indices
```

2.10.4.3 *Example: Finding multiple roots of functions*

The function f_xsin() created earlier has multiple roots. Let's locate and plot these roots using our Python code:

```
1  import autograd.numpy as anp              # Thinly-wrapped numpy
2  from autograd import grad
3
4  def f_xsin(x):                            # Define a chained composite function
5      f = x*anp.sin(0.8*x - 5.0)
6      return f
7
8  gf_xsin = grad(f_xsin)          # compute the gradient (derivative for 1D)
```

```
1  plt.figure(); plt.ioff()
2  plt.rcParams.update({'font.size': 5})
3  fig, ax = plt.subplots(1,1,figsize=(4,2))
4  xL, xR = 0., 50.
5  n_x = 200                                 # number of points on x axis
6  X = np.linspace(xL, xR, n_x)                            # for plot
7
8  # find the roots of the function, using Newton's method
9  fv, roots, _, indices = roots_finder(f_xsin, gf_xsin,
10                          xL, xR, n_x, limit=1000, atol=1e-6)
11
12  print(f'Total number of roots found: {len(roots)}; the first 10 are:')
13  gr.printl(roots[:10])
14  ax.set_xlabel('x')
15  ax.set_title('f(x)=x*sin(0.8*x - 5.0) and its roots')
16  ax.grid(color='r', linestyle=':', linewidth=0.5)
17  ax.plot(X, f_xsin(X), label="$f(x)$")
18  ax.scatter(roots,[0]*len(roots),c='r',s=10,label="Roots found for f(x)")
19  ax.legend()              #loc='center right', bbox_to_anchor=(1, 0.5))
20
21  plt.savefig('imagesDI/roots_xsinkx_b.png', dpi=500)
22  #plt.show()
```

```
Converged at 30th iteration to 1e-06
Total number of roots found: 14; the first 10 are:
[0.0, 2.323, 6.25, 10.177, 14.104, 18.031, 21.958, 25.885, 29.8119, 33.7389]
```

This code finds, in less than one second, all the 14 roots in the interval for the function, which is a composite of polynomials and the sine function, as shown in Fig. 2.16.

2.10.4.4 *Use of Python built-in method for roots finding*

NumPy includes built-in functions for root finding. As an example, we demonstrate using the following function, which is often encountered in structural mechanics problems:

```
1  def beamOnFoundation(bz):              # Deflection of a beam on foundation
2      return np.exp(-bz)*(np.sin(bz)+np.cos(bz))
3
4  def f(x):                      # a composite function whoes root to be found
5      return 2*beamOnFoundation(x/2)-1-beamOnFoundation(x)
```

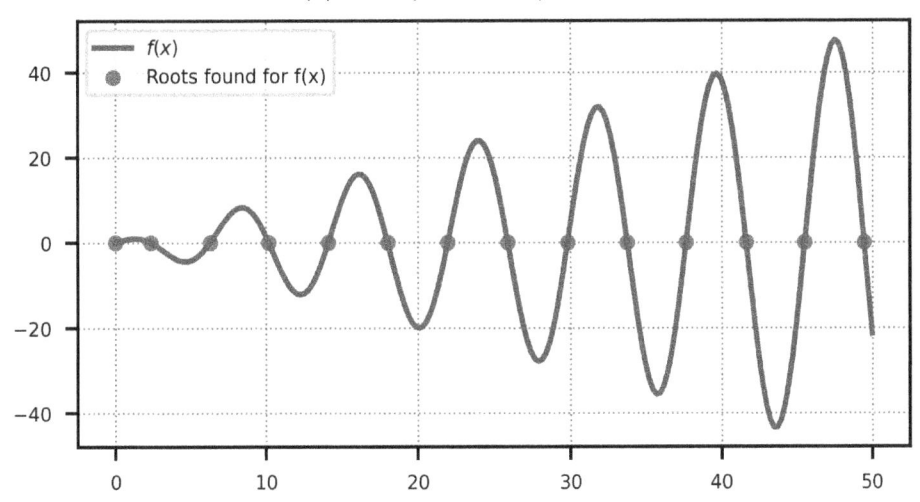

Figure 2.16. Roots found for a composite function using the present algorithm that uses dense sampling and sign-change detection for quality initial guesses and the Newton iteration.

```
1  from scipy.optimize import fsolve
2  root_guess = 5.                      # specify guess location of the root
3  x_root = fsolve(f, root_guess)
4  print(f"x_root={x_root}")
5  np.isclose(f(x_root), 0)             # check if f(x_root)=0.0.
```

```
x_root=[1.8585e+00]
```

```
array([ Truc])
```

As seen, `scipy.fsolve()` also requires an initial guess. Since we have an algorithm for quality initial guesses, it can be used together with `scipy.fsolve()` to find all roots for `f_xsin()`. We use the indices obtained for this function earlier as initial guesses.

```
1  x_root = [fsolve(f_xsin, x)[0] for x in X[indices]]
2  gr.print1(x_root[0:10], "The first 10 roots found for f_xsin():\n")
3  np.isclose(f_xsin(np.array(x_root)), 0)    # check if f(x_root)=0.0.
```

```
The first 10 roots found for f_xsin():
 [0.0, 2.323, 6.25, 10.177, 14.104, 18.031, 21.958, 25.885, 29.8119, 33.7389]
array([ True,  True,  True,  True,  True,  True,  True,  True,  True,
        True,  True,  True,  True,  True])
```

The roots found are exactly the same as we found earlier using our hand-crafted algorithm. This is largely because we have a quality guess algorithm.

2.10.5 *Finding vibration modes of structure components*

2.10.5.1 *Roots of characteristic equations*

The behavior of many systems in engineering is often controlled by characteristic equations. The derivation of such an equation depends on the discipline. To examine the behavior of a system, one needs to find the roots of the characteristic equation, which is a typical root-finding problem. The following example demonstrates finding the roots of the characteristic equation governing the free vibration of a beam, allowing us to obtain the natural frequency and the corresponding shape of each vibration mode.

2.10.5.2 *Example: Vibration modes of cantilever beams*

The normalized shapes of the vibration modes of a cantilever beam can be found in textbooks on vibrations or in the online article, http://emweb.unl.edu/Mechanics-Pages/Scott-Whitney/325hweb/Beams.htm, can be expressed as follows:

$$\Phi(x) = \frac{1}{2}\left[(\cos k_n x - \cosh k_n x) - \frac{\cos k_n l + \cosh k_n l}{\sin k_n l + \sinh k_n l}(\sin k_n x - \sinh k_n x)\right]$$
(2.73)

where $\Phi(x)$ is the function representing the shape of the modes, n $(1, 2, \ldots)$ stands for the nth mode, and $k_n l$ is the nth characteristic value that satisfies the following characteristic equation:

$$\cos k_n l + \cosh k_n l + 1 = 0$$
(2.74)

where k_n is the wave number [16]. It is determined by the bending stiffness and line density of the beam.

$$k_n = \sqrt[4]{\frac{\omega_n^2 \rho A}{EI}}$$
(2.75)

where E is the Young's modulus of the material of the beam, I is the second moment of area of the cross-section of the beam, and ω_n is the angular natural frequency of the beam. Once k_n is found as a root of Eq. (2.74), ω_n can be found using Eq. (2.75):

$$\omega_n = k_n^2 \sqrt{\frac{EI}{\rho A}}$$
(2.76)

The mode shape given in Eq. (2.73) is a combination of elementary functions, including sine, cosine, hyperbolic sine, and hyperbolic cosine functions, which we have studied in Ref. [1].

We proceed by writing code to plot the mode shape. First, we define the characteristic equation (2.74). To find all its roots using the Newton iteration algorithm, we also need its derivative. We will again use autograd to compute this derivative:

```
1  def beam_cf_ch(x):       # characteristic equation for clapped-free beams
2      return anp.cos(x)*anp.cosh(x)+1.
3
4  gf_x = grad(beam_cf_ch)  # for compute the gradient (derivative for 1D)
```

We are now ready to find the roots of the characteristic equation, which gives all the $k_n l$ in a specified domain:

```
1  # find the roots, using proposed algorithm with Newton iteration:
2  fv, kls, _, _ = roots_finder(beam_cf_ch, gf_x,
3                          1, 20, 100, limit=1000, atol=1e-6)
4  print(kls)
```

```
Converged at 22th iteration to 1e-06
[1.8751e+00 4.6941e+00 7.8548e+00 1.0996e+01 1.4137e+01 1.7279e+01]
```

Next, we define the function that computes the shape of the modes for a beam:

```
1  def beam_cf(kl, l, x):
2      '''Compute the mode shape of a cantilever (clampped-free) beam
3      '''
4      kx = x*kl/l
5      return (np.cos(kx)-np.cosh(kx))/2-(np.sin(kx)-np.sinh(kx))*\
6              (np.cos(kl)+np.cosh(kl))/2/(np.sin(kl)+np.sinh(kl))
```

Let's compute and plot all the mode shapes:

```
1  N = 6                                # compute & plot the first 6 mode shape
2  modes = [n for n in range(1,N+1)]
3  Labels=[str(n)+' mode' for n in modes]    # ['1st mode','2nd ',...,Nth]
4  xL, xR, dx = 0., 1., 0.01            # range along the beam axis, for plots
5  X = np.arange(xL, xR+dx, dx)
6  l = 1.0                                             # Length of the beam
7
8  plt.figure(); plt.ioff()
9  plt.rcParams.update({'font.size': 6})
10 fig_s = plt.figure(figsize=(6,4))
11
12 for i, n in enumerate(modes):
13     ax = fig_s.add_subplot(3,2,i+1)
14     ax.plot(X, beam_cf(kls[i],l,X), 'b', lw=1.2,label=Labels[i])
15     ax.plot(X,-beam_cf(kls[i],l,X),'b:')
16     if i==4 or i==5: ax.set_xlabel('x')
17     if i==0 or i==2 or i==4: ax.set_ylabel('$\Phi(x)$')
18     ax.legend()
19
20 plt.savefig('imagesDI/VibModeBeam.png',dpi=500,bbox_inches='tight')
21 #plt.show()
```

The first six vibration modes of the cantilever beam are computed and plotted in Fig. 2.17. This example is quite typical in structural mechanics. It demonstrates how straightforward it is to obtain solutions once we have a

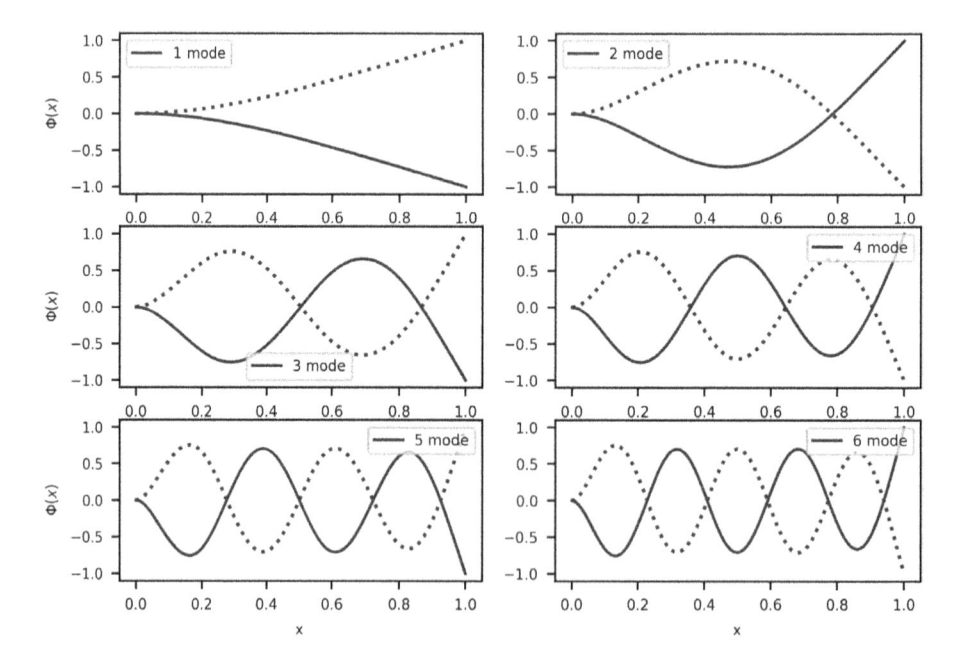

Figure 2.17.   Vibration modes of a beam clamped at its left end.

good understanding of functions and techniques for extracting their features. The tools available in Python and the algorithms provided here are both useful and effective.

2.10.6 *Finding extrema of functions*

2.10.6.1 *Stationary points of functions*

The derivative of a function is crucial for finding its extrema, as these points occur where the slope or tangent of the function is zero, known as **stationary points**. To find an extreme point, we look for roots of the derivative function, a process facilitated by the root-finding methods discussed in previous sections.

This process will be demonstrated clearly using the following example.

2.10.6.2 *Example: Derivative of trigonometric function and extrema*

The following code computes and plots a function that includes a sine function along with its first derivative. This function is differentiable across the entire real domain and exhibits multiple roots and extrema. It serves as an ideal example to illustrate the significant properties of function derivatives in relation to extreme values.

We will utilize autograd to compute both the first and second derivatives of the function:

```
1  # Define the function:
2  def fxsin(x):
3      f = x*anp.sin(x)/(2*np.pi)
4      return f
5
6  gfxsin = grad(fxsin)        # This is a new function of the 1st diff.
7  g2fxsin = grad(gfxsin)      # The diff of diff (the 2nd diff). It is
8                              # used to find the roots of 1st diff.
9  #gfxsin                     # Take a look at the 1st diff
```

The gfxsin obtained above is now the derivative function of the given function. Let us take a look at some of its values:

```
1  list_gf  = [gfxsin(x)  for x in np.linspace(0., 2.*np.pi, 4)]
2  list_g2f = [g2fxsin(x) for x in np.linspace(0., 2.*np.pi, 4)]
3  gr.printl(list_gf,  'd_fxsin/dx  values sampled at locations:')
4  gr.printl(list_g2f, 'd2_fxsin/dx values sampled at locations:')
```

```
d_fxsin/dx  values sampled at locations: [0.0, -0.0288, -0.4712, 1.0]
d2_fxsin/dx values sampled at locations: [0.3183, -0.4478, 0.4182, 0.3183]
```

Let's now use the following code to plot the curves of the function along with its first derivative, find roots and extreme values, and then mark these on the curves:

```
1  plt.figure(); plt.ioff()
2  plt.rcParams.update({'font.size': 9})
3  fig, ax = plt.subplots(1,1,figsize=(6,4.5))
4  xL, xR = 0.001, 5.6
5  n_x = 1000                              # number of points on x axis
6  X = np.linspace(xL, xR, n_x)
7
8  # compute the values of the functoin and its first derivative:
9  gfxsinX = np.array([gfxsin(x) for x in X])
10
11 # roots of df/dx via Newton iteration, hence 2nd derivative is needed:
12 fv, df_roots, _, indices = roots_finder(gfxsin, g2fxsin,
13                                 xL, xR, n_x, limit=1000, atol=1e-6)
14
15 # compute the values of f at the roots of df.
16 f_at_df_roots = np.array([fxsin(x) for x in df_roots])
17
18 ax.set_xlabel('x')
19 ax.grid(color='r', linestyle=':', linewidth=0.5)
20 ax.plot(X, fxsin(X), label="$f(x)$")
21 ax.plot(X, gfxsinX    , label="$df(x)/dx$")
22
23 ax.scatter(df_roots,[0]*len(df_roots),  c='r', s=20,
24            label="roots of $df(x)/dx$")
25 ax.scatter(df_roots, f_at_df_roots  ,   c='k', s=25, marker='*',
26            label="Extreme values of f(x)")
27
28 ax.axvline(x=0, c="k", lw=0.6); ax.axhline(y=0, c="k", lw=0.6)
29 ax.legend(loc='upper center', bbox_to_anchor=(0.68, 1.0))
30
31 plt.savefig('imagesDI/d_sin.png', dpi=500)
32 #plt.show()
```

Converged at 3rd iteration to 1e-06

As shown in Fig. 2.18, the derivative (orange curve) of the function has two roots in its domain. Each root corresponds to an extremum of the function: one maximum and one minimum (marked with star symbols). Readers can apply the same code to discover similar properties for other differentiable functions and identify all key values accordingly.

2.11 Higher derivatives of a function

As demonstrated in the examples above, the derivative of a function yields a new function. Naturally, we enquire about the derivative of this derivative.

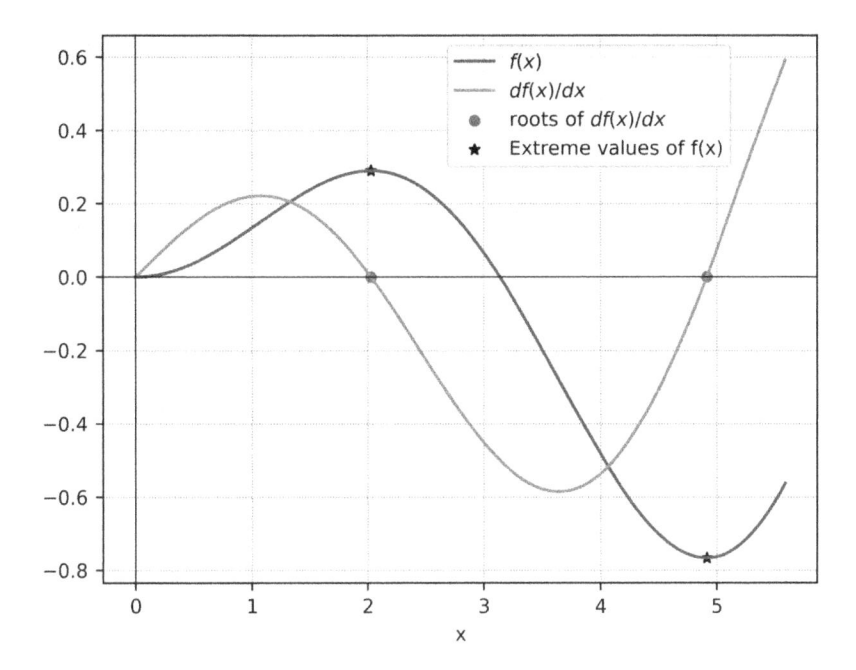

Figure 2.18. A given function and its derivative, roots and extrema.

In computational methods, understanding higher-order derivatives of a function, where applicable, is crucial. This knowledge allows us to explore local properties further using these higher derivatives.

The derivative of the derivative is termed the second derivative of the function. Similarly, the derivative of the second derivative is known as the third derivative, and so forth. All these higher derivatives can be obtained using Eq. (2.1), treating each lower derivative as a new function. Throughout this process, it's important to consider potential changes in the domain and codomain of these higher derivative functions.

2.11.1 *Example: Physical meaning of second derivative*

To deepen our understanding of the concept of higher derivatives, let's examine a familiar example involving the second-order derivative.

2.11.2 *A car accelerating at a constant rate*

Consider a car accelerating at a constant rate starting from time $t = 0$, at which speed $v = 0$ and distance $f = 0$. The acceleration is denoted as

$a \, [\text{m/s}^2]$. The distance that the car travels from start to time t should be

$$f(t) = \frac{1}{2}at^2 \tag{2.77}$$

where $f(t)$ is the function representing the distance that the car has traveled up to time t. The independent variable is time t. Following the definition in Eq. (2.1), its first-order derivative becomes

$$f'(t) = \frac{df(t)}{dt} = at \tag{2.78}$$

which is the speed v of the car at time t. It is a new function of t. Following, again, the definition in Eq. (2.1), we can find the derivative of the new function (the first-order derivative):

$$f''(t) = \frac{d}{dt}\left(\frac{df(t)}{dt}\right) = \frac{d}{dt}(at) = a \tag{2.79}$$

This is the second derivative of the distance function $f(t)$. It means that the second derivative of the distance function is the acceleration of the car.

Let $a = 4 \, \text{m/s}^2$, we can then draw the acceleration, speed, and the distance as functions of time t in Fig. 2.19.

In this case, due to the constant acceleration, the acceleration function $a(t)$ is a simple horizontal straight line, as shown in Fig. 2.19(a). The speed function $v(t)$ is an inclined straight line, as shown in Fig. 2.19(b). The distance function $f(t)$ is quadratic in t, as shown in Fig. 2.19(c). The distance traveled by the car in three seconds should be $f = \frac{1}{2}at^2 = 2 \times 9 = 18 \, [\text{m}]$.

This example shows the second derivative of a function can have a physical meaning. It depends on the type of the problem. Our discussion on higher-order derivatives is universal and shall apply to all differentiable functions that may be used in various disciplines.

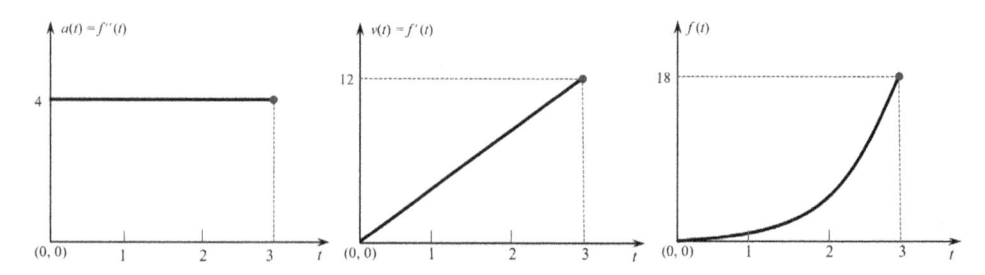

Figure 2.19. Constant acceleration, linear speed and quadratic distance function of a car.

2.11.3 *Example: Higher-order derivatives of quadratic functions*

Consider the quadratic function in the following general form:

$$f(x) = c_0 + c_1 x + c_2 x^2 \tag{2.80}$$

where c_0 and c_1 are given constants and c_2 is a non-zero constant. The unique and useful features of the quadratic function are listed as follows:

1. Its derivative is a linear function:

$$f'(x) = c_1 + 2c_2 x \tag{2.81}$$

   Since $f'(x)$ is linear, it must have one and only one root. This means that the quadratic function has only one unique extremum at the root of $f'(x) = c_1 + 2c_2 x = 0$, which gives $x = -\frac{c_1}{2c_2}$.

2. Its second derivative is a constant: $2c_2$.
3. At $x = -\frac{c_1}{2c_2}$, the slope of the quadratic function is zero. The second derivative of a function tells how the function curves. If $c_2 < 0$, implying that it curves downward (negative curvature), the extremum is a maximum. If $c_2 > 0$, it curves upward (positive curvature), and the extremum is a minimum.

If a problem, such as an optimization/minimization problem, can be cast in a quadratic function, the existence of extreme values and uniqueness are understood. Therefore, we always want to formulate problems as quadratic ones whenever possible. Even if it is not possible for the entire domain, we will still try to do so locally in an approximate manner, so that we can take advantage of quadratic functions.

One should not confuse problems of extrema-finding with root-finding. These are two different types of problems. *Finding extrema of a function requires root-finding for the derivative of the function.* The beauty of a quadratic function is that its derivative is linear and has one and only one root, implying that there is only one extremum. For third-order polynomials, extrema-finding becomes much more complicated because their derivatives are quadratic, whose roots can be complex-valued. We will discuss this in Section 2.14.3.

In the following example, we first use SymPy to derive the equations for the derivative of the quadratic function.

```
1  x, c0, c1, c2 = sp.symbols('x,c0, c1, c2')  # define symbolic variables
2  f_quad = c0 + c1*x + c2*x**2                      # define the function
3  display(Math(f" \\frac{{d}}{{dx}}({latex(f_quad)}) = \
4                         {latex(diff(f_quad,x))}"))
5  display(Math(f" \\frac{{d^2}}{{dx^2}}({latex(f_quad)}) = \
6                         {latex(f_quad.diff(x,2))}"))
```

$$\frac{d}{dx}\left(c_0 + c_1 x + c_2 x^2\right) = c_1 + 2c_2 x$$

$$\frac{d^2}{dx^2}\left(c_0 + c_1 x + c_2 x^2\right) = 2c_2$$

Using formulas from SymPy, we've written a NumPy code to compute the derivatives of a quadratic function and explicitly determine its extremum and location:

```
1  # define the general form of quadratic function:
2  def f_quad(x, c0, c1, c2):
3      f   = c2*x**2 + c1*x + c0        # function
4      df = 2*c2*x + c1                 # 1st derivative
5      d2f= 2*c2                        # 2nd derivative
6      xm = -c1/c2/2.                   # Location of the extremum
7      fm = c2*xm**2 + c1*xm + c0       # funtion extremum
8      return f, df, d2f, xm, fm
```

Let's plot a typical quadratic function using the following code. The plot is given in Fig. 2.20:

```
1  plt.figure(); plt.ioff()
2  plt.rcParams.update({'font.size': 8})
3  fig, ax = plt.subplots(1,1)#,figsize=(3,2))
4  X = np.arange(-7.5, 5, .01)
5  c0 = 8.; c1 = 5.; c2 = 2.
6  fq, dfq, d2fq, xm, fm = f_quad(X, c0, c1, c2)
7  print("The first derivatives = ", dfq[0], dfq[int(len(X)/2)], dfq[-1])
8  print(f"Extremum of f = {fm}, location = {xm}")
9
10 ax.set_xlabel('x')
11 ax.set_title('Quadratic function and its derivative')
12 ax.grid(color='r', linestyle=':', linewidth=0.5)
13 ax.plot(X, fq, label="$f(x)$")
14 ax.plot(X,dfq, label="$df(x)/dx$")
15 ax.scatter(xm, fm, color='m', s=9, label="Extreme")
16 xr = np.argmin(np.abs(dfq))
17
18 ax.scatter(X[xr], 0., color='k', s=9,label="Root of $df(x)/dx$")
19 ax.axvline(x=0, c="k", lw=0.6); ax.axhline(y=0, c="k", lw=0.6)
20 ax.legend(loc='upper center')
21
22 plt.savefig('imagesDI/f-quadextreme.png', dpi=500)
23 #plt.show()
```

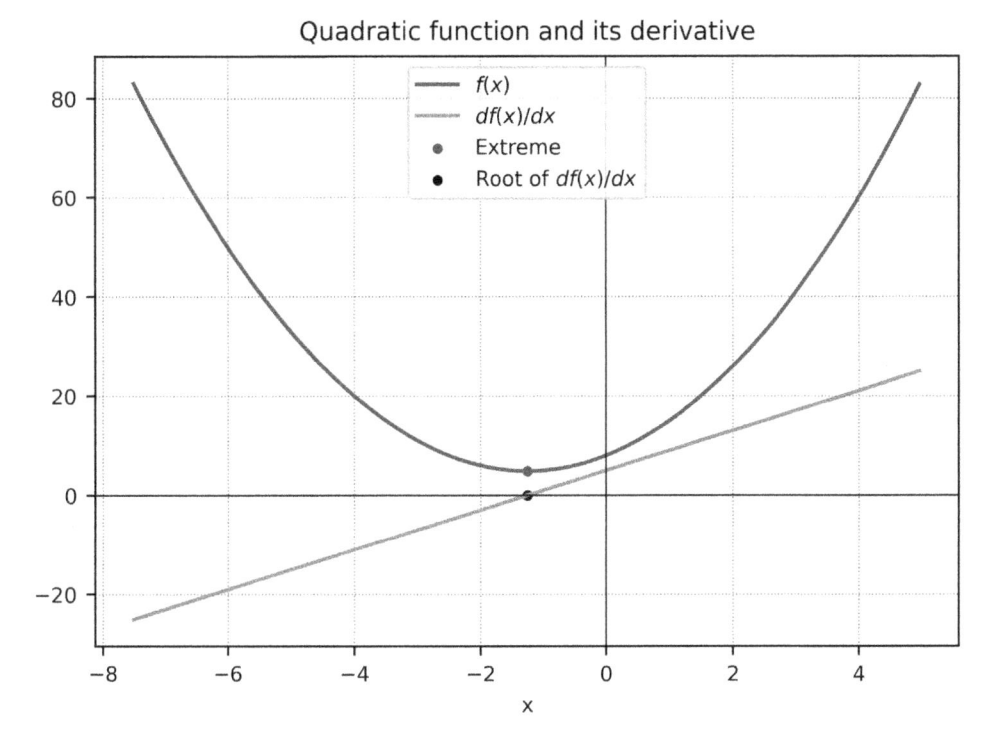

Figure 2.20. Quadratic function and its derivative function. At the only root of the derivative function, the function has one extremum (minimum in this case).

```
The first derivatives =   -25.0 -5.329070518200751e-13 24.959999999998935
Extremum of f = 4.875, location = -1.25
```

In conclusion, we shall make use of the properties of quadratic functions whenever it is possible.

2.11.4 *Example: Higher derivatives of the exponential function*

```
1  x = sp.symbols('x')
2  df_exp = sp.diff(sp.exp(x),x)      # define exponential function in sympy
3  print(" The derivative of exp(x)=", df_exp)
```

```
The derivative of exp(x)= exp(x)
```

The derivative of the (natural) exponential function is itself. This property holds true for any order of its derivatives.

```
1  dfn_exp=[sp.diff(sp.exp(x),(x, n)) for n in range(21)]  # 0th~20th diff
2  print("The higer derivatives of exp(x):")
3  dfn_exp                                          # Its higer derivatives
```

```
The higer derivatives of exp(x):
```

$$[\, e^x,\ e^x,\ e^x,\ e^x,\ e^x,\ e^x,\ e^x,\ e^x,\ e^x,\ e^x,\ e^x,\ e^x,\ e^x,\ e^x,\ e^x,\ e^x,\ e^x,\ e^x,\ e^x, e^x,\ e^x\,]$$

This property of higher derivatives implies that the exponential function not only grows rapidly, but all its derivatives also grow rapidly across the domain of $(-\infty, \infty)$. Its derivatives reveal the following characteristics:

1. It does not have any extremum because its first derivative is non-zero.
2. It is monotonic because its first derivative is always positive.
3. Due to item 2, it approaches zero only as $x \to -\infty$, but it never equals zero and thus has no roots.
4. Its curvature is never zero, always positive, and grows exponentially fast. It consistently curves upward, facilitating the transformation of values to positive ones in a monotonic fashion, which is advantageous in statistical analysis.
5. Since e^x grows extremely fast, e^{-x} decays extremely fast. This property is useful as weight functions to control the growth of certain functions, such as orthogonal Laguerre polynomials, ensuring convergence as $x \to \infty$.

For the exponential function, higher-order derivatives elucidate its global properties.

2.11.5 *Example: Derivatives of the logarithmic function*

The following code produces a list of derivatives of the logarithmic function from 0–10th order:

```
1  dfn_log=[sp.diff(sp.log(x),(x, n)) for n in range(11)]  # 0th~10th diff
2  print("The higer derivatives of log(x):")
3  dfn_log                                          # Its higer derivatives
```

```
The higer derivatives of log(x):
```

$$\left[\log{(x)},\ \frac{1}{x},\ -\frac{1}{x^2},\ \frac{2}{x^3},\ -\frac{6}{x^4},\ \frac{24}{x^5},\ -\frac{120}{x^6},\ \frac{720}{x^7},\ -\frac{5040}{x^8},\ \frac{40320}{x^9},\ -\frac{362880}{x^{10}}\right]$$

The logarithmic function has a domain of $(0, \infty)$, and its properties based on derivatives are as follows:

1. It has one root at $x = 1$ (a fundamental property of the function).
2. It never reaches an extremum and increases monotonically because its first derivative is always positive.
3. Due to item 2, the root at $x = 1$ is the only root of the logarithmic function.
4. It always curves downward because its second derivative is always negative.

These higher-order derivatives of the logarithmic function reveal its global properties.

2.11.6 *Example: Derivatives of the tanh function*

The hyperbolic tangent function, denoted as tanh, is widely utilized in various computational methods, including ML [5]. Defined by the natural exponential function, tanh is often differentiable infinitely. The following SymPy code computes up to the third-order derivatives of the tanh function:

```
1  x = sp.symbols('x')
2  dfn_tanh=[sp.tanh(x).diff(x,n) for n in range(4)]        # 0th~3rd diff
3  print("The higer derivatives of tanh(x):")
4  dfn_tanh
```

The higer derivatives of tanh(x):

$$
\left[\tanh(x),\; 1 - \tanh^2(x),\; 2\left(\tanh^2(x) - 1\right)\tanh(x),\right.
$$
$$
\left. -2\left(\tanh^2(x) - 1\right)\left(3\tanh^2(x) - 1\right)\right]
$$

The hyperbolic tangent function, denoted as tanh, has a domain of $(-\infty, \infty)$, and its properties based on derivatives are as follows:

1. It has one root at $x = 0$ (a fundamental property of the function).
2. It has no extremum and increases monotonically because its first derivative is always positive and lies within the interval $(0, 1)$ (noting that $-1 < \tanh(x) < 1$). This property makes tanh useful in signal processing and data normalization.
3. Due to item 2, the root at $x = 0$ is the only root of the hyperbolic tangent function.

4. It curves upward in $(-\infty, 0)$ because its second derivative is positive there, and it curves downward in $(0, \infty)$ because its second derivative is negative there. It has zero curvature at $x = 0$.

5. Since its first derivative is in $(0, 1)$, the first derivative of its inverse function is in $(1, \infty)$. This implies that the inverse function is real-valued in theory as long as the argument of tanh remains real. This property enables the construction of neural network models for both forward and inverse mechanics problems [17, 18].

For the tanh function, higher-order derivatives reveal its properties within its two sub-domains: $(-\infty, 0)$ and $[0, \infty)$.

The following NumPy code computes and plots the tanh function and its derivatives up to third order:

```python
 1  plt.figure(); plt.ioff()
 2  plt.rcParams.update({'font.size': 10})
 3  Label=['tanh(x)','1st diff.','2nd diff.','3rd diff.']
 4
 5  def f(x):                              # define a function
 6      return anp.tanh(x)         # Hyperbolic tangent often used in ML
 7      #return (np.exp(x)-np.exp(-x))/(np.exp(x)+np.exp(-x))
 8
 9  d_f = []                               # placeholder for derivative functions
10  d_f.append(f)                          # append the function to 0th diff.
11  n = 3                                  # compute nth diffs.
12  for i in range(1, n+1):
13      d_f.append(grad(d_f[i-1]))    # Obtain function diffs, put to a list
14
15  x = np.linspace(-5,5,200)              # create values of arguments, x
16  for k in range(n+1):                   # plot all the n function diffs
17      plt.plot(x, [d_f[k](xi) for xi in x], label = Label[k])
18
19  plt.grid(color='r', linestyle=':', linewidth=0.5)
20  plt.xlabel('x')
21  plt.legend()
22  plt.savefig('imagesDI/tanhd_x_.png', dpi=500)
23  #plt.show()
```

Figure 2.21 plots the hyperbolic tangent function and its derivatives. The function is bounded: $-1 < \tanh(x) < 1$ and monotonic across $(-\infty, \infty)$. Its first derivative is always positive. The second derivative is positive in $(-\infty, 0)$ and negative in $(0, \infty)$. It has zero curvature at $x = 0$. Tanh is an ideal activation function for ML models [5], enabling neural networks to address both forward and inverse problems [17, 18].

In summary, examining higher derivatives of functions reveals their properties and helps select the most suitable function for specific purposes. Readers can conduct similar analyses for other functions.

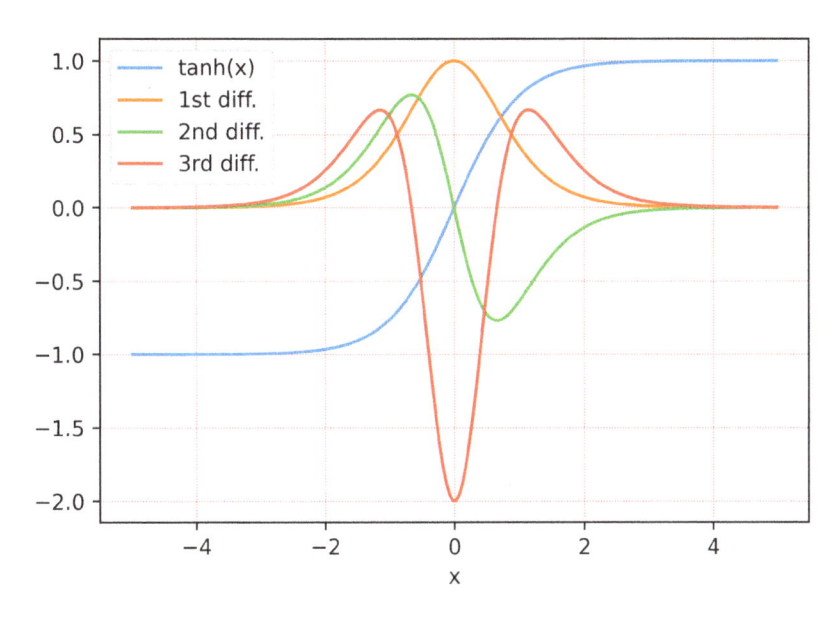

Figure 2.21. Hyperbolic tangent function and its derivatives.

2.11.7 *Derivatives of composite functions*

For a general function, finding these roots analytically can be challenging. Here, we employ the **root_finder()** function with Newton iteration, as previously presented, to analyze the first and second derivatives of the composite function. This approach helps us identify its roots, stationary points, and extrema:

```
 1  plt.figure(); plt.ioff()
 2  plt.rcParams.update({'font.size': 8})
 3  fig, ax = plt.subplots(1,1,figsize=(6,4.5))
 4
 5  g2f_xsin = grad(gf_xsin)      # 2nd gradient, gf_xsin is defined earlier
 6  xL, xR = 0., 25.
 7  n_x = 1000                               # number of points on x axis
 8  X = np.linspace(xL, xR, n_x)
 9
10  # find the roots of the 1st derivative of the function.
11  # Because Newton's method is used, we need the 2nd derivative of f.
12  gfv, roots, gfr2, _ = roots_finder(gf_xsin, g2f_xsin,
13                          xL, xR, n_x, limit=1000, atol=1e-6)
14  # Plot the curves:
15  ax.set_xlabel('x')
16  ax.set_title('Function, its derivatives, and roots')
17  ax.grid(color='r', linestyle=':', linewidth=0.5)
18  ax.plot(X, f_xsin(X), label="$f(x)$")
19  ax.plot(X, gfv      , label="$df(x)/dx$")
20  ax.plot(X, [g2f_xsin(xi) for xi in X], label="$d^2f(x)/dx^2$")
21  ax.scatter(roots,[0]*len(roots), c='r', s=15,
22              label="Roots of $df(x)/dx$")
```

```
23  ax.scatter(roots, gfr2,              c='m', s=8,
24            label="$d^2f(x)/dx^2$ at the roots")
25  ax.scatter(roots, f_xsin(roots), c='k', s=15, marker='*',
26            label="Extrema of f(x)")
27  ax.legend() #loc='center right', bbox_to_anchor=(1, 0.5))
28
29  plt.savefig('imagesDI/xsinkx_b.png', dpi=500)
30  #plt.show()
```

```
Converged at 14th iteration to 1e-06
```

In this example, we used **root_finder()** to locate the roots of the derivative of the given function. It was observed that at each of these roots, the function attained a local extremum, as depicted in Figure 2.22. We also computed the second derivative (shown in green) of the function. It is evident that when the extremum is a maximum, the second derivative is negative, and when the extremum is a minimum, the second derivative is positive. Higher-order derivatives thus reveal the local properties of the function.

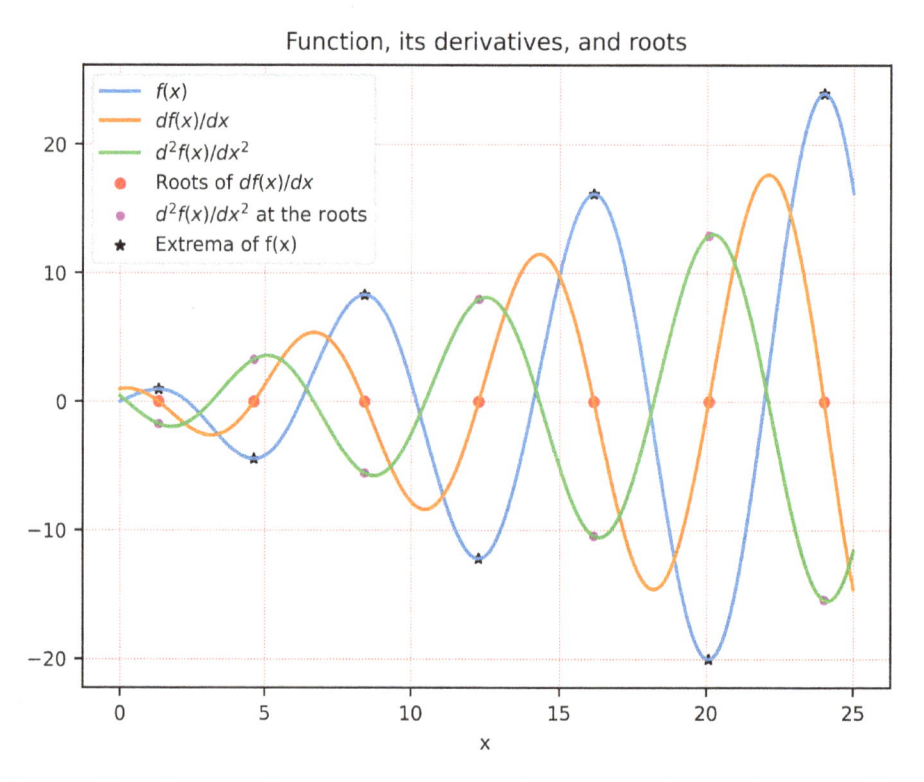

Figure 2.22. Roots, stationary points, extrema of a composite sinusoidal function. These key values are determined using the first and second derivatives of the function.

2.12 Taylor's expansion of functions

2.12.1 *Formulation*

Using derivatives, functions can be expanded using the Taylor series around a point in their domain where they are infinitely differentiable. Taylor's expansion for $f(x)$ at $x = a$ can be expressed as follows:

$$f(x) = f(a) + \left.\frac{df(x)}{dx}\right|_{x=a}(x-a) + \frac{1}{2!}\left.\frac{d^2 f(x)}{dx^2}\right|_{x=a}(x-a)^2 + \cdots$$
$$+ \frac{1}{k!}\left.\frac{d^k f(x)}{dx^2}\right|_{x=a}(x-a)^k + \mathcal{O}\big((x-a)^{k+1}\big) \tag{2.82}$$

Here, x can be on either the left or right side of a. If the series converges to the function value at x as the number of terms approaches infinity, we obtain the Taylor expansion for $f(x)$.

When using only the zeroth-order approximation, we have

$$f(x) = f(a) + \mathcal{O}\big((x-a)\big) \tag{2.83}$$

When using only the first derivative, we obtain the linearization (Eq. (2.68)) of a function with an error of $\mathcal{O}\big((x-a)^2\big)$. When using up to the second derivative, we get the quadratic approximation (Eq. (2.71)) of a function with an error of $\mathcal{O}\big((x-a)^3\big)$.

The coefficient $\frac{1}{k!}$ ensures that the kth derivatives of the approximation match those in the Taylor expansion, which can be verified by differentiating Eq. (2.82) k times.

Equation (2.82) signifies that the function can be locally expressed as a linear combination of monomial functions. This concept bears similarity to the fitting techniques discussed in Chapter 6 of Ref. [1]. However, there are differences:

1. The coefficients of the monomials in Taylor's expansion utilize the various orders of the function's derivatives. In contrast, the coefficients of fitted monomials are determined through a fitting process using function values at specified points within the fitting domain.
2. Taylor's expansion provides a good approximation near the point where derivatives are evaluated. In contrast, fitting techniques aim to provide a good approximation across the domain where the function is fitted.

2.12.2 *Example: Taylor's expansion of trigonometric functions*

Trigonometric functions are infinitely differentiable within their domains, allowing them to have Taylor expansions. The following examples demonstrate using SymPy to generate some of these expansions. Readers can easily modify the codes to find expansions for other functions.

```
1  #from mpmath import nprint, chop, taylor          # alternative module
2
3  #print("Coefficients of Taylor's expansion of sine near x=0:")
4  #nprint(chop(taylor(sp.sin, 0,7))) # use chop, expansion upto 7th order
5
6  x, a = sp.symbols("x, a", real=True)
7  a = 0                                    # expansion about a
8  n = 11                                   # expansion up to order n
9  sin_Taylor=sp.series(sp.sin(x), x,  x0=a,  n=n,  dir='+')   # use SymPy
10 #                              in x, at a, order, direction
11 # get the coefficients:
12 coefs = [sin_Taylor.coeff(x, i) for i in range(n)]
13 print("coefs=", coefs)
14 #coefs.reverse()    # reverse the order for np.poly1d() to be used later
15 #print("reversed coefs=",coefs)
16
17 print(f"\nTaylor's expansion of sin at vicinity of x=0:")
18 sin_Taylor
```

coefs= [0, 1, 0, -1/6, 0, 1/120, 0, -1/5040, 0, 1/362880, 0]

Taylor's expansion of sin at vicinity of x=0:

$$x - \frac{x^3}{6} + \frac{x^5}{120} - \frac{x^7}{5040} + \frac{x^9}{362880} + O\left(x^{11}\right)$$

Taylor's expansion of the sine function includes only odd orders because the sine function is antisymmetric. All its even-order derivatives at $x = 0$ are zero. Let's examine the cosine function:

```
1  sp.series(sp.cos(x), x, n=n)          # Taylor's expansion of cos near x=0
```

$$1 - \frac{x^2}{2} + \frac{x^4}{24} - \frac{x^6}{720} + \frac{x^8}{40320} - \frac{x^{10}}{3628800} + O\left(x^{11}\right)$$

Taylor's expansion of the cosine function has only even terms because it is symmetric. All its odd-order derivatives at $x = 0$ are zero. Let us look at the tangent function, which is antisymmetric:

```
1  sp.series(sp.tan(x), x, n=n)          # Taylor's expansion near x=0
```

$$x + \frac{x^3}{3} + \frac{2x^5}{15} + \frac{17x^7}{315} + \frac{62x^9}{2835} + O\left(x^{11}\right)$$

It has only odd terms.

Trigonometric inverse functions can be expressed as Taylor's series. The following gives just one example:

```
1  #import mpmath as mp                                    # alternative
2  #mp.nprint(mp.chop(mp.taylor(sp.asin,0,9)))  # arcsine: inverse of sine
3  asin_Taylor = sp.series(sp.asin(x),x, n=15,)
4  asin_Taylor                          # Taylor's expansion of arcsine near x=0
```

$$x + \frac{x^3}{6} + \frac{3x^5}{40} + \frac{5x^7}{112} + \frac{35x^9}{1152} + \frac{63x^{11}}{2816} + \frac{231x^{13}}{13312} + O\left(x^{15}\right)$$

Finally, let us get Taylor's expansion for the "strange" function $f(x) = x^x$ studied in Ref. [1]:

```
1  xx_Taylor = sp.series(x**x,x, n=6)
2  xx_Taylor                          # Taylor's expansion of x^x near x=0
```

$$1 + x \log\left(x\right) + \frac{x^2 \log\left(x\right)^2}{2} + \frac{x^3 \log\left(x\right)^3}{6} + \frac{x^4 \log\left(x\right)^4}{24} + \frac{x^5 \log\left(x\right)^5}{120}$$

$$+ O\left(x^6 \log\left(x\right)^6\right)$$

2.12.3 *Existence of Taylor's expansion*

Most continuous functions have Taylor series expansions around points where they are infinitely differentiable. However, not all functions have a Taylor expansion at a point even if the function is infinitely differentiable at that point. This is because we also require the Taylor expansion to converge to the function value at the point as the number of terms in the series approaches infinity. A typical example is the following function:

$$f(x) = \begin{cases} e^{-1/x^2} & \text{if } x \neq 0 \\ 0 & \text{if } x = 0 \end{cases} \tag{2.84}$$

This function is infinitely differentiable with zero derivatives when $x \to 0$:

```
1  fx = exp(-1/x**2)
2  [sp.limit(fx.diff(x,k),x,0) for k in [1, 2, 5, 10]]
```

[0, 0, 0, 0]

Therefore, its Taylor's expansion at $x = 0$ becomes zero regardless of how many terms are taken:

```
1  e_x2_Taylor = sp.series(fx,x, n=11)
2  display(Math(f" \\text{{Taylor's expansion of ${latex(fx)}$ = }}\
3                            {latex(e_x2_Taylor)}"))
```

Taylor's expansion of $e^{-\frac{1}{x^2}} = O\left(x^{11}\right)$

It has only the error term. However, the function around $x = 0$ is not zero (although it is small):

```
1  [exp(-1/x**2).subs(x,0.1) for xi in [-0.1, 0.1]]
```

$$\left[3.72007597602089 \cdot 10^{-44}, \ 3.72007597602089 \cdot 10^{-44}\right]$$

This means that the Taylor expansion at $x = 0$ does not converge to the function value regardless of the number of terms taken. This function has a Taylor expansion at any $x \neq 0$. This feature makes the function quite special.

In the following discussions, we will assume that the function has a Taylor expansion.

2.12.4 *Applications of Taylor expansion*

The Taylor expansion has a number of applications:

1. simplification of complicated functions in the vicinity of a point of interest;
2. comparison of the local behavior of functions, which is useful in finding the limit of some functions;
3. function approximation in the vicinity of a point of interest, which complements the fitting techniques discussed in Chapter 6 of Ref. [1], where the approximation is effective over a given domain.

2.12.4.1 *Simplification of complicated functions*

Taylor's expansion is a useful tool in theoretical studies because it transforms any smooth function into a polynomial form, providing an explicit view of the function's local behavior. Following are some examples of composite functions:

```
1  a, b, c = sp.symbols("a, b, c", real=True)
2  fx = sp.sin(a*x**2+b*x+c)              # A composite function
3  sp.series(fx, x, n=4)                  # Taylor's expansion near x=0
```

$$\sin(c) + x^2 \left(a\cos(c) - \frac{b^2 \sin(c)}{2} \right) + x^3 \left(-ab\sin(c) - \frac{b^3 \cos(c)}{6} \right)$$

$$+ \, bx\cos(c) + O\left(x^4\right)$$

For this composite function, the slope at x is simply $b\cos(c)$, and its curvature is $a\cos(c) - \frac{b^2 \sin(c)}{2}$.

Using SymPy, one can easily find such function behaviors. The following is another example that may require lengthy derivation by hand:

```
1  fx = sp.exp(c*x**2+b)*sp.sin(a*x**2+b*x+c)    # A composite function
2  sp.series(fx, x, n=3)                          # Taylor's expansion near x=0
```

$$e^b \sin(1.0c) + x^2 \left(-0.5b^2 e^b \sin(1.0c) + ce^b \sin(1.0c) + 4.5e^b \cos(1.0c) \right)$$

$$+ \, 1.0bxe^b \cos(1.0c) + O\left(x^3\right)$$

The intersection, slope and curvature at $x = 0$ is found explicitly by Taylor's expansion.

2.12.4.2 *Comparison of local behavior of functions*

Many equalities and inequalities can be derived or proven using Taylor expansion, which facilitates easy comparison between functions. For example, in proving the limits in Eq. (2.16), the use of Taylor expansion becomes useful. The term $\cos(\Delta x)$ in Eq. (2.16) can be expanded to

$$\cos(\Delta x) = 1 - \frac{(\Delta x)^2}{2} + O\left((\Delta x)^4\right) \tag{2.85}$$

This gives

$$\lim_{\Delta x \to 0} \frac{\cos(\Delta x) - 1}{\Delta x} = \lim_{\Delta x \to 0} \frac{1 - \frac{(\Delta x)\Delta x}{2} + O\left((\Delta x)^3 \Delta x\right) - 1}{\Delta x}$$

$$= \lim_{\Delta x \to 0} \left[\frac{\Delta x}{2} + O\left((\Delta x)^3\right) \right] = 0 \tag{2.86}$$

We observe that blue-cancellation is achieved effortlessly, leading to subsequent red-cancellation. Equation (2.85) essentially implies that $(\cos x - 1)$ behaves similar to x^2 near $x = 0$.

Readers can employ the same approach to demonstrate that $\sin x$ behaves similar to x near $x = 0$, hence $\lim_{\Delta x \to 0} \frac{\sin(\Delta x)}{\Delta x} = 1$, justifying its use in Eq. (2.16).

2.12.4.3 *Example: Limits of the ratio of a pair of functions*

Certainly, Taylor's expansions of functions can be combined with L'Hôpital's rule to evaluate the limit of the ratio of two functions. For instance, in Case II discussed in Section 2.6.9.2, we can expand $f(x)$ to obtain the following:

```
1  # Case II, f(x):
2  f = sp.sin(x) - x           # Taylor's expansion near x=0
3  display(Math(f" \\text{{{${latex(f)}$=}}}{latex(sp.series(f, x, n=5))}"))
```

$$-x + \sin(x) = -\frac{x^3}{6} + O\left(x^5\right)$$

Now, for $g(x) = x^2$, it is evident that $f(x)$ approaches zero one order faster than $g(x)$, hence we immediately have $\lim_{x \to 0} \frac{f(x)}{g(x)} = 0$.

If $g(x) = x^3$, we immediately have $\lim_{x \to 0} \frac{f(x)}{g(x)} = -\frac{1}{6}$.

If $g(x) = x^4$, we observe $\lim_{x \to 0} \frac{f(x)}{g(x)} = -\infty$.

We can conclude the following:

> Using Taylor's expansion, we can easily determine the limit of the ratio of any pair of functions that have Taylor expansions by comparing the rates of the lowest-order terms in their Taylor expansions.

2.12.4.4 *Function approximation using Taylor expansion*

Taylor expansion is valuable for approximating functions near x. This concept aligns with our discussions in Sections 2.9.1 and 2.9.2. Depending on the desired accuracy, one can select the order of terms to truncate. In this example, we use the sine function to illustrate how the order of terms taken from the Taylor expansion affects the accuracy of the approximation, specifically using orders $n = 1, 3, 5,$ and 7. For clarity, we present the plots showing the original function alongside its approximations at different orders:

```
1  x  = sp.symbols("x", real=True)
2  sin_Taylor=sp.series(sp.sin(x),x,n=11)   # SymPy, create Taylor of sine
3
4  coefs = [sin_Taylor.coeff(x, i) for i in range(11)]   # get the coeffs
5  coefs.reverse()   # reverse the order for np.poly1d() to be used later
6
7  n_order = [1, 3, 5, 7]                          # orders to be used
8  n = len(n_order)
9
10 # generate a list of n polynomials functions with n_orders:
11 f_taylor = [np.poly1d(coefs[-1-n_order[i]:]) for i in range(n)]
```

```
1  plt.figure(); plt.ioff()
2  plt.rcParams.update({'font.size': 8})
3  fig, ax = plt.subplots(1,1,figsize=(6,4.5))
4  colors = ['b', 'm','g', 'r', 'orange', 'y']
5  xL, xR = -1.2*np.pi, 1.2*np.pi
6  n_x = 1000                                # number of points on x axis
7  X = np.linspace(xL, xR, n_x)
8
9  # Plot the approximation curves:
10 ax.set_xlabel('x')
11 ax.grid(color='r', linestyle=':', linewidth=0.5)
12 ax.plot(X, np.sin(X), c="k", lw=2., label="$sin(x)$")
13
14 for i in range(len(n_order)): # plot approx. curves of different orders
15     ax.plot(X, f_taylor[i](X),c=colors[i],label=f"$n$={n_order[i]}")
16
17 ax.axvline(x=0, c="k", lw=0.6); ax.axhline(y=0, c="k", lw=0.6)
18 ax.legend(loc='center left', bbox_to_anchor=(.2, .8))
19 ax.set_ylim(-1.1, 1.1)
20
21 plt.savefig('imagesDI/Taylor_sin.png', dpi=500)
22 #plt.show()
```

The following can be observed from the plot shown in Fig. 2.23:

1. The approximation is accurate in the vicinity of the expansion point (in this case, around $x = 0$).
2. The inclusion of higher-order terms enhances the accuracy of the approximation and extends the range of validity. This improvement becomes clearer when examining the differences between the original sine function and its approximations.

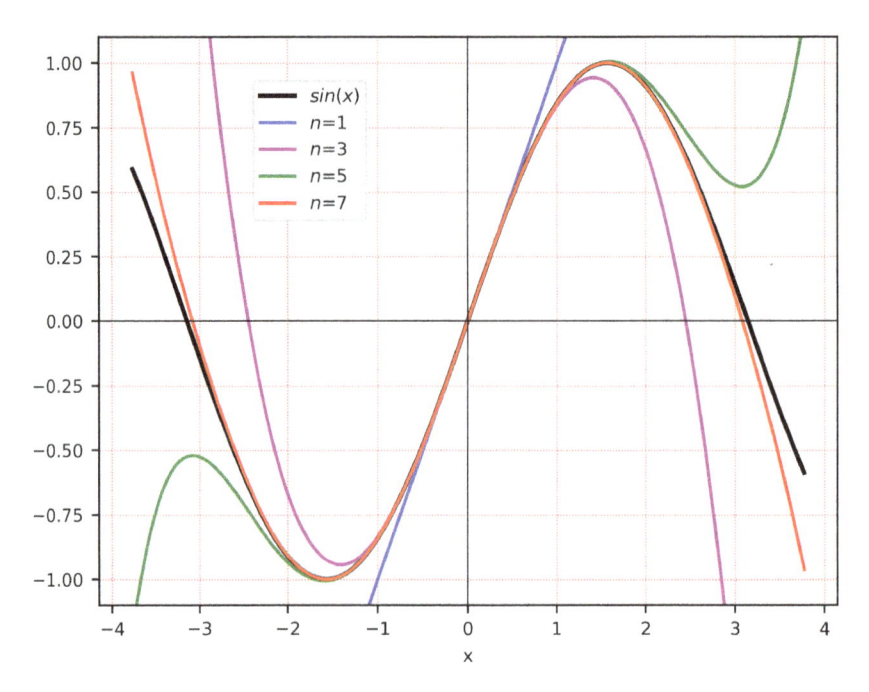

Figure 2.23. Taylor's expansions of the sine function using different order of derivatives at $x = 0$.

```
1   # Plot the curves of  the differences between original & approximated:
2   plt.figure(); plt.ioff()
3   plt.rcParams.update({'font.size': 8})
4   fig, ax = plt.subplots(1,1,figsize=(6,4))
5   ax.set_xlabel('x'); ax.grid(color='r',linestyle=':',linewidth=0.5)
6   ax.axvline(x=0, c="k", lw=0.6); ax.axhline(y=0, c="k", lw=0.6)
7
8   for i in range(len(n_order)):
9       f_diff = f_taylor[i](X)-np.sin(X)
10      ax.plot(X, f_diff, c=colors[i], label=f"$n$={n_order[i]}")
11
12  ax.legend(loc='center left', bbox_to_anchor=(.3, .8))
13  ax.set_ylim(-1.1, 1.1)
14  plt.savefig('imagesDI/Taylor_sin_diff.png', dpi=500)
15  #plt.show()
```

In Fig. 2.24, it is evident that the first-order (linear) approximation remains accurate only within approximately $[-0.3, 0.3]$, while the seventh-order approximation extends its accuracy to about $[-2.5, 2.5]$. Beyond these ranges, the approximations deviate significantly from the sine function.

In numerical computations, the use of limited terms is necessary due to the rapid growth of higher-order monomials [1]. Therefore, Taylor expansion

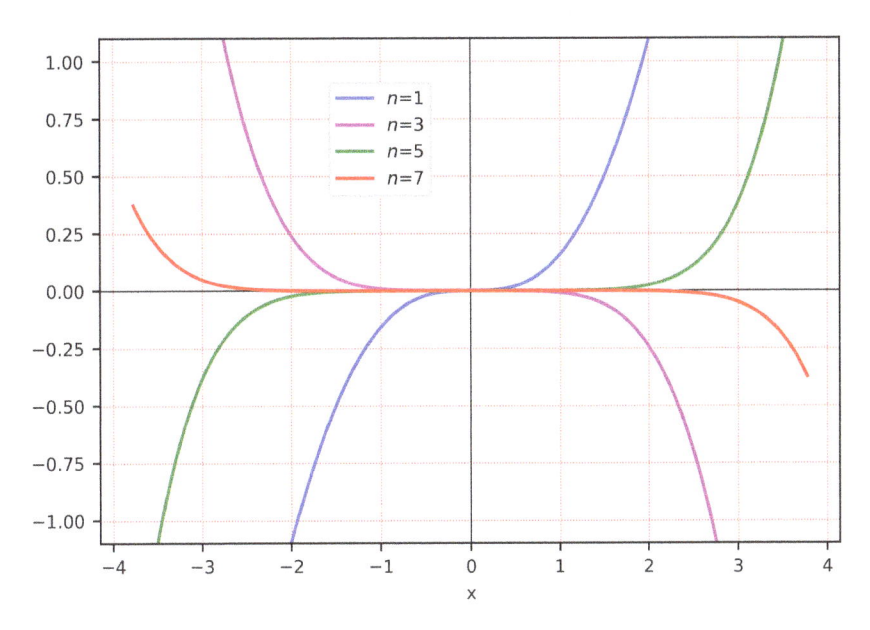

Figure 2.24. The differences between the sine function and its Taylor's expansions using different order of derivatives at $x = 0$.

is typically applied in the vicinity of x to avoid including too many high-order terms. This local approximation approach is a practical and effective strategy consistent with the discussion in Section 2.9.3.

2.12.4.5 *Computing Euler's number*

Taylor's expansion can be employed to compute special constants, such as Euler's number e and π.

From SymPy results, the power series for the real exponential function with argument $x \in \mathbb{R}$ is given by

$$e^x = \sum_{k=0}^{\infty} \frac{x^k}{k!} = 1 + x + \frac{x^2}{2} + \frac{x^3}{6} + \frac{x^4}{24} + \frac{x^5}{120} + \frac{x^6}{720} + \cdots \qquad (2.87)$$

Since $k!$ grows extremely fast with k, the series converges for all real arguments x. Therefore, Eq. (2.87) holds for $x \in \mathbb{R}$. Additionally, Euler's number e can be obtained by setting $x = 1$:

$$e = \exp(1) = \sum_{k=0}^{\infty} \frac{1}{k!} \qquad (2.88)$$

This formula is used to compute an approximate value of e. The accuracy depends on the terms used. The code is as follows:

```
1  e_approximated = 0.; k_factorial = 1
2  for k in range(10):
3      e_approximated += 1./k_factorial; k_factorial *= (k+1)
4
5  print(f"e_approximated={e_approximated}, e_in_Sympy={sp.E.evalf()}")
```

e_approximated=2.7182815255731922, e_in_Sympy=2.71828182845905

We observe that using only 10 terms, the approximate e is accurate to seven decimal places. The rapid convergence is attributed to the factorial growth of the denominator $k!$.

2.12.4.6 *Computing constant π*

To compute π, we may use Taylor's series expansion of $\arctan(x)$:

```
1  f = sp.atan(x)
2  atan_Taylor = sp.series(f, x, n=23)
3  display(Math(f" \\text{{${latex(f)}$ = }}{latex(atan_Taylor)}"))
```

$$\operatorname{atan}(x) = x - \frac{x^3}{3} + \frac{x^5}{5} - \frac{x^7}{7} + \frac{x^9}{9} - \frac{x^{11}}{11} + \frac{x^{13}}{13} - \frac{x^{15}}{15} + \frac{x^{17}}{17} - \frac{x^{19}}{19} + \frac{x^{21}}{21}$$
$$+ O\left(x^{23}\right)$$

This series with $x = 1$ is given as

$$\arctan(1) = \frac{\pi}{4} = \sum_{n=0}^{\infty} (-1)^n \frac{1}{2n+1} \tag{2.89}$$

The code snippet for computing π is as follows:

```
1  atan_Taylor100 = sp.series(f,x,n=100).removeO().subs(x,1)
2  print("π_approximated=",(atan_Taylor100.evalf(8)*4))
3  gr.printx("np.pi")
```

π_approximated= -0.63411606
np.pi = 3.141592653589793

It is observed that with an expansion up to the order of 100, the result has only two accurate decimal places due to slow convergence. Since the series alternates in sign, accuracy is primarily controlled by $\frac{1}{2n+1}$. To achieve an accuracy of 1%, approximately $n = \frac{100}{2} \approx 50$ terms are required.

In 1706, John Machin discovered a clever formula (https://en.wikipedia.org/wiki/Machin-like_formula):

$$\frac{\pi}{4} = 4 \arctan \frac{1}{5} - \arctan \frac{1}{239} \qquad (2.90)$$

Since the accuracy is now controlled by $\left(\frac{1}{5}\right)^{2n+1} \frac{1}{2n+1}$, the series converges much faster, allowing him to compute the value of π accurate to 100 decimal places.

Let's use the Python code to compute π using Machin's formula:

```
1  n = 20
2  atan_Taylor_n = sp.series(f,x,n=n).removeO()
3  π_4 = 4*atan_Taylor_n.subs(x,1/5)-atan_Taylor_n.subs(x,1/239)
4  print("π_approximated=",π_4.evalf(16)*4)
5  print("π_numpy=       ", np.pi)                    # for comparison
```

```
π_approximated= -0.02129065844587972
π_numpy=        3.141592653589793
```

With Machin's formula, we can obtain 15 accurate digits (to machine accuracy) using only 20 terms, which is a drastic improvement.

For readers' information, the mpmath module can compute the value of π accurately to as many digits as desired:

```
1  import mpmath
2  ndps = 50               # ndps decimal places + 1 for internal precision
3  mpmath.mp.dps = ndps
4  pi_mpmath = mpmath.pi                    # Compute pi using mpmath
5  print(f"π accurate to {ndps} digits:{pi_mpmath}")
```

```
π accurate to 50 digits:3.1415926535897932384626433832795028841971693993751
```

2.12.5 *Example: Taylor's expansion of other useful functions*

```
1  x = symbols('x')
2  f = 1/sp.sqrt(1+x)
3  f_Taylor = sp.series(f, x, n=9,)
4  display(Math(f" \\text{{${latex(f)}$ = }}{latex(f_Taylor)}"))
```

$$\frac{1}{\sqrt{x+1}} = 1 - \frac{x}{2} + \frac{3x^2}{8} - \frac{5x^3}{16} + \frac{35x^4}{128} - \frac{63x^5}{256} + \frac{231x^6}{1024} - \frac{429x^7}{2048} + \frac{6435x^8}{32768}$$

$$+ O\left(x^9\right)$$

This implies that $\frac{1}{\sqrt{x+1}}$ behaves similar to $1 - \frac{x}{2}$ in the vicinity of $x = 0$, a widely used approximation. It is a specific case of the binomial expansion used in Eq. (2.13). The complete expression for the binomial expansion is derived using the following:

```
1  n = symbols('n')
2  f =(1+x)**n
3  f_Taylor = sp.series(f, x, n=5)
4  display(Math(f" \\text{{${latex(f)}$ = }}{latex(f_Taylor)}"))
```

$$(x+1)^n = 1 + nx + \frac{nx^2\,(n-1)}{2} + \frac{nx^3\,(n-2)\,(n-1)}{6}$$

$$+ \frac{nx^4\,(n-3)\,(n-2)\,(n-1)}{24} + O\left(x^5\right)$$

Let us take a look at Taylor's expansion of $f(x) = \frac{\sin x}{x}$:

```
1  f = sp.sin(x)/x
2  f_Taylor = sp.series(f, x, n=14)
3  display(Math(f" \\text{{${latex(f)}$ = }}{latex(f_Taylor)}"))
```

$$\frac{\sin(x)}{x} = 1 - \frac{x^2}{6} + \frac{x^4}{120} - \frac{x^6}{5040} + \frac{x^8}{362880} - \frac{x^{10}}{39916800} + \frac{x^{12}}{6227020800} + O\left(x^{14}\right)$$

This implies that $\frac{\sin x}{x}$ behaves similar to 1, or $\sin x$ behaves similar to x in the vicinity of $x = 0$, which is also a widely used approximation. We utilized this in Eq. (2.16).

Note that we can also approximate $\frac{\sin x}{x}$ as $1 - \frac{x^2}{6}$ or even include higher-order terms for more accurate comparisons.

Let's consider a more complex composite function:

```
1  f = (x**2+2*x)*sp.sin(3*x**2+.8*x-5.0)*sp.exp(1.8*x)   # composite func.
2  f_Taylor  = sp.series(f,x, n=5)
3  f_Taylor5 = gr.printM(f_Taylor,n_dgt=5)              # limits its digit to 5
4  display(Math(f" \\text{{${latex(f)}$ = }}{latex(f_Taylor5)}"))
```

$$\left(x^2 + 2x\right) e^{1.8x} \sin(3x^2 + 0.8x - 5.0) = 1.9178x + 4.8649x^2 + 6.9651x^3$$
$$+ 2.4131x^4 + O\left(x^5\right)$$

This complicated function behaves similar to $1.9178x$ in the vicinity of $x = 0$.

As demonstrated by these examples, complex functions can be approximated by polynomials using Taylor's expansion, provided that the function is differentiable and the Taylor series converges. This approach is highly practical. However, caution is necessary to assess the accuracy of the approximation when x is far from the expansion point.

2.13 Gradient descent, minimization of functions

Minimizing a function is one of the most frequently used operations in computational methods. The gradient descent method is a fundamental technique applicable in various fields, including ML. It leverages the function's gradient to efficiently locate its minimizer. Here, we introduce the gradient descent method for 1D cases, referring to online materials available under the Apache License 2 at this GitHub repository (https://github.com/zackchase/mxnet-the-straight-dope/tree/master/chapter06_optimization) and to Ref. [5]. Our derivation in the following section utilizes the MVT discussed earlier.

2.13.1 *Formulation*

Consider a cost function $f(x)$ that is to be minimized. Assume that it has at least a first derivative in its domain. We can then apply the MVT in a local interval $[x, x + \Delta x]$ within the domain of $f(x)$:

$$f(x + \Delta x) = f(x) + f'(\xi)\Delta x \qquad (2.91)$$

where Δx is a small real number and $\xi \in (x, x+\Delta x)$. Since the function also has a first derivative at x and $x + \Delta x$, ξ can lie in $[x, x + \Delta x]$. Additionally, $f'(\xi)$ must be a finite number. Therefore, we choose a very small positive real number η as a scalar, such that $-\eta f'(\xi)$ is approximately equal to Δx. This implies that η should scale as $\frac{\Delta x}{f'(\xi)}$. Equation (2.91) can be rewritten as

$$f\left(x - \eta f'(\xi)\right) = f(x) - \eta\left(f'(\xi)\right)^2 \qquad (2.92)$$

Since η is chosen positive, we have

$$f\left(x - \eta f'(\xi)\right) \leq f(x) \qquad (2.93)$$

This means that an update to x with

$$x := x \underbrace{-\eta f'(\xi)}_{\Delta x} \tag{2.94}$$

would lead to a reduction of $f(x)$ value. The negative sign indicates that the update in x is in the negative (or downhill) direction of the gradient. This method is known as the **gradient descent method** for function minimization. The parameter η is commonly referred to as the learning rate.

Now, ξ can lie in $[x, x + \Delta x]$. Since we generally don't know its exact value and we know $f'(x)$, a straightforward assumption is $\xi = x$, given Δx is small. Equation (2.94) then becomes

$$x := x - \eta f'(x) \tag{2.95}$$

Using Eq. (2.95), regardless of which valley side x is on, the gradient descent makes a step closer to the minimum, assuming that the function is convex, as illustrated in Fig. 2.25.

If x is on the left hill, $f'(x)$ is negative, and Δx will be positive, implying that the update using Eq. (2.95) moves rightward to the minimizer x^*. If x is on the right hill, $f'(x)$ is positive (η is always positive), Δx will be negative, implying that the update using Eq. (2.95) moves leftward to the minimizer. In either case, we make progress getting closer to the minimizer with the gradient descent algorithm because the function is assumed to be convex. This is the essence of the gradient descent algorithm.

Note that we do not have to assume $\xi = x$. In fact, in the so-called Nesterov accelerated gradient method, a special ξ is chosen based on previous

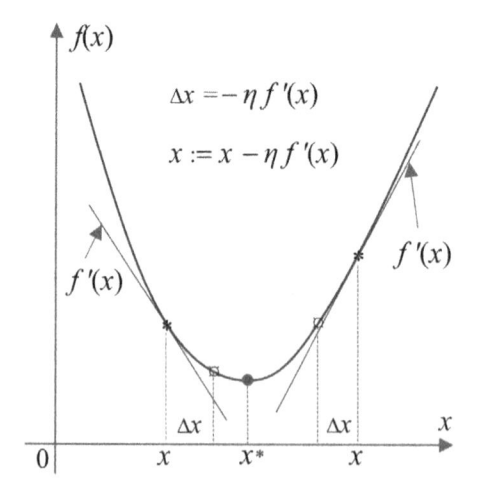

Figure 2.25.   Gradient descent in 1D case for a convex function.

step data to achieve faster convergence. Interested readers may take a look at Section 9.7 in Ref. [5].

2.13.2 *Key points on gradient descent*

Based on the above analysis, we note the following important points:

1. Updating in x is in the negative direction of the gradient, hence the name *gradient descent*.
2. The amount of reduction is proportional to the gradient of the function. The larger the gradient, the larger the reduction, provided that $\eta f'(x)$ is sufficiently small so that our analysis can still hold without overshooting.
3. Choosing η can be tricky [5]. It often requires some trial and error.
4. We must not take too big a η as it could undermine the basis of our analysis. A larger learning rate may lead to overshooting and hence oscillating behavior in the convergence process, which is often observed in practice.
5. If the derivative value $f'(x)$ at x is small, we make *very* small progress (one more order smaller) (note the $(f'(x))^2$ term). This implies that to speed up the process, one would need to access the higher-order derivatives of the function (such as higher-order methods like the method of Newton iteration discussed earlier).

2.13.3 *Python code for gradient descent*

We now write the following code for finding the minimum of a convex function, using only its first derivative:

```python
def primitive_gd(f_grad, x_init, η, limit=1000, atol=1.e-6):
    '''
        A primitive solver to find the minimum of a
        convex function f using the gradient descent algorithm.
        Input: f_grad: gradient function of f that can be in nD
               x_init--initial guess of x*
               η-learning rate
               limit--maximum iterations
               atol--tolerance for the error of the minimum
    '''
    i = 0; x = x_init
    while lg.norm(f_grad(x)) > atol and i < limit:
        x = x - η*f_grad(x)
        i += 1

    if lg.norm(f_grad(x)) > atol: print(f'Not yet converged:{atol}!!!')
    return x
```

2.13.4 *Example: Minimization of a simplified loss function*

Note that `primitive_gd()` works for multi-dimensional functions. This primitive code is simple, but many minimization algorithms used in ML models essentially use this algorithm with some modifications and improvements [5]. Here, we use it to find the minimizer of a simple 1D composite function that mimics a loss function in ML:

```
 1  import autograd.numpy as anp              # Thinly-wrapped numpy
 2  from autograd import grad
 3
 4  # define a composite function (mimic an ML model) in numpy
 5  def f_ML(x):
 6      f  = (anp.tanh(c1*x+c0) - y0)**2 + c2    # tanh: activiation func.
 7      return f
 8
 9  f_grad = grad(f_ML)          # gradient function (derivative for 1D)
10
11  y0 = 0.2                                        # target value
12  c0, c1, c2 = 5.0, -1.5, 0.2      # constants for weight, biase, error
13  x_init = 4.5 # 1.5               # one may use x_init = 1.5 and try
14  eta = 0.01; tol = 1e-7           # learning rate, error tolerance
15
16  x_star = primitive_gd(f_grad, x_init, eta, limit=1000, atol=1.e-6)
17  f_min  = f_ML(x_star)
18  df_min = f_grad(x_star)
19
20  print(f"Minimum value of f ={f_min:.4e}, at x={x_star:.4e}")
21  print(f"The slope of f ={df_min:.4e}, at x={x_star:.4e}")
```

```
Minimum value of f =2.0000e-01, at x=3.1982e+00
The slope of f =9.6340e-07, at x=3.1982e+00
```

Let us plot the function, its derivative, along with its minimum point found:

```
 1  plt.figure(); plt.ioff()
 2  plt.rcParams.update({'font.size': 5})
 3  fig, ax = plt.subplots(1,1,figsize=(4.5,3))
 4  X = np.arange(1, 5, 0.001)
 5  fq = f_ML(X)
 6  dfq = np.array([])
 7  for x in X:
 8      dfq = np.append(dfq, f_grad(x))
 9
10  ax.set_xlabel('x')
11  ax.set_title('Quadratic function and its derivative')
12  ax.grid(color='r', linestyle=':', linewidth=0.5)
```

```
13  ax.plot(X, fq, label="$f(x)$")
14  ax.plot(X,dfq, label="$df(x)/dx$")
15  ax.scatter(x_star, f_ML(x_star), c='g', s=9, label="Minimum")
16  ax.scatter(x_star, 0., c='r', s=6, label="Root of $df(x)/dx$")
17  ax.scatter(x_init, f_ML(x_init), c='m', s=9, label="Initial point")
18  #ax.axvline(x=0, c="k", lw=0.6); ax.axhline(y=0, c="k", lw=0.5)
19  ax.legend()#loc='upper center')
20
21  plt.savefig('imagesDI/f-ML-GDmin.png', dpi=500)
22  #plt.show()
```

The gradient descent algorithm starts at an initial point and finds the minimum of the composite function, as shown in Fig. 2.26. In the code given above, one may set x_init = 1.5 and run the code again. The code will still find the minimum.

For this kind of 1D function, one may use the Newton iteration method given earlier. For higher-dimensional problems, the gradient descent-based algorithm is often found to be more effective, robust, and easier to use. This is because it uses only the first derivative of the function. In the following chapter, we use primitive_gd() to find the minimum for a 1000-dimensional function.

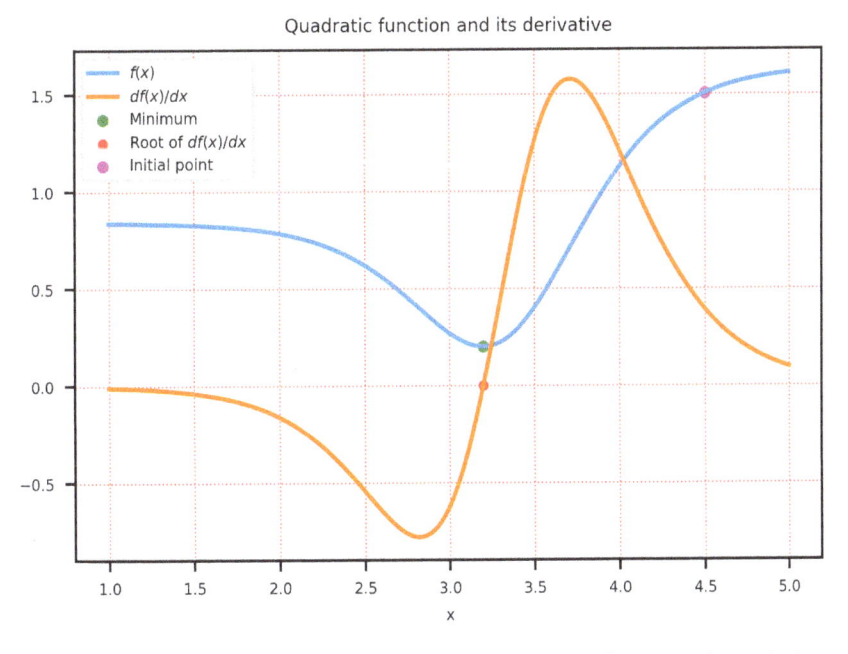

Figure 2.26. Finding the minimum of a simple composite function that mimics a loss function in an ML model via gradient descent algorithm.

2.13.5 *Case study: Conditions for roots of quadratic polynomials*

Consider a simple yet fundamental problem: determining the conditions for a quadratic polynomial with real coefficients to have different types of roots. This problem has been studied extensively in the past and is well known to many. These conditions can be derived from the discriminant in the quadratic formula [1]. The following process illustrates the use of the first and second derivatives of the polynomial, along with its geometry, to find these conditions.

Let us first acknowledge the following simple facts:

1. A constant polynomial $p(x) = c$, where c is a real constant, has either no root or infinitely many roots (when $c = 0$).
2. A first-order polynomial $p(x) = ax + c$, where $a \neq 0$, has one root $x^* = -c/a$.

Now, the quadratic polynomial can be expressed as

$$p_2(x) = ax^2 + bx + c \tag{2.96}$$

Here, we assume $a \neq 0$; otherwise, it is a first-order polynomial with one root $x^* = -c/a$.

The quadratic polynomial must have at most two roots, which can be of three types: two distinct real roots, one repeated root, or a pair of complex roots.

The first derivative of $p_2(x)$ is found as

$$p_2'(x) = 2ax + b \tag{2.97}$$

The stationary point, where $p_2'(x) = 0$, is found as

$$x^* = -\frac{b}{2a} \tag{2.98}$$

The value of the polynomial at the stationary point is computed as

$$p_2(x^*) = a\left(-\frac{b}{2a}\right)^2 - b\left(\frac{b}{2a}\right) + c = -\frac{b^2}{4a} + c \tag{2.99}$$

The second derivative of $p_2(x)$ is $p_2''(x) = 2a$. If $a > 0$, the polynomial curves upward, and if $a < 0$, it curves downward.

The following snippet plots all three quadratic polynomials with $a > 0$:

```
1  # Define a list of 2 roots for 3 2nd-order polynimials:
2  roots = [[2, 5], [3.8, 3.8], [5.5+2j, 5.5-2j]]
3
4  # Gerate coefficients for these 3 polys using roots:
5  p2_coeffs = [np.poly(roots[i]) for i in range(len(roots))]
6
7  # Produce 3 polynomials using these coeffs:
8  np_p2s = [np.poly1d(p2_coeffs[i])  for i in range(len(roots))]
9
10  x_start=[-p2_coeffs[i][1]/p2_coeffs[i][0]/2 for i in range(len(roots))]
11  p2_min =[np_p2s[i](x_start[i]) for i in range(len(roots))]
12  print(f"Stationary points, x*={x_start}")
13  print(f"                  p2(x*)={p2_min}")
```

```
Stationary points, x*=[3.5, 3.8, 5.5]
                  p2(x*)=[-2.25, 0.0, 4.0]
```

```
1  # Plot these polynomials:
2  #plt.figure(); plt.ioff()
3  plt.rcParams.update({'font.size': 12})
4  fig, ax = plt.subplots(1,1,figsize=(9,7))
5  colors = ['r', 'b', 'g']#, 'm'
6  labels = ['2 distinct real roots', '1 duplicated real root', \
7                                       '2 complex roots']
8  X = np.arange(1, 7, .1)
9
10  ax.set_xlabel('x'); ax.set_ylabel("2nd-order polynomials")
11  ax.grid(color='r', linestyle=':', linewidth=0.5)
12  for i in range(len(roots)):
13      ax.plot(X, np_p2s[i](X), colors[i], label=labels[i])
14
15  ax.scatter(x_start,p2_min, c=colors, marker="*")
16  ax.scatter(roots[0],[0,0], c=colors[0], label='real roots')
17  ax.scatter(roots[1],[0,0], c=colors[1], label='real roots')
18  ax.scatter((roots[2][0].real,roots[2][1].real), [0,0], c=colors[2], \
19                          label='real-part of complex roots')
20
21  ax.axvline(x=0, c="k", lw=0.6); ax.axhline(y=0, c="k", lw=0.6)
22  ax.legend() #loc='center right', bbox_to_anchor=(1, 0.5))
23
24  plt.savefig('imagesDI/threep2.png', dpi=500, bbox_inches='tight')
25  plt.show()
```

The stationary points and roots of the three quadratic polynomials are shown in Fig. 2.27. The red curve represents the polynomial with two real roots. Its stationary point is below the x-axis, causing the curve to cross the x-axis twice, producing two roots. The blue curve represents the polynomial with one repeated root. Its stationary point lies on the x-axis, producing a repeated root. The green curve represents the polynomial with two complex roots. Its stationary point is above the x-axis. Thus, the curve does not cross x-axis, and the entire curve remains in the upper plane. This means that it has two complex roots in the complex plane, with their real parts marked by the green dot.

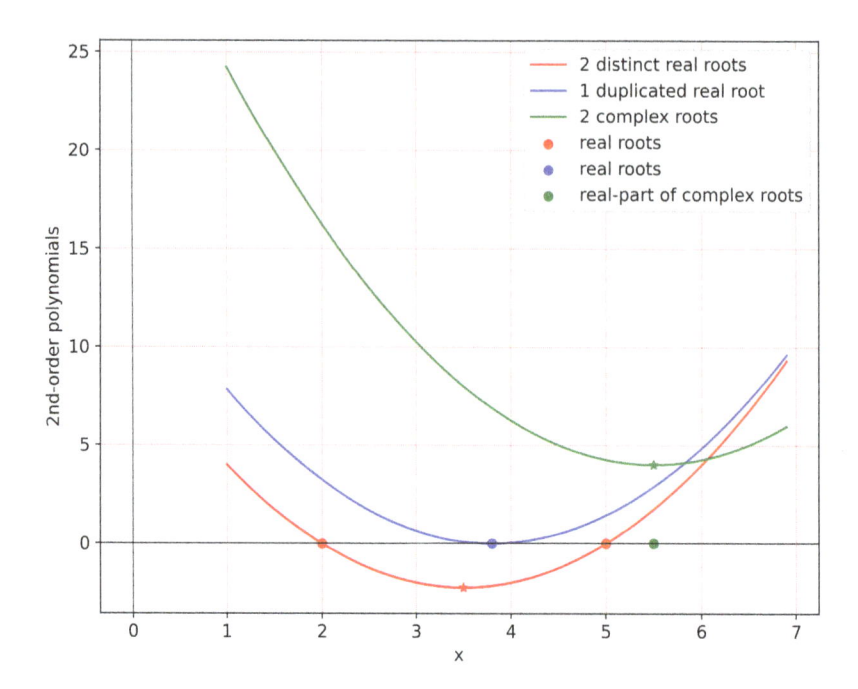

Figure 2.27.	Three typical quadratic polynomials with roots and stationary points.

Using these derivatives and the geometric features of these curves, we find the following:

1. If $p_2(x)$ has two distinct real roots, we must have $p_2(x^*) < 0$ when $a > 0$, so that the curve crosses the x-axis, producing two roots. Thus, Eq. (2.99) gives $b^2 - 4ac > 0$. If $a < 0$, the polynomial curves downward. Hence, $p_2(x^*) > 0$ must be true for the curve to cross the x-axis and produce two roots. In this case, Eq. (2.99) gives $-\frac{b^2}{4a} + c > 0$. When multiplying $4a$ on both sides of the inequality, the sign flips (because a is negative), which leads to the same condition $b^2 - 4ac > 0$.
2. If $p_2(x)$ has one repeated real root, we must have $p_2(x^*) = 0$, so that the stationary point lies on the curve, touching the x-axis to produce a repeated root. This gives the condition $b^2 - 4ac = 0$.
3. If $p_2(x)$ has a pair of complex roots, we must have $p_2(x^*) > 0$ if $a > 0$ or $p_2(x^*) < 0$ if $a < 0$, so that the curve does not cross the x-axis and thus has only complex roots. Both cases lead to the same condition $b^2 - 4ac < 0$, following the reasoning in item 1.

These conditions are summarized as follows:

$$b^2 - 4ac > 0 \text{ two real roots}$$
$$b^2 - 4ac = 0 \text{ one repeated root} \qquad (2.100)$$
$$b^2 - 4ac < 0 \text{ a pair of complex roots}$$

These are the same conditions that can be found using the quadratic formula [1]. It is a basic and important inequalty, which can also be used to prove many other inequalites, including the Cauchy–Schwarz inequality widely used in computational methods.

2.14 Higher-order derivatives of polynomials and applications

Polynomial functions are probably the most important functions in computational methods, as discussed in detail in Ref. [1]. We have also studied a few of these in this chapter. Computing their derivatives is particularly easy and is detailed in the following section.

2.14.1 *Formulation*

Consider a function that is (or approximated as) a polynomial of order n. It can be generally expressed as

$$f(x) := c_0 + c_1 x + c_2 x^2 + \cdots + c_n x^n = \sum_{k=0}^{n} c_k x^k \qquad (2.101)$$

where x is the independent variable, and c_k, $k = 0, 1, 2, \ldots, n$ are given coefficients. Using the power rule given in Eq. (2.14), the derivative of the polynomial function can be easily obtained as

$$\frac{df(x)}{dx} = c_1 + 2c_2 x + \cdots + n c_n x^{n-1} = \sum_{k=1}^{n} k c_k x^{k-1} \qquad (2.102)$$

It is still a polynomial, but of one order lower. Similarly, the second-order derivative is

$$\frac{d^2 f(x)}{dx^2} = 2c_2 + 6c_3 x + \cdots + n(n-1) c_n x^{n-2} = \sum_{k=2}^{n} k(k-1) c_k x^{k-2} \quad (2.103)$$

It is another polynomial of even lower order. The same process gives higher-order derivatives until it becomes zero. The general formula can be written as

$$\frac{d^p f(x)}{dx^p} = \sum_{k=p}^{n} \underbrace{k(k-1)\cdots(k-p+1)}_{\text{keep only } p \text{ consecutive terms}} c_k x^{k-p} \qquad (2.104)$$

Therefore, computing the derivatives of a polynomial function is particularly straightforward; they remain as polynomials. All we need to do is update the coefficients and reduce the terms accordingly. This simplicity is one reason why we often approximate complex functions with polynomials. Python offers a set of convenient tools for computing derivatives, utilizing the relation given in Eq. (2.104).

Care is needed when using a polynomial as an approximation of another complex function. The accuracy of derivatives can deteriorate even if the approximation accuracy at the function level is high.

The following section provides the examples using Python, including both NumPy and SymPy.

2.14.2 *Example: Python code for a polynomial and its derivatives*

This example demonstrates the use of Python to create, display, and differentiate polynomials. We utilize both NumPy and SymPy to derive formulas for derivatives at various orders. NumPy is efficient for numerical computations, while SymPy is valuable for symbolic derivation and LaTeX formatting. Both are effective in computing polynomial derivatives. We use the following polynomial as an example:

$$f(x) = 2x^4 + 5x^3 + 8x^2 + x + 6 \tag{2.105}$$

Since any polynomial can be completely defined using its coefficients, NumPy modules leverage this property when creating or storing any polynomial. Additionally, polynomials can be generated using their roots. In such cases, the generated polynomial is unique up to a constant factor, with the default being a unit constant for its highest-order term:

```
1  # Create polynomils, by providing its coefficients:
2  # f(x)=2*x**3+5*x**2+8*x**2+1*x+6:              # example polynomial
3
4  coeffs = [2, 5, 8, 1, 6]         # List of coeffs to define the poly.
5  np_f = np.poly1d(coeffs, variable='x')     # Generate the poly in numpy
6                                             # using the coeffs
7  np_f
```

```
poly1d([2, 5, 8, 1, 6])
```

As seen, the output consists only of the coefficients of the polynomial, as they are sufficient to fully determine it.

To produce a LaTeX display, we generate a SymPy polynomial using the same coefficients:

```
1  sp_f = sp.Poly.from_list(coeffs, sp.symbols('x'))
2  gr.printM(sp_f, 'In sympy (produced using np_f):')
```

```
In sympy (produced using np_f):
```

$$\text{Poly}\left(2x^4 + 5x^3 + 8x^2 + x + 6, x, domain = \mathbb{Z}\right)$$

Note that in SymPy, the domain for polynomials is defined in the complex domain by default. This is because the complex domain is closed under polynomial roots (see Ref. [1]).

Once the polynomial is created, its coefficients can be accessed using the following:

```
1  print(f"The coefficients in the numpy poly: {np_f.c}")
2  print(f"The coefficients in the sympy poly: {sp_f.coeffs()}")
```

```
The coefficients in the numpy poly: [2 5 8 1 6]
The coefficients in the sympy poly: [2, 5, 8, 1, 6]
```

We can also easily obtain the roots of the polynomial simply using the following:

```
1  print(f"The roots of the numpy poly: \n{np_f.r}")
2
3  x = sp.symbols('x')
4  sp_roots = [r.evalf(4) for r in sp.roots(sp_f).keys()]
5  print(f"The roots of the sympy poly:\n{sp_roots}")
```

```
The roots of the numpy poly:
[-1.4238+1.5044j -1.4238-1.5044j  0.1738+0.8179j  0.1738-0.8179j]
The roots of the sympy poly:
[-1.424 - 1.504*I, -1.424 + 1.504*I, 0.1738 - 0.8179*I, 0.1738 + 0.8179*I]
```

We observe that both NumPy and SymPy produce the same roots, all of which are complex-valued. Additional code is required for SymPy because it produces formulas for these roots by default in a dictionary format. To compute numerical values, we need to access each value in the dictionary and evaluate it. Readers interested in viewing the raw formulas for these roots produced by SymPy can use `sp.roots(sp_f)`.

Let us compute the derivative of the polynomial, and the roots of the derivative:

```
1  np_df = np_f.deriv()          # produce its derivative function in numpy
2  print(f"The derivative of the polynomial: \n{np_df}")
3  print(f"The roots of the derivative of the polynomial: \n{np_df.r}")
```

```
The derivative of the polynomial:
   3      2
8 x + 15 x + 16 x + 1
The roots of the derivative of the polynomial:
[-0.9043+1.0306j -0.9043-1.0306j -0.0665+0.j    ]
```

Note that the number of roots of the derivative is reduced by one, and one of the roots is real. Since the derivative is of third order, it must have at least one real root. In general, it is also possible for all three roots to be real even if the original polynomial has all complex roots.

Let's compute the derivative using SymPy:

```
1  sp_df = sp_f.diff(x)
2  display(Math(f" \\text{{Derivative of the polynomial:}}{{latex(sp_df)}}"))
3  sp_roots = [r.evalf(4) for r in sp.roots(sp_df).keys()]
4  print(f"The roots of the derivative of the polynomial:\n{sp_roots}")
```

Derivative of the polynomial: $\text{Poly} \left(8x^3 + 15x^2 + 16x + 1, x, domain = \mathbb{Z}\right)$

```
The roots of the derivative of the polynomial:
[-0.06650, -0.9043 + 1.031*I, -0.9043 - 1.031*I]
```

In a similar way, we can obtain the second derivatives of the polynomial:

```
1  np_d2f= np_df.deriv()                # produce 2nd derivative function
2  print(f"The 2nd derivative of the polynomial (numpy): \n{np_d2f}")
3
4  sp_d2f = sp_df.diff(x)
5  display(Math(f" \\text{{The 2nd derivative of the polynomial (sympy):}}\
6                                        {{latex(sp_d2f)}}"))
```

```
The 2nd derivative of the polynomial (numpy):
    2
24 x + 30 x + 16
```

The second derivative of the polynomial (sympy): $\text{Poly} \left(24x^2 + 30x + 16, x,\right.$ $\left. domain = \mathbb{Z}\right)$

This process can be continued to obtain derivatives of all orders of the polynomial. Since the original polynomial is of fourth order, the fifth derivative must be zero:

```
1  np_d5f = np_f.deriv().deriv().deriv().deriv().deriv()
2  print(f"The 5th derivative of the polynomial (numpy): {np_d5f}")
3
4  sp_d5f = sp_f.diff(x).diff(x).diff(x).diff(x).diff(x)
5  display(Math(f" \\text{{5th derivative of the polynomial (sympy):}}\
6                                        {{latex(sp_d5f)}}"))
```

```
The 5th derivative of the polynomial (numpy):
0
```

5th derivative of the polynomial (sympy): $\text{Poly} \left(0, x, domain = \mathbb{Z}\right)$

To obtain only the polynomial part from a SymPy expression, we can query the arguments in the expression:

```
1  sp_f.args
```

$$\left(2x^4 + 5x^3 + 8x^2 + x + 6, \ x\right)$$

The expression can be obtained by simply using the following:

```
1  sp_polyf = sp_f.args[0]
2  display(sp_polyf)
3  display(sp_f.as_expr())                    # Alterntively, use as_expr()
```

$$2x^4 + 5x^3 + 8x^2 + x + 6$$

$$2x^4 + 5x^3 + 8x^2 + x + 6$$

Both NumPy and SymPy offer rich and useful tools for handling polynomials. Generally, NumPy's tools are more efficient for numerical computations, while SymPy's are well suited for theoretical study and formula derivation. The examples given above cover only a few of these capabilities.

2.14.3 *Case study: Condition for a third-order polynomial having three real roots*

Let's delve into a simple yet fundamental problem: determining the condition under which a third-order polynomial with real coefficients has three real roots. This problem has been studied extensively in the past. The following process, developed by the author using SymPy, illustrates the use of the first and second derivatives of a polynomial to derive important characteristics.

We know that any third-order polynomial must have at least one real root. The other two roots may be either real or form a complex conjugate pair, depending on the coefficients of the polynomial. To analyze this scenario, we can express the third-order polynomial as

$$ax^3 + bx^2 + cx + d = 0 \tag{2.106}$$

We assume $a \neq 0$; otherwise, it is a second- or first-order polynomial, and their roots are well understood. The condition for two real roots is $c^2 - 4bd \geq 0$ [1]. We can divide by a for all terms in Eq. (2.106). Therefore, Eq. (2.106) can be rewritten as follows without loss of generality:

$$\text{poly3}(x) = x^3 + bx^2 + cx + d = 0 \tag{2.107}$$

For convenience in discussion, we give the polynomial a name, poly3. After obtaining the results, one can reintroduce a (if desired) through simple substitutions:

$$b := b \times a; \quad c := c \times a; \quad d := d \times a \tag{2.108}$$

We can now study the derivatives of poly3 using the procedure described in the following section.

2.14.3.1 *Define the general third-order polynomial*

```
1  x, b, c, d = sp.symbols('x, b, c, d') #  the general form of polynomial
2  poly3 = x**3 + b*x**2 + c*x + d
3
4  d_poly3 = poly3.diff(x)                        # Its 1st derivative
5  display(Math(f" \\frac{{d}}{{dx}}({latex(poly3)}) = {latex(d_poly3)}"))
6
7  d2_poly3 = poly3.diff(x, 2)                    # Its 2nd derivative
8  display(Math(f" \\frac{{d}}{{dx}}({latex(poly3)}) = {latex(d2_poly3)}"))
```

$$\frac{d}{dx}(bx^2 + cx + d + x^3) = 2bx + c + 3x^2$$

$$\frac{d}{dx}(bx^2 + cx + d + x^3) = 2(b + 3x)$$

Therefore, at $x = -\frac{b}{3}$, the polynomial has zero curvature, implying that this point is a saddle point. If $b \geq 0$, the polynomial curves upward to the left of $x = -\frac{b}{3}$ and downward to the right. If $b < 0$, the curvature reverses. In either case, the polynomial intersects $y = 0$ and must have at least one real root.

2.14.3.2 *Find the roots of the derivative of the polynomial*

The existence of the other two roots depends on the vertical positions of the stationary points on the curve relative to $y = 0$. Therefore, we need to find these stationary points by computing the roots of its first derivative. The code is as follows:

```
1  derivative_roots = sp.solve(d_poly3, x)
2  derivative_roots
```

$$\left[-\frac{b}{3} - \frac{\sqrt{b^2 - 3c}}{3}, \ -\frac{b}{3} + \frac{\sqrt{b^2 - 3c}}{3} \right]$$

These two roots are denoted as follows for later reference:

$$x_L = -\frac{b}{3} - \frac{\sqrt{b^2 - 3c}}{3}$$

$$x_R = -\frac{b}{3} + \frac{\sqrt{b^2 - 3c}}{3}$$

(2.109)

2.14.3.3 *Vertical positions of the polynomial at these two roots*

At these two roots of the derivative, we compute the function values. At the left root, we have the following:

```
1  y_xL = poly3.subs(x, derivative_roots[0])    # y-value at the left root
2  y_xL.simplify()
```

$$\frac{2b^3}{27} + \frac{2b^2\sqrt{b^2 - 3c}}{27} - \frac{bc}{3} - \frac{2c\sqrt{b^2 - 3c}}{9} + d$$

We shall do the same for the root on the right:

```
1  y_xR = poly3.subs(x,derivative_roots[1]).simplify()
2  y_xR                                          # y-value at the right root
```

$$\frac{2b^3}{27} - \frac{2b^2\sqrt{b^2 - 3c}}{27} - \frac{bc}{3} + \frac{2c\sqrt{b^2 - 3c}}{9} + d$$

2.14.3.4 *Condition 1: Stationary point in real space*

To ensure these stationary points exist in real space, the expression under the square roots must be non-negative. We use the following code to determine this condition:

```
1  condition1 = derivative_roots[0].args[1].args[1].args[0]
2  condition1                       # condition1 must be >=0; for later use
```

$$b^2 - 3c$$

We have identified the first condition for a third-order polynomial to have three real roots:

$$b^2 - 3c \geq 0$$

(2.110)

If Eq. (2.110) is satisfied, both x_L and x_R, and hence y_{xL} and y_{xR}, will all be real.

Furthermore, if $y_{xL} \cdot y_{xR} \leq 0$, indicating that the polynomial crosses the $y = 0$ line between x_L and x_R, it must have two additional real roots on the real axis. To illustrate this clearly, we provide the following code to plot four typical polynomials, which may have either one or three real roots.

First, let's generate four different polynomials:

```python
1  # Define a list of 3 roots for 4 3rd-order polynimials:
2  roots = [[1, 5, 8], [1.5, 4, 9], [2, 6+2j, 6-2j], [2.5, 7.5+5j, 7.5-5j]]
3
4  # Gerate coefficients for 4 polys using roots:
5  p3_coeffs = [np.poly(roots[i]) for i in range(len(roots))]
6
7  # Produce 4 polynomials using these coeffs:
8  np_p3s = [np.poly1d(p3_coeffs[i])  for i in range(len(roots))]
```

```python
1  # Plot these polynomials:
2
3  plt.figure(); plt.ioff()
4  plt.rcParams.update({'font.size': 12})
5  fig, ax = plt.subplots(1,1,figsize=(8,6))
6  colors = ['r', 'b', 'g', 'm']
7  labels = ['3 real roots', '3 real roots', '1 real root', '1 real root']
8  factors = [1,-1, 0.5, 0.25]      # scales the funcs. to fit in the plot
9  X = np.arange(0, 10, .1)
10
11 ax.set_xlabel('x'); ax.set_ylabel("3rd-order polynomials")
12 ax.grid(color='r', linestyle=':', linewidth=0.5)
13 for i in range(len(roots)):
14     ax.plot(X,factors[i]*np_p3s[i](X), colors[i], label=labels[i])
15
16 ax.scatter(roots[0],[0,0,0], c=colors[0], label='real (part) of roots')
17 ax.scatter(roots[1],[0,0,0], c=colors[1])
18 ax.scatter((roots[2][0],roots[2][1].real), [0,0], c=colors[2])
19 ax.scatter((roots[3][0],roots[3][1].real), [0,0], c=colors[3])
20
21 ax.axvline(x=0, c="k", lw=0.6); ax.axhline(y=0, c="k", lw=0.6)
22 ax.legend() #loc='center right', bbox_to_anchor=(1, 0.5))
23
24 plt.savefig('imagesDI/fourp3.png', dpi=500, bbox_inches='tight')
25 #plt.show()
```

Figure 2.28 shows four typical third-order polynomials. The blue and red curves represent polynomials with three real roots. Both have two stationary points, and the product of the function values at these points is negative,

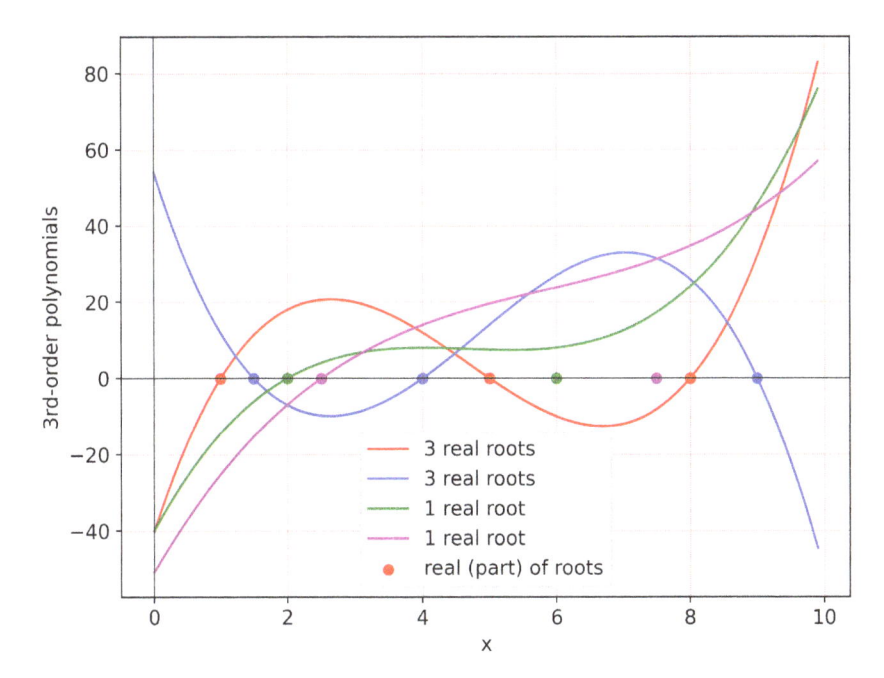

Figure 2.28. Four typical third-order polynomials: Two of these have three real roots, and other two have only one real root (but two complex roots that cannot be shown here).

indicating that the curve intersects the x-axis, producing two additional roots.

The green curve has one real root and two stationary points, but the product of the function values at these points is positive. This means the function values stay on the same side of the x-axis, making it impossible to have two more real roots.

The magenta curve has one real root and does not even have a stationary point in the real domain. It increases monotonically toward infinity, making it impossible to have two more real roots.

Note that the expressions for both y_{xL} and y_{xR} can be simplified since we are interested only in their relative vertical positions. We remove the denominator by multiplying through by 27. The second condition then becomes

```
1  condition2 = (y_xL * y_xR).simplify()*27   # multiply 27 to simplify it
2  condition2
```

$$4b^3d - b^2c^2 - 18bcd + 4c^3 + 27d^2$$

We have identified the second condition for a third-order polynomial to have three real roots:

$$4b^3d - b^2c^2 - 18bcd + 4c^3 + 27d^2 \leq 0 \qquad (2.111)$$

Conditions in Eqs. (2.110) and (2.111) together provide the answer to the question posed at the beginning.

Now, let us examine the equalities in these conditions. When $b^2 = 3c$, the two roots for the derivative given in Eq. (2.109) become identical: $x_L = x_R = -\frac{b}{3}$. This implies that these two roots coincide. Therefore, if we exclude the equality $b^2 = 3c$, Eqs. (2.110) and (2.111) become the conditions for distinct roots.

Let's validate these conditions using the following code. The strategy for testing is as follows:

1. Generate random coefficients for third-order polynomials.
2. Compute conditions in Eqs. (2.110) and (2.111).
3. If both conditions are met, compute the roots of the polynomial and check if there are any complex roots. If found, print them.
4. If the conditions are not met, compute the roots of the polynomial and check if they are all real. If they are, print them.

```python
1  # Python code for testing these conditions:
2  n_tests = 1_000                        # take about 10s on author's laptop
3
4  for _ in range(n_tests):
5      poly3_cs = np.random.randn(3)          # generate random coefficients
6      poly3_cs = np.append(1, poly3_cs)      # form coefficients for poly3
7
8      cond1 = poly3_cs[1]**2 - 3.*poly3_cs[2]              # set condition 1
9      yL = y_xL.subs({b:poly3_cs[1],c:poly3_cs[2],d:poly3_cs[3]}).evalf(4)
10     yR = y_xR.subs({b:poly3_cs[1],c:poly3_cs[2],d:poly3_cs[3]}).evalf(4)
11     np_poly3 = np.poly1d(poly3_cs, variable='x')
12
13     if cond1 >= 0:                                  # check condition 1
14         if yL*yR <= 0:                              # check condition 2
15             if np.any(np.iscomplex(np_poly3.r)):
16                 print(yL, yR, cond1, np_poly3.r)
17
18     elif not np.any(np.iscomplex(np_poly3.r)):
19             print(yL, yR, cond1, np_poly3.r)
```

We ran the code multiple times, with each run conducting 1000 random tests, all of which passed without exception. Readers are welcome to conduct their own tests using the provided code.

If necessary, a can be reintroduced into these two conditions using the following code snippets:

```
1  # For condition 1:
2  a = sp.symbols('a')
3  dict_ = {b:b/a, c:c/a, d:d/a}
4  condition1.subs(dict_).simplify()
```

$$\frac{-3ac + b^2}{a^2}$$

Since $a \neq 0$, Eq. (2.110) becomes

$$b^2 - 3ac \geq 0 \tag{2.112}$$

```
1  # For condition 2:
2  condition2.subs(dict_).simplify()
```

$$\frac{27a^2 d^2 + 2ac\left(-9bd + 2c^2\right) + b^2 \cdot \left(4bd - c^2\right)}{a^4}$$

Since $a \neq 0$, Eq. (2.111) becomes

$$27a^2 d^2 + 2ac(-9bd + 2c^2) + b^2 \cdot (4bd - c^2) \leq 0 \tag{2.113}$$

The conditions in Eqs. (2.112) and (2.113) are the ones found in the literature, which are identical to the simplified ones, Eqs. (2.110) and (2.111).

Note that when Eq. (2.110) is satisfied, the third-order polynomial has two real stationary points x_L and x_R, as given in Eq. (2.109). These are known as branching points, where the inverse function of the third-order polynomial gives rise to two complex branches [1].

2.14.4 *Case study: Extrema and their locations of a polynomial*

Since the derivative polynomial of a polynomial can have complex roots, as shown in the previous example, studying the extreme values of the polynomial can be complicated. However, if a polynomial has at least two real roots within a domain, we can assert that the function must have at least one extreme value. This is a direct consequence of Rolle's theorem. This example explores the extreme values and their locations for a polynomial that has at least two real roots.

2.14.4.1 *Setting of the problem*

Consider a third-order polynomial, called simply p3, which has three real roots: 1, 5, 8. Determine the maximum and minimum values of the polynomial function p3 within the domain of interest $x \in [0, 10]$.

2.14.4.2 *The general procedure to solve this problem*

The general procedure to find the maximum and minimum values of a function is outlined as follows:

1. Compute the first derivative of the function.
2. Find the roots of the first derivative.
3. Evaluate the second derivative of the function at these roots.
4. Determine whether the extrema are maxima or minima by examining the sign of the second derivative at these points. A negative second derivative indicates a maximum, a positive second derivative indicates a minimum, and a zero second derivative suggests a saddle point.
5. Calculate the values of the function at the boundaries of the problem domain.
6. Compare the boundary values with the extrema to determine the maximum and minimum values of the function within the problem domain.

This procedure is applicable to all types of functions. For polynomial functions, these computations are straightforward, as demonstrated in the previous examples. Let's illustrate this process with code snippets.

2.14.4.3 *Create the polynomial function*

Let us first create a polynomial by giving these roots.

```
1  roots = [1, 5, 8]        # a list of 3 roots, hence for a 3rd-order poly.
2
3  p3_coeff = np.poly(roots)      # get coefficients of the poly using roots
4
5  np_p3 = np.poly1d(p3_coeff)   # produce polynomial function using coeffs
6
7  print(f"The roots:{np.roots(np_p3)}")        # find and print its roots
```

```
The roots:[8.0000e+00 5.0000e+00 1.0000e+00]
```

We have created p3 in NumPy and confirmed that it does have three roots same as those used to create it. Let us do the same in SymPy.

```
1  x = sp.symbols('x')
2  sp_p3 = sp.Poly.from_list(p3_coeff, sp.symbols('x'))      # in sympy
3  sp_p3
```

$$\text{Poly}\left(1.0x^3 - 14.0x^2 + 53.0x - 40.0, x, domain = \mathbb{R}\right)$$

We created the p3 with the leading coefficient being 1 because it is created using its roots:

```
1  np_dp3 = np_p3.deriv()          # produce its derivative function in numpy
2  print(np_dp3)
3  sp_dp3 = sp_p3.diff(x)          # produce its derivative function in sympy
4  sp_dp3
```

```
   2
3 x - 28 x + 53
```

$$\text{Poly}\left(3.0x^2 - 28.0x + 53.0, x, domain = \mathbb{R}\right)$$

```
1  np_d2p3= np_dp3.deriv()         # produce 2nd derivative function in numpy
2  sp_d2p3 = sp_dp3.diff(x)        # produce 2nd derivative function in sympy
3  sp_d2p3
```

$$\text{Poly}\left(6.0x - 28.0, x, domain = \mathbb{R}\right)$$

2.14.4.4 *Roots of the derivative function and the extrema*

The following code finds the roots of the derivative of p3, using sp.root():

```
1  #           roots  poly_value_at_roots  roots
2  extremes = [(r,    sp_p3.subs(x,r)) for r in sp.roots(sp_dp3)]
3  extremes
```

$$[(2.63907915657,\ 20.745349157)\,,\ (6.69425417677,\ -12.597201009)]$$

We have successfully identified two roots of the derivative of $p3$ and obtained the corresponding values of $p3$ at these points, as expected. Given that $p3$ has three roots: 1, 5, and 8, being a third-order polynomial, we anticipate two stationary points: one between 1 and 5, and another between 5 and 8 due to Rolle's theorem.

The following code demonstrates how to filter the necessary roots within the domain of interest $[0, 10]$. While in this example all roots fall within $[0, 10]$, this may not always be the case in general scenarios. Let's proceed with the demonstration:

```
1  # Filter the roots in the problem domain
2  domain = [0, 10]                        # define the problem domain
3
4  extremes = list(filter(lambda x: x[0] >= domain[0] and
5                              x[0] <= domain[1], extremes))
6
7  gr.printt(extremes[0])    # 1st extremum: [root, poly_value_at_root]
8  gr.printt(extremes[1])    # 2nd extremum: [root, poly_value_at_root]
```

```
[2.6391, 20.7453]
[6.6943, -12.5972]
```

Now, we determine whether these extrema are maxima or minima by evaluating the second-order derivative of $p3$:

```
1  sp_d2p3 = sp_dp3.diff(x)                     # 2nd derivative
2  x1 = extremes[0][0]              # 1st location of the stationary point
3  x2 = extremes[1][0]              # 2nd location of the stationary point
4
5  print(f'x1={x1:.4f}, x2={x2:.4f}')
6  print(f'p3(x1)={sp_p3.subs(x, x1):.4f}, p3(x2)={sp_p3.subs(x, x2):.4f}')
7  print(f'sp_d2p3(x1)={sp_d2p3.subs(x, x1):.4f}, '
8         f'sp_d2p3(x2)={sp_d2p3.subs(x, x2):.4f}')
```

```
x1=2.6391, x2=6.6943
p3(x1)=20.7453, p3(x2)=-12.5972
sp_d2p3(x1)=-12.1655, sp_d2p3(x2)=12.1655
```

The second derivative at x_1 is negative, and hence $f(x_1) = 20.745$ is a maximum. The second derivative at x_2 is positive, and hence $f(x_2) = -12.597$ is a minimum.

2.14.4.5 *Values at the domain boundaries*

It's important to note that the global maximum and minimum values may also occur at the boundaries of the domain of interest. Computing these values at the domain boundaries is straightforward:

```
1  bxL = domain[0]          # x value at the left boundary of the domain
2  bxR = domain[1]          # x value at the right boundary of the domain
3
4  print(f'bxL = {bxL:.4f}, bxR = {bxR:.4f}')
5  print(f'p3(bxL)={sp_p3.subs(x,bxL)}, p3(bxR)={sp_p3.subs(x,bxR)}')
```

```
bxL = 0.0000, bxR = 10.0000
p3(bxL)=-40.0000000000000, p3(bxR)=90.0000000000000
```

For this problem, it's evident that the maximum function value of 90.0 occurs at the right boundary of the domain, while the minimum function value of -40.0 occurs at the left boundary. However, this outcome may vary if the problem domain is adjusted.

2.14.4.6 *Plot the function with roots and extrema*

The following code snippet plots a polynomial function, its derivative function, roots, the extrema, and boundary values. The plot is shown in Fig. 2.29.

```
 1  plt.figure(); plt.ioff()
 2  plt.rcParams.update({'font.size': 12})
 3  fig, ax = plt.subplots(1,1,figsize=(10,5))
 4  X = np.arange(0, 10, .1)
 5
 6  ax.set_xlabel('x')        #ax.set_ylabel("$f(x)$")
 7  #ax.set_title('3rd polynomial and its derivative')
 8  ax.grid(color='r', linestyle=':', linewidth=0.5)
 9  ax.plot(X,np_p3(X), label="$f(x)$")                        # use numpy
10  ax.plot(X,np_dp3(X), label="$df(x)/dx$")
11
12  ax.scatter(roots,[0,0,0], c='r', label="Roots of f(x)")
13  ax.scatter([extremes[0][0],extremes[1][0]],[0,0], c='m', marker='x',
14              label="Roots of $df(x)/dx$")
15  ax.scatter([extremes[0][0],extremes[1][0]],
16              [extremes[0][1],extremes[1][1]], c='m',label="Extremes")
17  ax.scatter([X[0], X[-1]], [np_p3(X[0]), np_p3(X[-1])], c='k',\
18              label="Boundary")
19
20  ax.axvline(x=0, c="k", lw=0.6); ax.axhline(y=0, c="k", lw=0.6)
21  ax.legend() #loc='center right', bbox_to_anchor=(1, 0.5))
22
23  plt.savefig('imagesDI/p3rd.png', dpi=500)
24  #plt.show()
```

2.15 Remarks

This chapter discussed theories, principles, techniques, and applications of derivatives for 1D real functions. We conclude with the following remarks:

1. The derivative of a function provides its local properties such as slope, tangent, or gradient at a point. SymPy is invaluable for computing derivatives of elementary functions, while Autograd is useful for numerical automatic differentiation of chained elementary functions.
2. Differentiable functions are necessarily continuous, but not all continuous functions are differentiable. Discontinuities like jumps, singularities, or kinks can prevent differentiability.

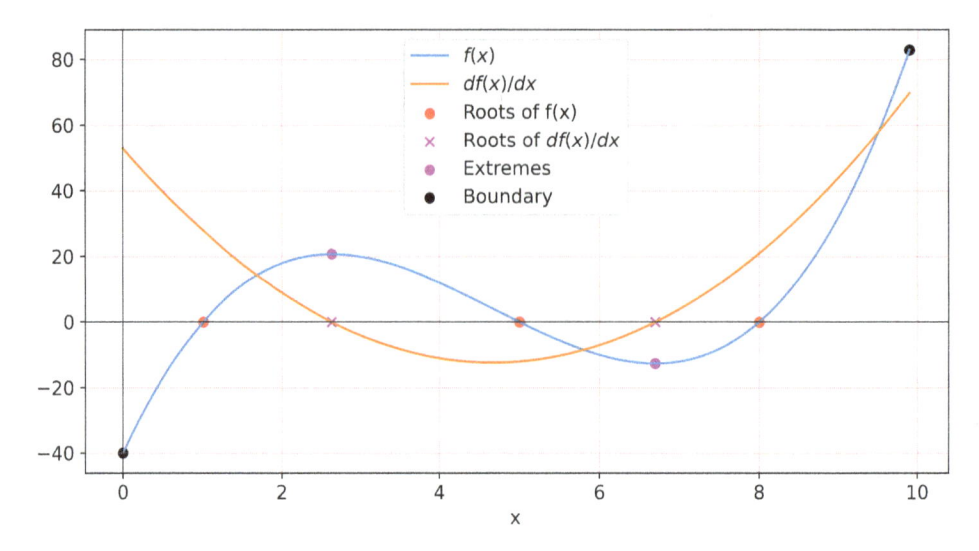

Figure 2.29. Roots, maximum and minimum values of a third-order polynomial function.

3. The chapter presented valuable theories, rules, techniques, and practical coding examples for handling various types of functions.
4. Higher-order derivatives extend the concept of derivatives, revealing important function properties like stationary points, extrema, and monotonicity.
5. Taylor series expansions utilize derivatives to approximate functions near a given point, aiding in function approximation and comparison.
6. Gradient descent and Newton's method are effective tools for locating extrema in functions. Gradient descent uses the first derivative for optimization, while Newton's method leverages both the first and second derivatives for faster convergence. Both methods require initial points to initiate optimization.

The following chapter will explore partial derivatives and gradients of multi-dimensional functions.

References

[1] G.R. Liu, *Numbers and Functions: Theory, Formulation, and Python Codes*, World Scientific, 2024.
[2] Y.C. Fan, *Advanced Mathematics*, 1979.
[3] L.X. Yang and B.Y. Bi, *Analytic Mathematics Practices*, Boris Demidovich, 2005.

[4]   G. Strang, *Calculus*, 1991. Available online (http://ocw.mit.edu/OcwWeb/resources/RES-18-001Spring-2005/ResourceHome/index.htm).

[5]   G.R. Liu, *Machine Learning with Python: Theory and Applications*, World Scientific, 2023.

[6]   G.R. Liu and S.S. Quek, *The Finite Element Method: A Practical Course*, Butterworth–Heinemann, New York, 2013.

[7]   G.R. Liu and X. Xu George, A gradient smoothing method (GSM) for fluid dynamics problems, *International Journal for Numerical Methods in Fluids*, 58(10), 1101–1133, 2008.

[8]   G.R. Liu and Zirui Mao, *Gradient Smoothing Methods with Programming: Applications to Fluids and Landslides*, 2023.

[9]   G.R. Liu, G.Y. Zhang, K.Y. Dai *et al.*, A linearly conforming point interpolation method (LCPIM) for 2D solid mechanics problems, *International Journal of Computational Methods*, 2(4), 645–665, 2005.

[10] G.R. Liu and T.T. Nguyen, *Smoothed Finite Element Methods*, Taylor and Francis Group, New York, 2010.

[11] G.R. Liu, A generalized gradient smoothing technique and the smoothed bilinear form for Galerkin formulation of a wide class of computational methods, *International Journal of Computational Methods*, 5(2), 199–236, 2008.

[12] G.R. Liu, On G space theory, *International Journal of Computational Methods*, 6(2), 257–289, 2009.

[13] G.R. Liu, Y. Li, K.Y. Dai *et al.*, A linearly conforming radial point interpolation method for solid mechanics problems, *International Journal of Computational Methods*, 3(4), 401–428, 2006.

[14] G.R. Liu, *Mesh Free Methods: Moving Beyond the Finite Element Method*, Taylor and Francis Group, New York, 2010.

[15] G.R. Liu and X. Han, *Computational Inverse Techniques in Nondestructive Evaluation*, Taylor and Francis Group, New York, 2003.

[16] G.R. Liu and Z.C. Xi, *Elastic Waves in Anisotropic Laminates*, 2001.

[17] G.R. Liu, FEA-AI and AI-AI: Two-way deepnets for real-time computations for both forward and inverse mechanics problems, *International Journal of Computational Methods*, 16(8), 1950045, 2019.

[18] S. Duan, L. Wang, F. Wang *et al.*, *A Technique for Inversely Identifying Joint Stiffnesses of Robot Arms via Two-way TubeNets*, 2021.

Chapter 3

Partial Derivatives of Functions

```
1  # Import necessary dependences.
2  import sys
3  sys.path.append('../grbin/')
4  from commonImports import *              # import dependences
5  import grcodes as gr                     # import own modules
6  importlib.reload(gr)                     # when grcodes is modified
7  np.set_printoptions(precision=4,suppress=True,    # Digits in print-outs
8                     formatter={'float_kind': '{:.4e}'.format})
```

Discussions on one-dimensional (1D) scalar functions have been provided in detail in Ref. [1], including the concept of functions, domain and codomain of functions, types of functions, behavior of functions, role of basis functions, and techniques for function approximation. Chapter 2 studies the theory, formulation, rules, property, techniques, and applications of the derivatives of 1D functions.

This chapter presents the functions in multi-dimensions (nD) and their partial derivatives and gradients. The function to be studied has several independent variables. An nD function is sometimes also termed as a multivariate function.

nD functions are, in fact, more common in computational methods because most of the real-life problems are in multiple dimensions. For example, the temperature in a room is usually different from one location to another. This means that the function of the room temperature depends on the three coordinate variables, x, y, and z. Similarly, functions of pressure in a room also depends on the three coordinates. Such a scalar function has a three-dimensional domain, but its codomain is still 1D. A loss function in a machine learning (ML) model [2] may have extremely high dimensions (millions and billions), and its codomain is also typically 1D. This chapter examines such a scalar function in nD, which is often encountered in computational methods for solving various problems in sciences and engineering. We will also introduce examples in structural mechanics to demonstrate how nD functions and their partial derivatives are actually used in real-life problems.

This chapter is written in reference to textbooks in Refs. [1, 3, 4]. Both NumPy and SymPy are used in the development of code for the demonstration examples. Wikipedia pages on partial_derivative (https://en.wikipedia.org/wiki/Partial_derivative) and function of several real variables (https://en.wikipedia.org/wiki/Function_of_several_real_variables) serve as valuable additional references. "Discussions" with ChatGPT, Gemini, and Bing have also helped greatly in coding and preparation of this chapter.

Our discussions are built upon the 1D scalar functions discussed in Ref. [1] and in Chapter 2 because many basic concepts, theorems, rules, and techniques are very similar. The major difference is caused by the dimension extension, which will be our focus of discussions. This chapter will also be more example-based.

Any problem in nD becomes more complex in terms of formulation and calculation. The key idea is often to convert it eventually to a 1D problem, which we are familiar with solving. Readers should keep this in mind while studying this chapter to avoid the confusion caused by the complex formulations.

3.1 Definition of functions in multi-dimensions

For functions defined in multi-dimensional domains (or spaces), their **domain** $\mathbb{X}$ is determined by more than one independent variable. The function produces a scalar in the **codomain** $\mathbb{Y}$. We consider functions in nD and their derivatives or gradients. For clearer elaboration, we often use two-dimensional (2D) functions as examples so that we can visualize the function graphically. Extensions will be made to three-dimensional (3D), and even 1000D, functions to show how the formulations and techniques can be extended to higher dimensions.

3.1.1 *Scalar functions in multi-dimensions*

Let us denote a 2D function in the following form:

$$f(\mathbf{x}) \equiv f(x, y) \equiv f(x_1, x_2) \tag{3.1}$$

where x and y are two real axes of the Cartesian coordinate in $\mathbb{R}^2$. In many texts, x and y are written, respectively, as x_1 and x_2 in the indicial notation. We can write the independent variables in a concise vector form:

$$\mathbf{x} \equiv \begin{bmatrix} x_1 \\ x_2 \end{bmatrix} \equiv \begin{bmatrix} x \\ y \end{bmatrix} \tag{3.2}$$

Using this notation convention, one can easily write an nD function in a general form of $f(\mathbf{x})$. for 2D, $\boldsymbol{x}$ has coordinates of x and y or x_1 and x_2; for 3D, x has x, y, z or x_1, x_2, x_3; for nD, $\mathbf{x}$ has $x_1, x_2, \ldots, x_n$, each is one coordinate of the domain $\mathbb{X} \in \mathbb{R}^n$.

As the convention in this book, we use **boldface** font representing a vector or matrix that have multiple components. This is convenient in the formulation when dealing with nD functions. Readers may need to pay special attention when seeing a boldface variable. It contains multiple variables.

In general, the codomain of a function defined in an nD domain can be $\mathbb{R}^m$ where m is an integer. This means that the output of the function is in $\mathbb{R}^m$. In such cases, the function its self is a vector with function components. If $m = 1$, $f(\mathbf{x})$ is a scalar function with a codomain $\mathbb{R}$, and the function is a mapping $f \colon \mathbb{R}^n \to \mathbb{R}$. Our discussion is only for scalar functions defined in the nD domain. Therefore, the discussion applies to one of the component functions of a vector function. Studies on vector functions will be presented in Ref. [13].

Note that the real space $\mathbb{R}$ can be naturally extended to the complex domain $\mathbb{C}$ which is, in fact, more general and enclosed, as discussed in Ref. [1]. In our discussion, we confine ourselves in $\mathbb{R}$ by default, unless specifically mentioned.

Figure 3.1 gives a graphic representations of three continuous functions $f(x_1, x_2, \ldots, x_n)$ in the space $\mathbb{R}^n$.

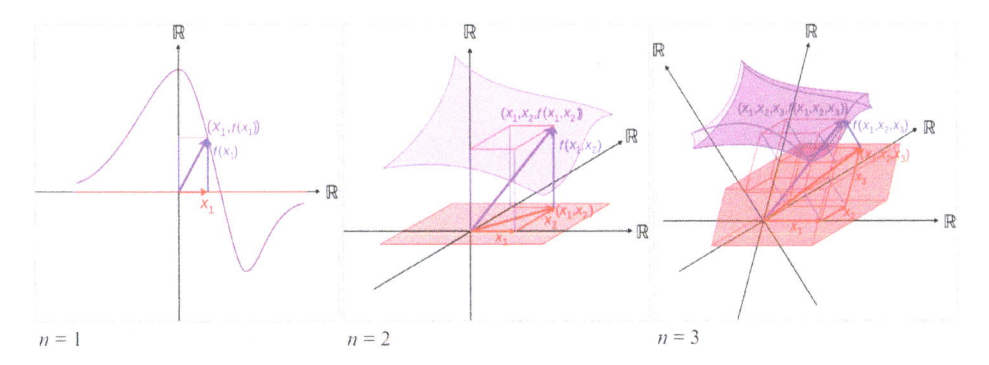

Figure 3.1. Graphic representations of functions $f(x_1, x_2, \ldots, x_n)$ in the space $\mathbb{R}^n$. $n = 1$: the function is a curve; $n = 2$: it is a surface; $n = 3$: a volume.

Source: Images from the wiki commons by Maschen under CC0 1.0 license.

The definition of an nD function is quite similar to that of a 1D function discussed in Ref. [1]. The following is an example function in $\mathbb{R}^2$:

$$f(x_1, x_2) = ax_1x_2 + bx_1 + c\sin(x_2) + d \qquad (3.3)$$

where $a, b, c,$ and d are given constants, each of which is in $\mathbb{R}$, which are independent of coordinate variables x_1 and x_2. Discussions on properties of various 1D functions (elementary functions, basis functions, and function approximation techniques using basis functions, etc.) in Ref. [1] are largely applicable to nD functions if we simply look at one of the coordinate variables and keep all the remaining variables fixed.

Operations, such as limits, differentiation, and integration, can also be performed on an nD function. Naturally, these need to be done with respect to one of the coordinate variables, typically one at a time. Proper formulations are needed to study the overall behavior and effects.

3.1.2 *Comparison of differential operations: 1D and 2D*

For differentiations of nD functions, many of the concepts studied in the previous chapter will be similar. The following provides a list of comparisons of operations for 1D and 2D functions, which gives an overall idea before we delve into each one in detail:

Comparison of differential operations

Curve: $y = f(x)$	Surface: $z = f(\mathbf{x}) = f(x, y)$
x	$\mathbf{x} \equiv \begin{bmatrix} x & y \end{bmatrix}^T$
$\dfrac{d}{dx}$	$\nabla \equiv \begin{bmatrix} \frac{\partial}{\partial x} \\ \frac{\partial}{\partial y} \end{bmatrix}$
$\dfrac{df}{dx}$	$\nabla f = \begin{bmatrix} \frac{\partial}{\partial x} \\ \frac{\partial}{\partial y} \end{bmatrix} f$
$\dfrac{df}{dx} = 0$	$\nabla f = \mathbf{0} = \begin{bmatrix} 0 \\ 0 \end{bmatrix}$
$\dfrac{d^2 f}{dx^2}$	$\nabla \nabla^T f = \begin{bmatrix} \frac{\partial^2}{\partial x^2} & \frac{\partial^2}{\partial x \partial y} \\ \frac{\partial^2}{\partial y \partial x} & \frac{\partial^2}{\partial y^2} \end{bmatrix} f$
$\Delta f \approx \dfrac{df}{dx}\Delta x$	$\Delta f \approx \Delta \mathbf{x} \nabla f = \dfrac{\partial f}{\partial x}\Delta x + \dfrac{\partial f}{\partial y}\Delta y$
$\dfrac{dy}{dt} = \dfrac{dy}{dx}\dfrac{dx}{dt}$	$\dfrac{dz}{dt} = \dfrac{\partial z}{\partial x}\dfrac{dx}{dt} + \dfrac{\partial z}{\partial y}\dfrac{dy}{dt}$ (3.4)

Using the matrix formulation, operations for an nD function have the same form as those for a 2D function. With this in mind, we can start to describe each of the operations for nD functions. Since an nD function depends on multiple independent variables, and the differentiation is done with respect to one variable at a time, we must use the so-called partial differential operation denoted conventionally by ∂. In Eq. (3.4), ∇ is called the **gradient operator**. It is a vector (of partial differential operators), and ∇f is called the gradient of f, which is also a vector. In other words, the gradient of a scalar function becomes a vector of functions. Each component in the vector function is a partial derivative of the function with respect to one of the multiple independent variables. Therefore, the shape of ∇f is the same as that of **x**.

From these simple comparisons, we note the following for an nD function:

1. It will have n first-order derivatives (a vector, a type of first-order tensor).
2. It will have $n \times n$ second-order derivatives (a matrix, a type of second-order tensor), and so on.

3.2 Partial derivatives of an nD function

3.2.1 *Definition*

Since an nD function depends on more than one independent variable, we should naturally perform differentiations with respect to **each of these variables**. A differentiation with respect to one of these variables gives a **partial derivative** of the function. The definition of a partial derivative is similar to the derivative of 1D functions. It measures how fast the function changes with the change of one independent variable, say x_1, at a point of x_1 and x_2, when only x_1 has an infinitely small change, while holding all the other independent variables unchanged. The partial derivative with respect to x_1 is expressed as

$$\frac{\partial f(x_1, x_2)}{\partial x_1} = \lim_{\Delta x_1 \to 0} \frac{f(x_1 + \Delta x_1, x_2) - f(x_1, x_2)}{\Delta x_1} \tag{3.5}$$

where Δx_1 is a small change in x_1 along x_1, implying that Δx_1 can be positive and negative.

In this entire process, x_2 is held unchanged as if it is a constant. The existence of a partial derivative requires the foregoing limit to be finite and gives the same value regardless of whether Δx is taken as positive or negative. Equation (3.5) is the basic formula for deriving a partial derivative of a function. Similarly, the partial derivative with respect to example x_2 is expressed as

$$\frac{\partial f(x_1, x_2)}{\partial x_2} = \lim_{\Delta x_2 \to 0} \frac{f(x_1, x_2 + \Delta x_2) - f(x_1, x_2)}{\Delta x_2} \tag{3.6}$$

where Δx_2 is a small change in x_2 along x_2, implying that Δx_2 can be positive and negative. In this entire process, x_1 is held unchanged as if it is a constant.

3.2.2 An example

Consider the function defined as

$$f(x_1, x_2) = ax_1 + b\sin(x_2) + c \tag{3.7}$$

By the definition given in Eq. (3.5), the partial derivative of the function with respect to x_1 becomes

$$\frac{\partial f(x_1, x_2)}{\partial x_1} = \lim_{\Delta x_1 \to 0} \frac{[a(x_1 + \Delta x_1) + b\sin(x_2) + c] - (ax_1 + b\sin(x_2) + c)}{\Delta x_1}$$

$$= \lim_{\Delta x \to 0} \frac{a\Delta x_1}{\Delta x_1} = \lim_{\Delta x_1 \to 0} a = a \tag{3.8}$$

Note that the limit for Δx_1 is taken at the final step in the same way as for the 1D functions. The outcome in Eq. (3.8) means that the partial derivative of this function along x_1 is a constant, which is the slope of the plane along the x_1-axis. Similarly, the partial derivative with respect to x_2 is evaluated as

$$\frac{\partial f(x_1, x_2)}{\partial x_2} = \lim_{\Delta x_2 \to 0} \frac{[ax_1 + b\sin(x_2 + \Delta x_2) + c] - (ax_1 + b\sin(x_2) + c)}{\Delta x_2}$$

$$= b \lim_{\Delta x_2 \to 0} \frac{\sin(x_2 + \Delta x_2) - \sin(x_2)}{\Delta x_2}$$

$$= b \lim_{\Delta x_2 \to 0} \frac{\sin(x_2)\cos(\Delta x_2) + \cos(x_2)\sin(\Delta x_2) - \sin(x_2)}{\Delta x_2}$$

$$= b \underbrace{\lim_{\Delta x_2 \to 0} \frac{\sin(x_2)(\cos(\Delta x_2) - 1)}{\Delta x_2}}_{\to 0} + b \lim_{\Delta x_2 \to 0} \frac{\cos(x_2)\sin(\Delta x_2)}{\Delta x_2}$$

$$= b\cos(x_2) \lim_{\Delta x_2 \to 0} \frac{\sin(\Delta x_2)}{\Delta x_2} = b\cos(x_2) \tag{3.9}$$

In the foregoing derivation, we used the Taylor expansion for both $\cos(\Delta x_2)$ and $\sin(\Delta x_2)$ as we did multiple times in Chapter 2.

Therefore, performing partial differentiations is the same as the differentiation we did for 1D functions discussed in Chapter 2. The only difference is that we shall hold other variables unchanged when doing it with respect to one of these variables. All the rules, techniques, and findings, including differentiability and higher derivatives, discussed in Chapter 2 apply. We can also conveniently obtain the partial derivatives of an nD function using SymPy. The only difference is a need to specify the independent variable to which the differentiation is performed.

The code for computing the partial derivatives for functions given in Eq. (3.7) is as follows:

```
1  x1, x2, a, b, c = sp.symbols('x1, x2, a, b, c')
2  f = a*x1 + b *sp.sin(x2) + c
3  print(f"df/dx1 = {f.diff(x1)}")          # differentiation w.r.t x1
4  print(f"df/dx1 = {diff(f, x1)}")                    # alternative
5  print(f"df/dx2 = {f.diff(x2)}")          # differentiation w.r.t x2
6  print(f"df/dx2 = {diff(f, x2)}")                    # alternative
```

```
df/dx1 = a
df/dx1 = a
df/dx2 = b*cos(x2)
df/dx2 = b*cos(x2)
```

3.2.3 *Rules for partial differentiation*

Based on the definition of partial differentiation, the rules used in the differentiation discussed in Chapter 2 largely apply, including the linear combination rule, product rule, power rule, reciprocal rule, and quotient rule. There may, however, be some minor notation subtleties. For example, the **chain rule** of partial differentiation shall have the following form.

Consider a simple composite function in 2D defined as

$$f\big(g(x,y)\big) \tag{3.10}$$

where $f(g)$ and $g(x,y)$ are all differentiable. The partial derivative of the composite function f with respect to x becomes

$$\frac{\partial f}{\partial x} = \frac{df}{dg}\frac{\partial g}{\partial x} \tag{3.11}$$

where we used differentiation (not partial) of f with respect to g because f depends only on one local variable g. When the differentiation is shifted to g, we shall use partial because g depends on both x and y. When $\frac{\partial f}{\partial x}$ is required, we need only $\frac{\partial g}{\partial x}$.

On the other hand, if the composite function is defined as

$$f\big(g(x,y), h(x,y)\big) \tag{3.12}$$

Implying that f depends on both $g(x,y)$ and $h(x,y)$. When computing $\frac{\partial f}{\partial x}$, the formulation becomes

$$\frac{\partial f}{\partial x} = \frac{\partial f}{\partial g}\frac{\partial g}{\partial x} + \frac{\partial f}{\partial h}\frac{\partial h}{\partial x} \tag{3.13}$$

In this case, all the differentiations are partial. In addition, we have now two terms because f depends on both $g(x,y)$ and $h(x,y)$. Their partials with respect to x must be added up, so that change in f resulting from the change in x can all be accounted for.

All these can be demonstrated using SymPy:

```
1  # Chain rule of partial differentiation
2  x, y, a, b, c = sp.symbols('x, y, a, b, c')
3  g = Function('g')(x, y)                              # 2D functions
4  h = Function('h')(x, y)
5  f = Function('f')(g, h)
6  display(Math(f" \\frac{{∂f}}{{∂x}} = {latex(f.diff(x))}"))     # wrt x
```

$$\frac{\partial f}{\partial x} = \frac{\partial}{\partial g(x,y)} f(g(x,y), h(x,y)) \frac{\partial}{\partial x} g(x,y) + \frac{\partial}{\partial h(x,y)} f(g(x,y), h(x,y)) \frac{\partial}{\partial x} h(x,y)$$

Let us use the following SymPy code to test out all these rules:

```
1  # Linear combination rule:
2  a, b = symbols('a, b', real=True)          # define symbolic variables
3  display(Math(f" \\frac{{∂}}{{∂x}}\\big({latex(a*g+b*h)}\\big) = \
4                              {latex((a*g+b*h).diff(x))}"))
```

$$\frac{\partial}{\partial x}\big(ag(x,y) + bh(x,y)\big) = a\frac{\partial}{\partial x}g(x,y) + b\frac{\partial}{\partial x}h(x,y)$$

```
1  # Product rule:
2  display(Math(f" \\frac{{∂}}{{∂x}}\\big({latex(g*h)}\\big) = \
3                              {latex((g*h).diff(x))}"))
```

$$\frac{\partial}{\partial x}\big(g(x,y)h(x,y)\big) = g(x,y)\frac{\partial}{\partial x}h(x,y) + h(x,y)\frac{\partial}{\partial x}g(x,y)$$

```
1  # Power rule for ∂[g(x)^n]/∂x:
2  n = symbols('n', integer=True)              # define symbolic variables
3  display(Math(f" \\frac{{∂}}{{∂x}}\\big({latex(g**n)}\\big) = \
4                        {latex((g**n).diff(x).simplify())}"))
```

$$\frac{\partial}{\partial x}\left(g^n(x,y)\right) = ng^{n-1}(x,y)\frac{\partial}{\partial x}g(x,y)$$

```
1  #  Reciprocal rule:
2  display(Math(f" \\frac{{∂}}{{∂x}}\\big({latex(1/g)}\\big) = \
3                        {latex((1/g).diff(x).simplify())}"))
```

$$\frac{\partial}{\partial x}\left(\frac{1}{g(x,y)}\right) = -\frac{\frac{\partial}{\partial x}g(x,y)}{g^2(x,y)}$$

```
1  #  Reciprocal rule for the ratio of two functions:
2  display(Math(f" \\frac{{∂}}{{∂x}}\\big({latex(h/g)}\\big) = \
3                        {latex((h/g).diff(x).simplify())}"))
```

$$\frac{\partial}{\partial x}\left(\frac{h(x,y)}{g(x,y)}\right) = \frac{g(x,y)\frac{\partial}{\partial x}h(x,y) - h(x,y)\frac{\partial}{\partial x}g(x,y)}{g^2(x,y)}$$

3.2.4 *Higher-order partial derivatives of functions*

Similar to 1D functions, an nD function can have higher-order partial derivatives. The definition is very similar. Each time, a higher-order partial derivative of its immediate lower-order derivative function is computed with respect to one of the independent variables, while all the other independent variables are fixed. The following sections present the examples of how higher-order partial derivatives of various functions are computed and used.

3.3 Linear functions: The most widely used functions

We introduce the simple but the most widely used linear scalar function. In the Cartesian coordinates, it can be expressed as

$$f(\mathbf{x}) = c_0 + c_1 x_1 + c_2 x_2 + \cdots + c_n x_n \tag{3.14}$$

where $x_i(i = 1, 2, \ldots, n)$ are the independent variables each in $\mathbb{R}$, $c_i(i = 0, 1, 2, \ldots, n)$ in $\mathbb{R}$ are constants (independent of any x_i in $\mathbf{x}$) called coefficients. Function $f(\mathbf{x})$ is also a scalar in $\mathbb{R}$ depending on $\mathbf{x}$. Its domain $\mathbb{X}$ is $(-\infty, \infty)^n$ or $\mathbb{R}^n$ and codomain $\mathbb{Y}$ is $(-\infty, \infty)$ or $\mathbb{R}$.

3.3.1 *Example: Partial derivatives of 2D linear functions*

The simplest 2D function is a linear function in 2D space, which is a plane. It is expressed as

$$f(\mathbf{x}) = ax_1 + bx_2 + c \qquad (3.15)$$

where x_1 and x_2, each in $\mathbb{R}$, are the independent variables and $a, b, c \in \mathbb{R}$ are constants (independent of x_1 and x_2). Function $f(\mathbf{x})$ is a scalar in $\mathbb{R}$ depending on $\mathbf{x}$. Its domain $\mathbb{X}$ is $(-\infty, \infty)^2$ or $\mathbb{R}^2$ and codomain $\mathbb{Y}$ is $(-\infty, \infty)$ or $\mathbb{R}$.

Let us use SymPy to find out these partial derivatives of the linear function:

```
1  x1, x2, a, b, c = symbols('x1, x2, a, b, c')       # symbolic variables
2  f  = a*x1 + b*x2 + c
3
4  display(Math(f" \\frac{{∂}}{{∂x_1}}({latex(f)}) = {latex(f.diff(x1))}"))
5  display(Math(f" \\frac{{∂}}{{∂x_2}}({latex(f)}) = {latex(f.diff(x2))}"))
6  display(Math(f" \\frac{{∂^2}}{{∂x_1∂x_2}}({latex(f)}) = \
7                  {latex(f.diff(x1).diff(x2))}"))
8  display(Math(f" \\frac{{∂^2}}{{∂x_2∂x_1}}({latex(f)}) = \
9                  {latex(f.diff(x2).diff(x1))}"))
```

$$\frac{\partial}{\partial x_1}(ax_1 + bx_2 + c) = a$$

$$\frac{\partial}{\partial x_2}(ax_1 + bx_2 + c) = b$$

$$\frac{\partial^2}{\partial x_1 \partial x_2}(ax_1 + bx_2 + c) = 0$$

$$\frac{\partial^2}{\partial x_2 \partial x_1}(ax_1 + bx_2 + c) = 0$$

It is seen that the partial derivative of the linear function with respect to any one of its independent variables is a constant, specifically the coefficient of the corresponding coordinate variable. This coefficient represents the slope of the function along the corresponding coordinate. The cross (also called mixed) partial derivatives $\frac{\partial^2 f}{\partial x_2 \partial x_1}$ or $\frac{\partial^2 f}{\partial x_1 \partial x_2}$ are all zero.

For given values of a, b, and c, we plot a linear function using the following Python code:

```
1  # Define a linear function in 2D in numpy
2  def f_linear(x1, x2, a, b, c):
3      f  = a*x1 + b*x2 + c
4      dfx1= a              # partial derivative for linear function w.r.t x1
5      dfx2= b                                        # w.r.t x2
6      return f, dfx1, dfx2
```

```
1  plt.figure(); plt.ioff()
2  x1 = np.linspace(.0 ,3.5,100)
3  x2 = np.linspace(-1.,4.0,100)
4  lx1, lx2 = len(x1), len(x2)
5  print('length of x1=',lx1,'length of x2=',lx2)
6
7  X1, X2 = np.meshgrid(x1, x2)
8  a, b, c = 1.0, 0.5, 2.0
9  Z, dfx1, dfx2 = f_linear(X1, X2, a, b, c)
10
11 fig_s = plt.figure()#figsize=(4,2.5))
12 ax = fig_s.add_subplot(1,1,1,projection='3d')
13
14 ax.set_xlabel('$x_1$', labelpad=-7, fontsize=6)
15 ax.set_ylabel('$x_2$', labelpad=-9, fontsize=6)
16 ax.set_zlabel('$f(x,y)$', labelpad=-9, fontsize=6)
17 ax.tick_params(axis='x', pad=-5)
18 ax.tick_params(axis='y', pad=-5)
19 ax.tick_params(axis='z', pad=-2)
20
21 ax.plot_surface(X1,X2,Z, rstride=1, cstride=1, shade=False,cmap="jet")
22
23 #plt.title('Linear function: a='+str(a)+', b=' +str(b)+' c=' +str(c))
24 fig_s.tight_layout()            # otherwise right y-label is clipped
25 plt.savefig('imagesDI/linear2D.png',dpi=500,bbox_inches='tight')
26 #plt.show()
```

```
length of x1= 100 length of x2= 100
```

A linear function in 2D is a plane, as shown in Fig. (3.2):

```
1  # print out the results of partial derivatives of the linear function:
2  print(f'Coefficient of x1 = {a};            Coefficient of x2 = {b}')
3  print(f'Partial derivative ∂f/∂x1 = {dfx1};   '\
4        f'Partial derivative ∂f/∂x2 = {dfx2}')
```

```
Coefficient of x1 = 1.0;            Coefficient of x2 = 0.5
Partial derivative ∂f/∂x1 = 1.0;   Partial derivative ∂f/∂x2 = 0.5
```

The slope of the linear function shown in Fig. 3.2 along an axis is determined by the coefficient with respect to that axis:

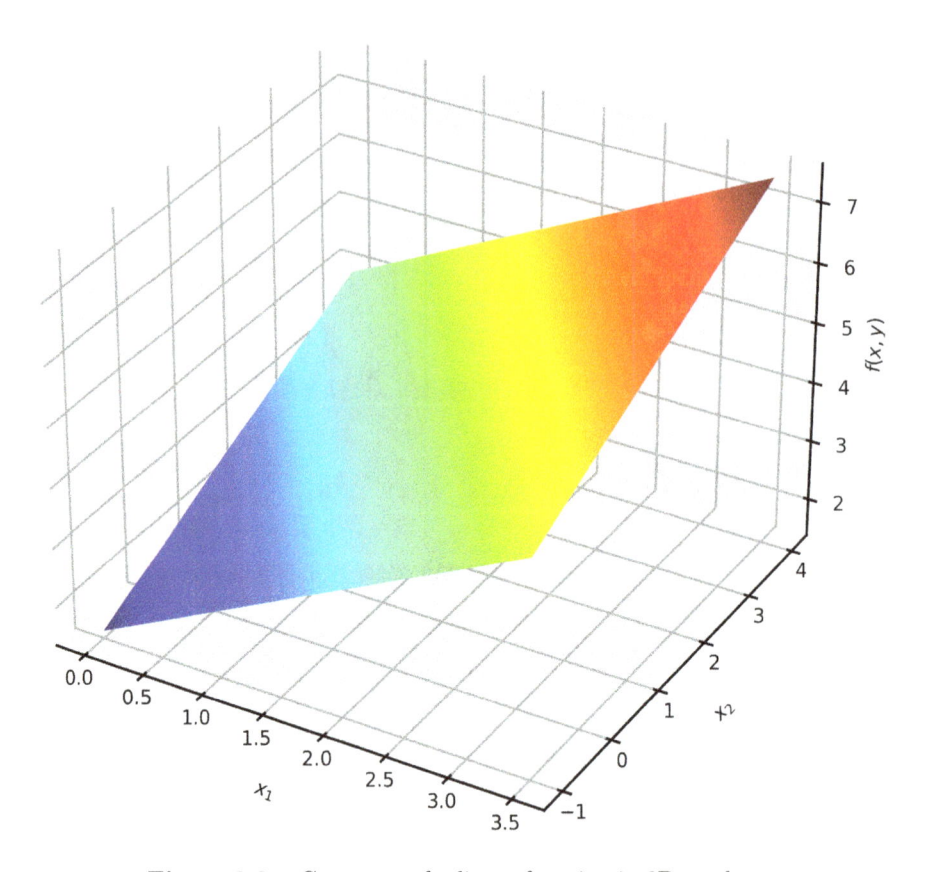

Figure 3.2.   Geometry of a linear function in 2D: a plane.

Equation (3.15) is perhaps the simplest nD function, yet it is widely used to construct linear elements in computational methods for approximating complicated functions, such as in the finite element method (FEM) [5], smoothed finite element method (SFEM) [6], and meshfree methods [7]. These coefficients a, b, c in Eq. (3.15) can be determined or converted to other types of coefficients depending on the application area, using conditions that $f(\mathbf{x})$ needs to satisfy. With only three constants, we can use triangular mesh for domain discretization, which is highly versatile for complex geometries.

A linear function in higher dimensions is a **hyperplane**. The surface of a complicated function is represented using a large number of local hyperplanes. Such an approximation can drastically simplify problems with complicated domains.

3.3.2 Example: Partial derivatives of 2D bilinear functions

Consider a function in 2D space defined as

$$f(x, y) = ax + by + cxy + d \qquad (3.16)$$

where a, b, c, and d are constants. A bilinear function varies linearly in x if y is fixed and linearly in y if x is fixed. Otherwise, it varies quadratically. This means that it is linear only in two directions and hence the name of **bilinear function**.

Let us use SymPy to compute its partial derivatives:

```
1  x, y, a, b, c, d = symbols('x, y, a, b, c, d')      # symbolic variables
2  f = sp.Function('f')(x,y)
3  f = a*x + b*y + c*x*y + d                            # define a function
4
5  display(Math(f" \\frac{{∂}}{{∂x}}({latex(f)}) = {latex(f.diff(x))}"))
6  display(Math(f" \\frac{{∂}}{{∂y}}({latex(f)}) = {latex(f.diff(y))}"))
7
8  display(Math(f" \\frac{{∂^2}}{{∂x∂y}}({latex(f)}) = \
9                              {latex(f.diff(x).diff(y))}"))
10 display(Math(f" \\frac{{∂^2}}{{∂y∂x}}({latex(f)}) = \
11                              {latex(f.diff(y).diff(x))}"))
12
13 display(Math(f" \\frac{{∂^2}}{{∂x^2}}({latex(f)}) = \
14                              {latex(f.diff(x,2))}"))
15 display(Math(f" \\frac{{∂^2}}{{∂y^2}}({latex(f)}) = \
16                              {latex(f.diff(y,2))}"))
```

$$\frac{\partial}{\partial x}(ax + by + cxy + d) = a + cy$$

$$\frac{\partial}{\partial y}(ax + by + cxy + d) = b + cx$$

$$\frac{\partial^2}{\partial x \partial y}(ax + by + cxy + d) = c$$

$$\frac{\partial^2}{\partial y \partial x}(ax + by + cxy + d) = c$$

$$\frac{\partial^2}{\partial x^2}(ax + by + cxy + d) = 0$$

$$\frac{\partial^2}{\partial y^2}(ax + by + cxy + d) = 0$$

As shown, the partial derivatives of the bilinear function are linear functions. The cross partial derivatives are no longer zero and give $\frac{\partial^2 f}{\partial x \partial y} = \frac{\partial^2 f}{\partial y \partial x} = c$. The second-order partial derivatives $\frac{\partial^2 f}{\partial x^2}$ and $\frac{\partial^2 f}{\partial y^2}$ are all zero, implying that the bilinear function is not a complete second-order polynomial.

Due to its partial quadratic behavior, the accuracy of bilinear functions in function approximation is higher than that of linear functions. For this reason, they are preferred in many cases. Bilinear functions are widely used in quadrilateral elements in FEM [5] and SFEM [6]. Since there are four constants involved, one needs to use a quadrilateral mesh for domain discretization, which is suitable for 2D domains. However, for higher dimensions, mesh generation can be challenging, especially for domains with complicated geometries. To overcome mesh difficulties, one may use a mixed mesh consisting of triangular elements for linear functions and quadrilateral elements for bilinear functions.

3.4 Partial derivatives of trigonometric functions

Trigonometric functions detailed in Chapter 4 of Ref. [1] for a 1D domain can be extended to multi-dimensions. ND trigonometric functions have applications in many fields in computational methods. Here, we introduce a few examples from structural mechanics.

3.4.1 *Case study: Deflection and moments of a plate subjected to sinusoidal load*

In engineering, plate structures are frequently used. When a plate is loaded, it deflects (albeit very slightly). Designers need to determine the amount of such deflection, which is a 2D function of x and y over the plane of the plate. From the deflection function, one can further derive the moments and stresses in the plate for safety design purposes. Various formulations can be found in textbooks, such as Refs. [8–10], although notations may differ.

The deflection function, denoted as $w(x, y)$, of a rectangular plate is found as

$$w(x, y) = w_0 \sin \frac{\pi x}{a} \sin \frac{\pi y}{b} \tag{3.17}$$

where a and b are constants standing for the length (in the x-direction) and width (in the y-direction) of the plate and w_0 is the amplitude of the deflection of the plate, which is a constant given by

$$w_0 = \frac{q_0 ab}{\pi^4 D \sqrt{a^2 + b^2}} \tag{3.18}$$

where D is a constant representing the bending rigidity of the plate defined as

$$D = \frac{Et^3}{12(1 - \nu^2)} \tag{3.19}$$

where E is the Young modulus, ν is the Poisson ratio of the material of the plate, and t is the thickness of the plate. In Eq. (3.18), q_0 is the amplitude of the sinusoidally distributed load on the top surface of the plate, which is given in the following equation:

$$q(x, y) = q_0 \sin \frac{\pi x}{a} \sin \frac{\pi y}{b} \tag{3.20}$$

Note that the domain of $w(x, y)$ is $\mathbb{R}^2$ and codomain is $\mathbb{R}$.

The components of the internal moments in the plate can be computed using the partial derivatives of the deflection function [8]:

$$M_{xx} = -D \left(\frac{\partial^2 w}{\partial x^2} + \nu \frac{\partial^2 w}{\partial y^2} \right)$$

$$M_{yy} = -D \left(\frac{\partial^2 w}{\partial y^2} + \nu \frac{\partial^2 w}{\partial x^2} \right)$$

$$M_{xy} = -(1 - \nu) D \frac{\partial^2 w}{\partial x \partial y} \tag{3.21}$$

where the moment M_{xy} relates to the cross partial derivatives of the deflection. Let us plot it using the following code. Readers may print out other moment components by uncommenting the corresponding code line:

```
1  x, y, a, b, wD, v = symbols('x, y, a, b, w_D, v ') # symbolic variables
2  w = wD*sp.sin(sp.pi*x/a)*sp.sin(sp.pi*y/b)              # wD = w0*D
3  #Mij = -(w.diff(x,2) + v*w.diff(y,2))                  # to plot Mxx
4  #Mij = -(w.diff(y,2) + v*w.diff(x,2))                  # to plot Myy
5  Mij  = -(1-v)*w.diff(x).diff(y)    # Mxy, cross partial derivatives of w
6  display(Math(f" \\text{{$M_{{xy}}$ = }} {latex(Mij)}"))
```

$$M_{xy} = \frac{\pi^2 w_D (\nu - 1) \cos \left(\frac{\pi x}{a} \right) \cos \left(\frac{\pi y}{b} \right)}{ab}$$

```
 1  # This code plots Mij for plates, can also be used to plot deflection.
 2  a_=2.; b_=1.; t_=0.05; D_=1.; q0_=-100.; v_ =1/3
 3  w0 =   q0_*a_*b_/D_/np.sqrt(a_**2+b_**2)/np.pi**4
 4  x_ = np.linspace(0., a_, 100)
 5  y_ = np.linspace(0., b_, 100)
 6  X, Y = np.meshgrid(x_, y_)
 7
 8  # for plotting deflection w:
 9  #Z = w0*np.sin(np.pi*X/a_)*np.sin(np.pi*Y/b_)              # deflection w
10
11  # for plotting Mxx or Myy or Mxy:
12  dic = {a:a_, b:b_, v:v_, wD:w0}
13  np_Mij = lambdify((x, y), Mij.subs(dic), "numpy")          # moment Mij
14  Z = np_Mij(X, Y)/5.    # scale down 5 times the maginitude for plotting
15
16  plt.figure(); plt.ioff()
17  fig_s = plt.figure(figsize=(6.,4.))
18  ax = fig_s.add_subplot(1,1,1,projection='3d')
19  ax.set_xlabel('$x$', labelpad=-6, fontsize=6)
20  ax.set_ylabel('$y$', labelpad=-7, fontsize=6)
21  ax.set_zlabel('$Mxy(x,y)$', labelpad=-8, fontsize=6)
22  ax.tick_params(axis='x', pad=-5)   # pad tick labels
23  ax.tick_params(axis='y', pad=-5)
24  ax.tick_params(axis='z', pad=-2)
25  ax.set_zticks([0, -0.5, -1.])             # tick labels
26  ax.set_box_aspect((1, 1, 0.2))
27  ax.set_zlim(-1.5, 0)             # set limits of z-axis
28  ax.plot_surface(X,Y,Z, rstride=1,cstride=1, shade=False,cmap="jet")
29
30  fig_s.tight_layout()       # otherwise right y-label is clipped
31  plt.savefig('imagesDI/plateMxy.png',dpi=500,bbox_inches='tight')
32  #plt.show()
```

The moment M_{xy} on the cross-section of the rectangular plate loaded with the sinusoidally distributed force has the maximum values at the four corners of the plate, as shown in Fig. (3.3).

3.4.2 *Case study: Series solution of a plate subjected to uniform load*

When a rectangular plate is subjected to an arbitrarily distributed load, the deflection formula can become much more complicated. For a uniformly distributed load, the deflection function can be expressed in the form of an infinite series, where each term is a combination of exponential and trigonometric functions in 2D [8].

Consider a simply supported square plate with edge length a. Its center is at $(\frac{a}{2}, 0)$. The plate is loaded with a uniformly distributed load.

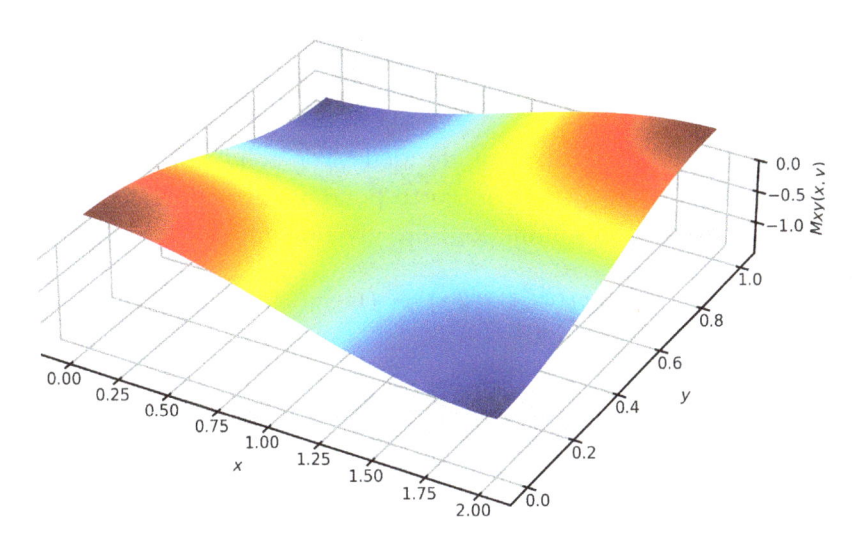

Figure 3.3. The moment M_{xy} on the cross-section of a rectangular plate subjected to a sinusoidally distributed load on the top surface. It reaches its maximum values at the four corners of the plate.

The deflection function can be written as follows:

$$w(x,y) = \frac{q_0 a^4}{D} \sum_{m=1,3,5,\dots}^{\infty} \left(A_m \cosh \frac{m\pi y}{a} + B_m \frac{m\pi y}{a} \sinh \frac{m\pi y}{a} + C_m \right) \sin \frac{m\pi x}{a}$$

(3.22)

where

$$A_m = -\frac{2\left(\alpha_m \tanh \alpha_m + 2\right)}{\pi^5 m^5 \cosh \alpha_m}$$

$$B_m = \frac{2}{\pi^5 m^5 \cosh \alpha_m}$$

$$C_m = \frac{4}{\pi^5 m^5}$$

$$\alpha_m = \frac{m\pi b}{2a}$$

(3.23)

Deriving partial derivatives of functions of this type can be quite tedious and prone to errors. This can be done easily using SymPy. Let's proceed with this for the term within the summation:

```
1  # Define the function:
2  x, y, m, a, α = symbols('x, y, m, a, α')           # symbolic variables
3  wm, Am, Bm, Cm = symbols('w_m, A_m, B_m, C_m')
4  y_ = m*sp.pi*y/a
5  x_ = m*sp.pi*x/a
6  wm = (Am*sp.cosh(y_) + Bm*y_*sp.sinh(y_) + Cm)*sp.sin(x_)
7  wm
```

$$\left(A_m \cosh\left(\frac{\pi m y}{a}\right) + \frac{\pi B_m m y \sinh\left(\frac{\pi m y}{a}\right)}{a} + C_m \right) \sin\left(\frac{\pi m x}{a}\right)$$

The first partial derivatives of the deflection relate to the slope of the loaded plate with respect to the corresponding directions. The formulas (for the term within the summation) are given as follows:

```
1  display(Math(f" \\frac{{∂w_m}}{{∂x}} = {latex(wm.diff(x))}"))
```

$$\frac{\partial w_m}{\partial x} = \frac{\pi m \left(A_m \cosh\left(\frac{\pi m y}{a}\right) + \frac{\pi B_m m y \sinh\left(\frac{\pi m y}{a}\right)}{a} + C_m \right) \cos\left(\frac{\pi m x}{a}\right)}{a}$$

```
1  display(Math(f" \\frac{{∂w_m}}{{∂y}} = {latex(wm.diff(y))}"))
```

$$\frac{\partial w_m}{\partial y} = \left(\frac{\pi A_m m \sinh\left(\frac{\pi m y}{a}\right)}{a} + \frac{\pi B_m m \sinh\left(\frac{\pi m y}{a}\right)}{a} + \frac{\pi^2 B_m m^2 y \cosh\left(\frac{\pi m y}{a}\right)}{a^2} \right) \sin\left(\frac{\pi m x}{a}\right)$$

The second partial derivatives of the deflection relate to the moments (hence the normal stresses) in the loaded plate. The formulas for the term within the summation are given as follows:

```
1  display(Math(f" \\frac{{∂^2w_m}}{{∂x^2}} = {latex(wm.diff(x, 2))}"))
```

$$\frac{\partial^2 w_m}{\partial x^2} = -\frac{\pi^2 m^2 \left(A_m \cosh\left(\frac{\pi m y}{a}\right) + \frac{\pi B_m m y \sinh\left(\frac{\pi m y}{a}\right)}{a} + C_m \right) \sin\left(\frac{\pi m x}{a}\right)}{a^2}$$

```
1  display(Math(f" \\frac{{∂^2w_m}}{{∂y^2}} = {latex(wm.diff(y, 2))}"))
```

$$\frac{\partial^2 w_m}{\partial y^2} = \frac{\pi^2 m^2 \left(A_m \cosh\left(\frac{\pi m y}{a}\right) + 2 B_m \cosh\left(\frac{\pi m y}{a}\right) + \frac{\pi B_m m y \sinh\left(\frac{\pi m y}{a}\right)}{a} \right) \sin\left(\frac{\pi m x}{a}\right)}{a^2}$$

```
1  display(Math(f" \\frac{{∂^2w_m}}{{∂x∂y}}={latex(wm.diff(x).diff(y))}"))
```

$$\frac{\partial^2 w_m}{\partial x \partial y} = \frac{\pi m \left(\frac{\pi A_m m \sinh\left(\frac{\pi m y}{a}\right)}{a} + \frac{\pi B_m m \sinh\left(\frac{\pi m y}{a}\right)}{a} + \frac{\pi^2 B_m m^2 y \cosh\left(\frac{\pi m y}{a}\right)}{a^2} \right) \cos\left(\frac{\pi m x}{a}\right)}{a}$$

The third partial derivative of the deflection relate to the shear stresses in the loaded plate. The formulas are given as follows. We print out only one of these:

```
1  #gr.printM(wm.diff(x, 3), "Partial derivative d3wm/dx3:")
2  #gr.printM(wm.diff(y, 3), "Partial derivative d3wm/dy3:")
3  #gr.printM(wm.diff(x, 2).diff(y), "Partial derivative d3wm/dx2dy:")
4  display(Math(f" \\frac{{∂^3w_m}}{{∂x∂y^2}}=\
5                 {latex(wm.diff(x).diff(y,2))}"))
```

$$\frac{\partial^3 w_m}{\partial x \partial y^2} = \frac{\pi^3 m^3 \left(A_m \cosh\left(\frac{\pi m y}{a}\right) + 2B_m \cosh\left(\frac{\pi m y}{a}\right) + \frac{\pi B_m m y \sinh\left(\frac{\pi m y}{a}\right)}{a} \right) \cos\left(\frac{\pi m x}{a}\right)}{a^3}$$

Derivations for the equations given above can be found in Ref. [8] and the Wikipedia page on bending of plates (https://en.wikipedia.org/wiki/Bending_of_plates). It requires some background in mechanics to understand. Using the derivatives of the deflection, one can compute the internal moments in the plate similar to the previous example. The point of presenting these functions here is that using elementary analytical functions to represent real-life problems in multiple dimensions can be very complicated. SymPy is a valuable tool for performing the heavy lifting in deriving these formulas.

3.5 Gradient of nD functions

3.5.1 *Definition*

Consider an nD function $f(\mathbf{x})$ that is differentiable at point $\mathbf{x}_0$, implying that all its partial derivatives exist in the neighborhood of $\mathbf{x}_0$ and are continuous at $\mathbf{x}_0$. Its partial derivatives with respect to all the independent variables $x_1, x_2, \ldots, x_n$ are denoted as $\frac{\partial f(\mathbf{x})}{\partial x_i}, i = 1, 2, \ldots, n$. Assume that all these independent variables at $\mathbf{x}_0$ have independent infinitely small changes: $dx_1, dx_2, \ldots, dx_n$. Based on the definition of partial derivatives, the change in $f(\mathbf{x})$ corresponding to dx_i, df_{x_i} at $\mathbf{x}_0$ should be

$$df_{x_i} = \frac{\partial f(\mathbf{x}_0)}{\partial x_i} dx_i \tag{3.24}$$

The total change in $f(\mathbf{x})$ at $\mathbf{x}_0$, $df_{\mathbf{x}_0}$, resulting from all these infinitely small changes in all $x_i (i = 1, 2, \ldots, n)$, should be the sum of these df_{x_i}:

$$df(\mathbf{x}_0) = \sum_{i=1}^{n} \frac{\partial f(\mathbf{x}_0)}{\partial x_i} dx_i = \frac{\partial f(\mathbf{x}_0)}{\partial x_1} dx_1 + \frac{\partial f(\mathbf{x}_0)}{\partial x_2} dx_2 + \cdots + \frac{\partial f(\mathbf{x}_0)}{\partial x_n} dx_n$$

$$= \underbrace{\left[\frac{\partial f(\mathbf{x}_0)}{\partial x_1} \quad \frac{\partial f(\mathbf{x}_0)}{\partial x_2} \quad \cdots \quad \frac{\partial f(\mathbf{x}_0)}{\partial x_n} \right]}_{\frac{df(\mathbf{x}_0)}{d\mathbf{x}}} \underbrace{\begin{bmatrix} dx_1 \\ dx_2 \\ \vdots \\ dx_n \end{bmatrix}}_{d\mathbf{x}} = \frac{df(\mathbf{x}_0)}{d\mathbf{x}} \cdot d\mathbf{x} \tag{3.25}$$

where $\frac{df(\mathbf{x}_0)}{d\mathbf{x}}$ is called the **derivative** of the nD function $f(\mathbf{x})$ at at $\mathbf{x}_0$. It is a row vector.

The **gradient** of a function, $\nabla f(\mathbf{x})$, may be more often used in computational methods. It is the transpose of the derivative of the function $\frac{df(\mathbf{x}_0)}{d\mathbf{x}}$ shown in Eq. (3.25). Hence, the gradient is a column vector with the same shape of $\mathbf{x}$. It gives a vector field and is expressed as

$$\nabla f(\mathbf{x}) = \begin{bmatrix} \frac{\partial f(\mathbf{x})}{\partial x_1} \\ \frac{\partial f(\mathbf{x})}{\partial x_2} \\ \vdots \\ \frac{\partial f(\mathbf{x})}{\partial x_n} \end{bmatrix} \qquad (3.26)$$

where ∇ is called the gradient operator defined as

$$\nabla = \begin{bmatrix} \frac{\partial}{\partial x_1} \\ \frac{\partial}{\partial x_2} \\ \vdots \\ \frac{\partial}{\partial x_n} \end{bmatrix} \qquad (3.27)$$

The gradient vector describes how a function changes in all directions at a point in space. The direction of the gradient vector indicates the direction of the fastest increase of the function at that point. The magnitude of the gradient vector represents the rate of increase in that direction. At a stationary point of a function, the gradient is the zero vector. Figure 3.4 shows an example function with its gradient field.

Gradient plays a fundamental role in optimization algorithms. It is used to minimize a function via gradient descent (or maximize via gradient ascent). It is widely used in a number of applications, including ML, where we always want to find the direction in which the function changes the most. In the autograd algorithms, the gradient is usually stored together with $\mathbf{x}$ (they have the same shape) in ML models [2].

3.5.2 *Directional derivative*

The **dot product** of $\nabla f(\mathbf{x})$ with an arbitrary vector $\mathbf{v} = [v_1, v_2, \ldots, v_n]^\top$ in $\mathbb{R}^n$ becomes

$$D_{\mathbf{v}} f = \nabla f(\mathbf{x}) \cdot \mathbf{v} = \frac{\partial f(\mathbf{x})}{\partial x_1} v_1 + \frac{\partial f(\mathbf{x})}{\partial x_2} v_2 + \cdots + \frac{\partial f(\mathbf{x})}{\partial x_n} v_n \qquad (3.28)$$

where $D_{\mathbf{v}} f$ denotes the **directional derivative** of f. It is a projection of the gradient vector on vector $\mathbf{v}$. Hence, it is the directional derivative of $f(\mathbf{x})$

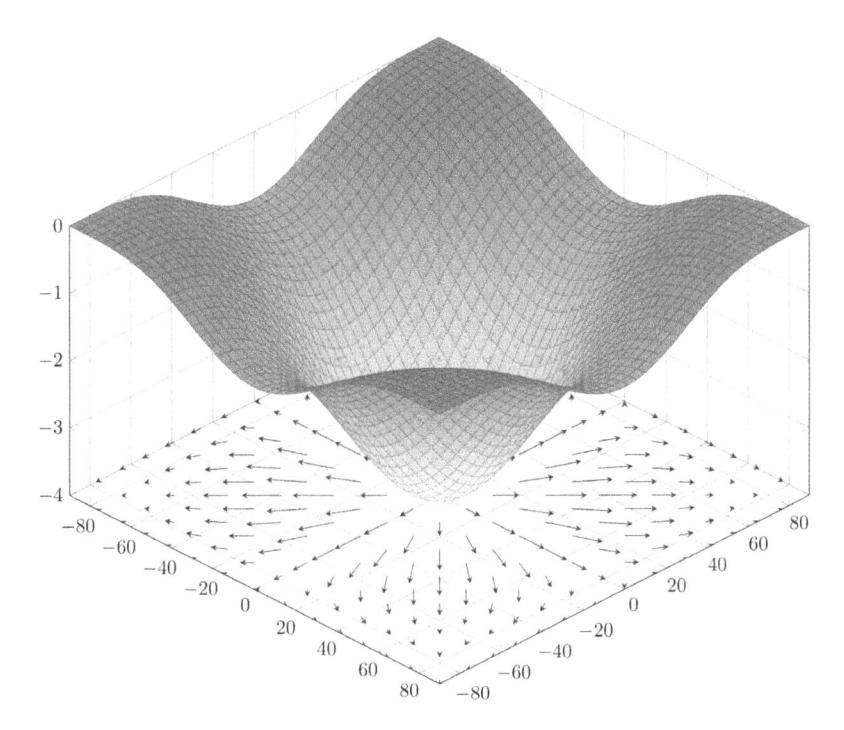

Figure 3.4. Function $f(x, y) = -(\cos^2 x + \cos^2 y)^2$ and its gradient vector field in $[-90°, 90°] \times [-90°, 90°]$. The gradient vector is shown as projected vectors on the bottom plane.

Source: Images from the wiki commons by MartinThoma under the CC0 1.0 license.

in the direction of **v**. It becomes a scalar and is the rate of change of the function in the direction of vector **v**.

3.5.3 *Gradient descent algorithm*

1. Assume that **v** is orthogonal to $\nabla f(\mathbf{x})$; their dot product is zero, and hence there is no change in the function $f(\mathbf{x})$ in the direction of **v**. In other words, **v** is tangent to a contour line of $f(\mathbf{x})$. The function value does not change when moving along such **v**.
2. If **v** is in the same direction as $\nabla f(\mathbf{x})$, the dot product is maximized, and hence the rate of change of $f(\mathbf{x})$ is the largest in the direction of **v**.
3. If **v** is in the opposite direction of $\nabla f(\mathbf{x})$, the dot product is minimized, and hence the rate of change of $f(\mathbf{x})$ is the largest but in the negative direction of **v**.

   This discussion essentially proves the gradient descent algorithm. Explicit derivations of the gradient descent algorithm with detailed formulas are given in Chapter 2 for 1D cases. For a more detailed discussion, proofs, and Python code examples, refer to Section 9.4 in Ref. [2].

The same analysis can be done using $\frac{df(\mathbf{x})}{d\mathbf{x}}$ (row vector) **acting** on the (column) vector. We use the gradient concept more often.

3.5.4 *Example: Minima of a 2D function, gradient descent*

We use the primitive_gd() developed in Chapter 2 to find a minimum of the following 2D function:

$$f(x_1, x_2) = -\big(\cos^2(x_1) + \cos^2(x_2)\big) \tag{3.29}$$

First, define the function using autograd.numpy, and then create the gradient function of the function:

```
1  import autograd.numpy as anp              # Thinly-wrapped numpy
2  from autograd import grad
3
4  def f_cos22(x):                  # Define a chained composite function
5      f1 = anp.cos(x[0])**2 + anp.cos(x[1])**2        # a 2D function
6      f = -f1**2                          # Functions be chained
7      return f
8
9  gf_cos22 = grad(f_cos22)             # generate the gradient function
```

Next, let us use the following code to plot the function:

```
1  a = 2.8 # np.pi/2
2  xL = -a; xR = a; x_points = 200
3  yL = -a; yR = a; y_points = 200
4  x1 = np.linspace(xL, xR, x_points)
5  x2 = np.linspace(yL, yR, y_points)
6
7  X1, X2 = np.meshgrid(x1, x2)
8  Z = f_cos22([X1, X2])
9
10 plt.figure(); plt.ioff()
11 plt.rcParams.update({'font.size': 4})
12 fig_s = plt.figure()  #figsize=(6,4))
13 ax = fig_s.add_subplot(1,1,1, projection='3d')
14
15 ax.set_xlabel('$x_1$', labelpad=-11, fontsize=6)
16 ax.set_ylabel('$x_2$', labelpad=-11, fontsize=6)
17 ax.tick_params(axis='x', pad=-5)
18 ax.tick_params(axis='y', pad=-3)
19 ax.tick_params(axis='z', pad=-5)
20 ax.plot_surface(X1,X2,Z, rstride=1,cstride=1,shade=False,cmap='jet')
21
22 plt.savefig('imagesDI/cos22jet.png',dpi=500,bbox_inches='tight')
23 #plt.show()
```

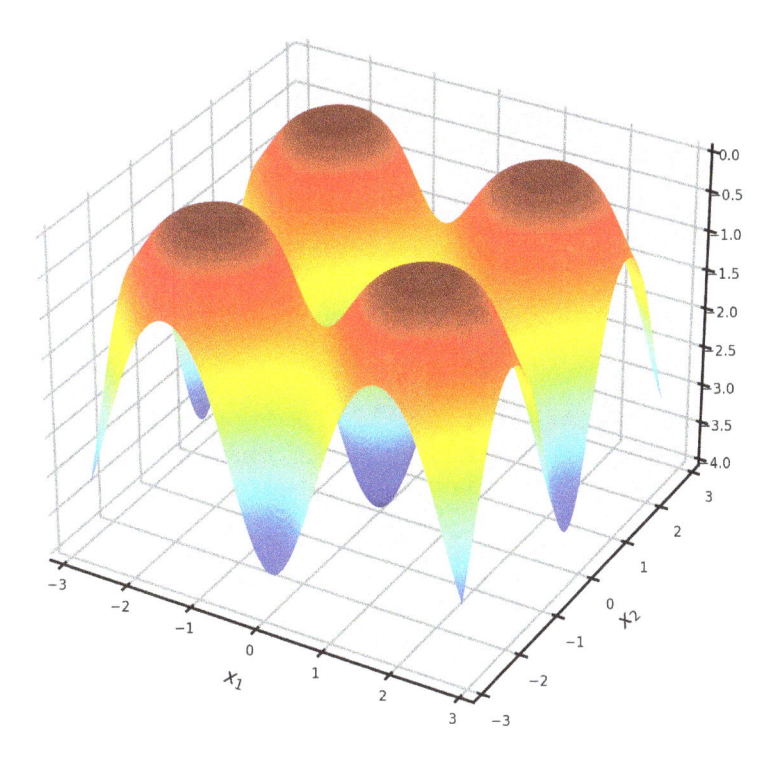

Figure 3.5. Surface plot of a 2D function made of sine and cosine.

The surface plot of the function defined in Eq. (3.29) is obtained using the code given above and is shown in Fig. 3.5.

As shown, this function is periodic and should have an infinite number of minima within its domain. Which one is found depends on the initial point used similar to the discussion for 1D functions in Chapter 2. The following code finds the minimum of the function for a given initial guess. It uses `gr.primitive_gd()`, which was developed in Chapter 2 and used for 1D problems:

```python
1  x_init = np.array([1., 1.])              # provide a x_init, may try others
2  η = 0.01; tol = 1e-6
3
4  x_star = gr.primitive_gd(gf_cos22, x_init, η, limit=1000, atol=tol)
5  f_min  = f_cos22(x_star)
6  df_min = gf_cos22(x_star)
7
8  print(f"Minimum value of f ={f_min},   at x={x_star}")
9  print(f"The slope of f ={df_min}, at x={x_star}")
```

```
Minimum value of f =-3.999999999999943,   at x=[8.4851e-08 8.4851e-08]
The slope of f =[6.7881e-07 6.7881e-07], at x=[8.4851e-08 8.4851e-08]
```

The minimizer we found is $(0,0)$, and the minimum value of the function is -4.0 when using an initial guess of $(1,1)$. The slopes in both directions on the surface defined by $f(x,y)$ are all very close to zero, and hence the minimum is a stationary point. If one changes the initial guess, the code may find another minimizer. Readers may play with initial guesses and see what the code finds.

3.5.5 *Case study: A mimic ML model with high dimensions*

Using autograd and gradient descent algorithms, the function can exist in extremely high dimensions. These two algorithms are essentially the keys that enable an ML model to be built with an extremely large number of learning parameters. The GPT-3 is known to have 175 billion parameters. This example mimics such an ML model, but drastically simplified with 1000 parameters, arranged in a vector $\mathbf{w} = [w_0, w_0, \ldots, w_{999}]^\top$. Thus, the number of dimensions becomes $\mathrm{nD} = 1000$. The loss function is simplified as

$$f(\mathbf{w}) = \tanh([\mathbf{w} - \mathbf{y}_0]^\top [\mathbf{w} - \mathbf{y}_0]) \qquad (3.30)$$

where tanh is the activation function and $\mathbf{y}_0$ is the vector of the target values similar to the label vector in an ML model. In this idealized example, we set it with all 1s. Vector $\mathbf{w}$ contains all the independent variables for the loss function. The loss function $f(\mathbf{w})$ is the dot product of the difference vector between the learning parameters and the target values, and hence it is a scalar function of n dimensions. Our task is to numerically find a $\mathbf{w}^*$ (a vector) that minimizes the loss function. Clearly, $\mathbf{w}^*$ must equal $\mathbf{y}_0$ for $f(\mathbf{w})$ to be at its minimum.

This example is kept simple, but captures the essences of an ML model using the so-called L2 loss functions [2]. The code is as follows. First, the loss function is defined with arbitrary dimensions. The gradient function of the loss function is then generated:

```
1  import autograd.numpy as anp              # Thinly-wrapped numpy
2  from autograd import grad
3
4  def loss_f(w):                             # Define the loss function
5      f1 = sum((w-y0)*(w-y0))                # a dot-product of error
6      f = anp.tanh(f1)            # use tanh as the activation function
7      return f         # a scalar function, nD depends on the length of x
8
9  g_loss_f = grad(loss_f)    # generate the gradient function of the loss
```

Next, we set the parameters for the model and initial guess for $\mathbf{w}^*$ using randomly generated values:

```
 1  nD = 1000                              # number of dimensions
 2  w_init = np.random.random(nD)      # set a w_init with random numbers
 3  y0 = np.full_like(w_init, 1.)             # set all in y0 as 1.
 4
 5  η = 0.01; tol = 1e-6         # define learning rate and error tolerance
 6
 7  # find the minimizers w*:
 8  w_star = gr.primitive_gd(g_loss_f, w_init, η, limit=10000, atol=tol)
 9  f_min  = loss_f(w_star)
10  df_min = g_loss_f(w_star)
11
12  print(f"Minimum of f ={f_min}\n at x={w_star[:4]}...{w_star.shape}")
13  print(f"Slope of f = {df_min[:4]}...{df_min.shape}\n at x={w_star[:4]}")
```

```
Minimum of f =1.0
 at x=[2.3575e-02 9.1352e-01 1.6973e-01 6.5880e-02]...(1000,)
Slope of f = [-3.9548e-303 -3.5028e-304 -3.3628e-303 -3.7834e-303]...(1000,)
 at x=[2.3575e-02 9.1352e-01 1.6973e-01 6.5880e-02]
```

As seen, we successfully found the minimizer w star that minimizes the loss function. The minimum value of the function is 1, as expected. It took microseconds for this 1000-dimensional problem. For a real ML model, the setting will be a lot more complicated. The minimizer in general can be multiple. Interested readers are directed to Ref. [2]. We stop here.

3.6 Case study: Derivatives and extrema of functions

In Chapter 2, we discussed in detail finding and examining extreme values of 1D functions. The same process for nD functions is much more complicated, but the theory and procedure are largely the same. We shall discuss these using an example, through which the basic procedure and necessary techniques and codes will be provided for effectively studying similar problems.

Before delving into detailed analysis, we should be able to anticipate our conclusions. For a 1D function, we know from Chapter 2 that the second derivative of the function determines the nature of its extrema. For an nD function, the features of the extrema should also depend on the second-order partial derivatives. Since there are $n \times n$ of them, we must have a matrix of second-order derivatives. Yes, we will have such a matrix, and it is called the Hessian matrix. The matrix characteristics, specifically its eigenvalues, determine the features of the extrema of the nD function.

Let us see how we can arrive at this conclusion.

3.6.1 *Setting of the example*

Assume a non-linear temperature distribution function defined over a 2D domain as

$$T_f(x, y) = \sin^2(ax) + \cos(by) + cx + dy \tag{3.31}$$

We would like to examine the features of the temperature distribution and find out where the heat sinks (where heat is being removed) and heat sources (where heat is being supplied) are located, as well as possible saddle points in the temperature field. Therefore, we need to compute higher-order partial derivatives to reveal all these properties. Clearly, a lot more detailed analysis is required.

Based on Fourier's law, the local heat flux at a point in the domain is proportional to the negative gradient of the temperature function. Therefore, our analysis shall start by computing the partial derivatives of the temperature. These can be evaluated using the following code:

```
1  x, y, a, b, c, d, e = symbols('x, y, a, b, c, d, e')      # variables
2  Tf = sp.sin(a*x)**2 + sp.cos(b*y) + c*x + d*y + e*x*y     # the function
3  df_dx = Tf.diff(x);  df_dy = Tf.diff(y) # the first partial derivatives
4  display(Math(f" \\frac{{∂}}{{∂x}}\\big({latex(Tf)}\\big) = \
5                                    {latex(df_dx)}"))
6  display(Math(f" \\frac{{∂}}{{∂y}}\\big({latex(Tf)}\\big) = \
7                                    {latex(df_dy)}"))
```

$$\frac{\partial}{\partial x}\left(cx + dy + exy + \sin^2(ax) + \cos(by)\right) = 2a\sin(ax)\cos(ax) + c + ey$$

$$\frac{\partial}{\partial y}\left(cx + dy + exy + \sin^2(ax) + \cos(by)\right) = -b\sin(by) + d + ex$$

3.6.2 *3D surface plot*

It is helpful if we can view the temperature distribution. We thus write the following code to plot the surface of the temperature function:

```
 1  xL = .5; xR = 5.; x_points = 200
 2  yL =1.5; yR = 5.; y_points = 200
 3  x1 = np.linspace(xL, xR, x_points)
 4  x2 = np.linspace(yL, yR, y_points)
 5  lx1, lx2 = len(x1), len(x2)
 6  print('length of x1=',lx1,'length of x2=',lx2)
 7
 8  X1, X2 = np.meshgrid(x1, x2)
 9  dic = {a:.8, b:1.5, c:0.1, d:0.1, e:0.1}
10  np_f = lambdify((x,y), Tf.subs(dic), "numpy")
11  Z = np_f(X1, X2)
12
13  plt.figure(); plt.ioff()
14  plt.rcParams.update({'font.size': 6})
15  fig_s = plt.figure()#figsize=(6,4))
16  ax = fig_s.add_subplot(1,1,1, projection='3d')
17
18  ax.set_xlabel('$x_1$', labelpad=-11, fontsize=6)
19  ax.set_ylabel('$x_2$', labelpad=-11, fontsize=6)
20  ax.set_zlabel('$Temperature$', labelpad=-12, fontsize=6)
21  ax.tick_params(axis='x', pad=-5)        # pad tick labels
22  ax.tick_params(axis='y', pad=-3)
23  ax.tick_params(axis='z', pad=-5)
24
25  ax.plot_surface(X1,X2,Z, rstride=1,cstride=1, shade=False,cmap="jet")
26
27  #fig_s.tight_layout()  # otherwise right y-label is clipped
28  plt.savefig('imagesDI/temperature.png',dpi=500,bbox_inches='tight')
29  #plt.show()
```

```
length of x1= 200 length of x2= 200
```

The surface of the temperature function is plotted in Fig. 3.6.

3.6.3 *Contour plot for 2D functions*

Three-dimensional effects can sometimes be better viewed using a contour plot, which can be produced using the following code:

```
 1  plt.figure(); plt.ioff()
 2  fig_s = plt.figure()#figsize=(2.5,2.5))              # start a new figure
 3  plt.rcParams.update({'font.size': 8})
 4  ax = fig_s.add_subplot(1,1,1)
 5  ax.set_xlabel('$x_1$'); ax.set_ylabel('$x_2$')
 6
 7  ax.contour(X1, X2, Z, levels=80, linewidths=0.8)
 8
 9  plt.axis('equal')
10  plt.savefig('imagesDI/temperatureContour.png',dpi=500,\
11                                  bbox_inches='tight')
12  #plt.show()
```

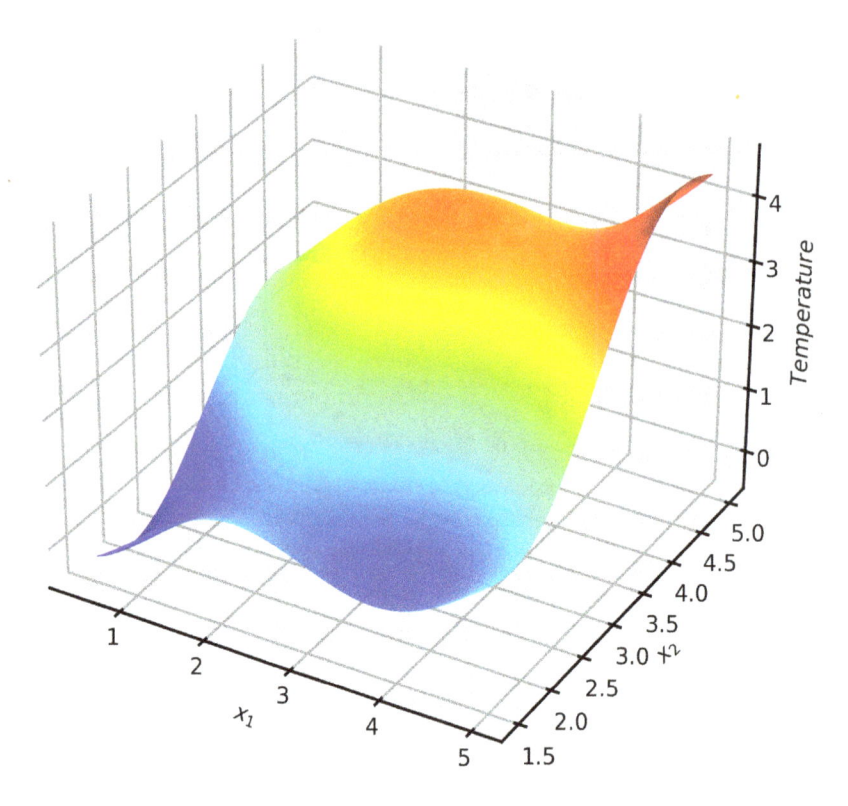

Figure 3.6. Surface plot of the temperature function that is non-linear in 2D domain.

The contour of the temperature function produced using the code given above is shown in Fig. 3.7.

Since the function value is the same on a contour curve, it provides a good estimation of possible stationary points. Inside any enclosed curve, there must be a stationary point corresponding to either a maximum or minimum. In Fig. 3.7, we can see two of these. A point near the cross point of an "X" pattern should be a stationary point corresponding to a saddle point. In Fig. 3.7, we can also find two of these. The following analysis aims to find the precise locations.

3.6.4 *Curve plot along one axis*

The contour plot shows that the temperature function exhibits some hills and valleys. We can plot the function by fixing, for example, $x_2 = 2.0$, and identify the hills and valleys along $x_2 = 2.0$. This reduces the problem to a 1D scenario, where we can write a Python function to quickly find

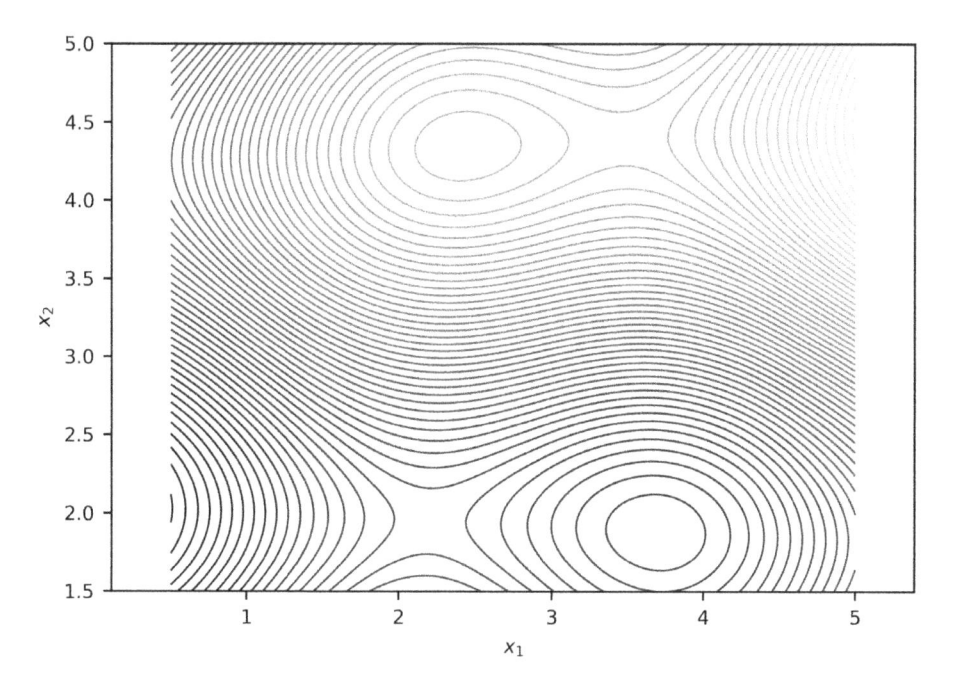

Figure 3.7. Contour plot of the temperature function.

the locations of the roots of a given SymPy function defined in a domain using `sp.nsolve()`:

```
 1  def find_roots(sp_f, xL, xR, n_division = 200):
 2      '''Find the roots of a 1D sympy function, sp_f, within the domain
 3          of (xL, xR)
 4      '''
 5      dx = (xR-xL)/n_division
 6      X = np.arange(xL, xR+dx, dx)              # discretize the domain
 7      np_f = lambdify(x, sp_f, "numpy")         # convert sp_f to np_f
 8
 9      # compute the sign of np_f
10      signs = np.sign(np_f(X))
11      # find the indices where sign changes: gives guesses of the roots.
12      indices = np.where(np.diff(signs))[0] + 1
13      guesses = X[indices]                      # guesses of the roots
14      roots = [sp.nsolve(sp_f, x, guess) for guess in guesses]
15      return roots
```

Let us plot the curve along the x-axis for $y = 2.0$ and use `find_roots()` to locate the roots of the derivative of the temperature function on this curve.

```
 1  plt.figure(); plt.ioff()
 2  plt.rcParams.update({'font.size': 7})
 3  fig, axis = plt.subplots(1,1)#,figsize=(3,2))
 4  X = np.arange(xL, xR, .01)
 5  dic = {a:.8, b:1.5, c:0.1, d:0.1, e:0.1}
 6
 7  y0 = 2.0                          # take a look at the distribution long y=y0
 8  Tfx = Tf.subs(dic).subs({y:y0})
 9  Tdf_dx = df_dx.subs(dic).subs({y:y0})
10  np_f = lambdify(x, Tfx, "numpy")
11  np_dfdx = lambdify(x, Tdf_dx, "numpy")
12
13  axis.set_xlabel('x')
14  axis.grid(color='r', linestyle=':', linewidth=0.5)
15  axis.plot(X, np_f(X),    label="$f(x)$")
16  axis.plot(X, np_dfdx(X), label="$∂f(x, y)/∂x$")
17
18  # plot the extreme points of Tf, by finding the root of Tdf_dx
19  roots = find_roots(Tdf_dx, xL, xR)
20  print(f" Roots of Tdf_dx = {roots}")
21
22  axis.scatter(roots, np_f(np.array(roots,dtype=float)), c='r', s=8)
23  axis.scatter(roots, np_dfdx((np.array(roots,dtype=float))), c='b',s=8)
24  axis.axvline(x=0, c="k", lw=0.6); axis.axhline(y=0, c="k", lw=0.6)
25  axis.legend(loc='upper center')#, bbox_to_anchor=(1, 0.5))
26  plt.savefig('imagesDI/Tfxy_2.png', dpi=500)
27  #plt.show()
```

```
Roots of Tdf_dx = [2.20374339255339, 3.68674283292747]
```

It is found that at $x = 2.0418$, there is a maximum, and at $x = 3.8487$, a minimum (along $y = 2.0$), as shown in Fig. 3.8. These extreme points correspond to the roots of the partial derivative of the function. Note that these extreme points are not the global extremes; they are only for the sliced 1D function. This provides us with a better understanding of the problem. More work is needed to find the global extrema.

3.6.5 *Plot of gradient field and stationary points*

The gradient vector function of the temperature function, often called (negated) heat flux, can be computed as follows:

```
1  X = sp.Matrix([x, y])
2  flux = gr.grad_f(Tf.subs(dic), X)    # independent variable is a vector
3  flux                                  # The flux will also be a vector
```

$$\begin{bmatrix} 0.1y + 1.6\sin{(0.8x)}\cos{(0.8x)} + 0.1 \\ 0.1x - 1.5\sin{(1.5y)} + 0.1 \end{bmatrix}$$

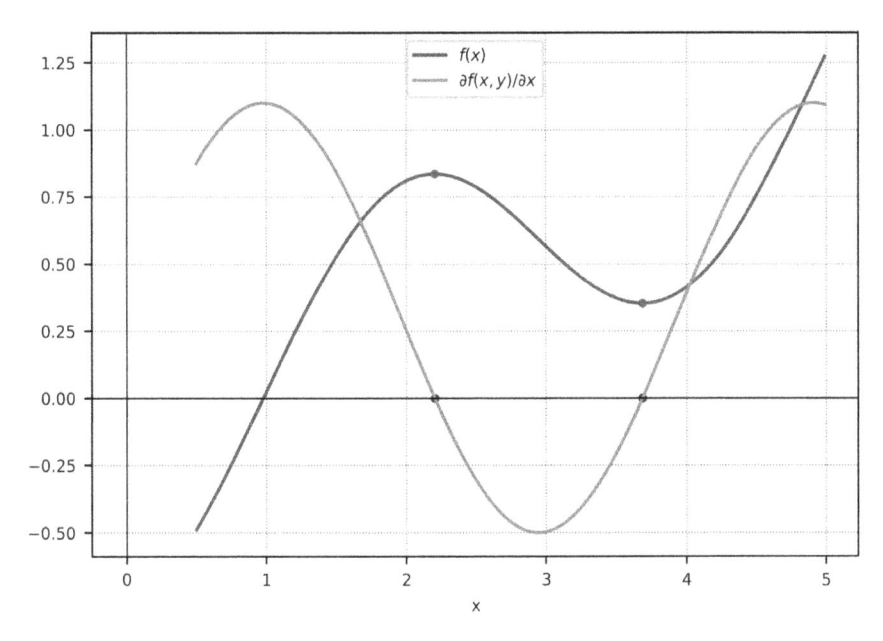

Figure 3.8. The curve of a sliced 1D function of the temperature function along x-axis at $y = 2.0$, the extreme values on it, and the roots of the derivative of the sliced 1D function.

From Figs. 3.6 and 3.7, we can see that the surface has four possible stationary points: one valley, one summit, and two saddle points. To find the precise locations of these stationary points, we need to provide initial guesses, which can be done by observing Fig. 3.7:

```
1  # Define initial guesses:
2  guesses = [[2.2, 2], [3.8, 2.], [3.5, 4.4], [2.5, 4.3]]
3              # point-1    point-2    point-3     point-4 # counterclockwise
```

The following code finds the precise locations of these stationary points:

```
1  # Solve the flux equation for x and y, where temperature is stationary:
2  slns = [sp.nsolve(flux, [x, y], guess).evalf(8) for guess in guesses]
3  slns
```

$$\left[\begin{bmatrix} 2.1996268 \\ 1.9510884 \end{bmatrix}, \begin{bmatrix} 3.6966488 \\ 1.8820845 \end{bmatrix}, \begin{bmatrix} 3.4649158 \\ 4.3902847 \end{bmatrix}, \begin{bmatrix} 2.4205012 \\ 4.3421618 \end{bmatrix} \right]$$

We can now write the following code to plot the gradient vector field marked with these stationary points at the precise locations:

```
 1  from matplotlib import cm
 2  plt.figure(); plt.ioff()
 3  plt.rcParams['figure.dpi'] = 500
 4  xL = .5; xR = 5.
 5  yL =1.5; yR = 5.
 6
 7  x1 = np.linspace(xL, xR, 30)
 8  x2 = np.linspace(yL, yR, 30)
 9  X1, X2 = np.meshgrid(x1, x2)
10
11  flux_x = lambdify((x,y), flux[0], "numpy")
12  flux_y = lambdify((x,y), flux[1], "numpy")
13
14  Amplidute = np.sqrt(flux_x(X1,X2)**2 + (flux_y(X1,X2)**2))
15  plt.quiver(X1, X2, flux_x(X1,X2), flux_y(X1,X2), Amplidute, scale=20, \
16              cmap='viridis')
17
18  plt.scatter(slns[0][0], slns[0][1], c='k')
19  plt.scatter(slns[1][0], slns[1][1], c='b')
20  plt.scatter(slns[2][0], slns[2][1], c='k')
21  plt.scatter(slns[3][0], slns[3][1], c='r')
22
23  plt.axis('equal')  #plt.colorbar()
24
25  plt.savefig('imagesDI/SaddlePoint.png', dpi=500)
26  #plt.show()
```

The gradient vector field of the temperature function is plotted in Fig. 3.9. It has one source point marked red, one sink point marked blue, and two saddle points marked black.

3.6.6 *An intuitive use of second derivatives*

As shown in Fig. 3.9, there is one heat source (red dot) in the domain, from which the heat flows out (heat flux at a point is proportional to the negative gradient). There is one heat sink (blue dot), into which the heat flows. There are two points marked in black, where the heat flows in some directions but out in others. These are saddle points. These sources, sinks, and saddle points can be identified by computing the second partial derivatives at these points. The intuitive analysis process that is analogous to 1D functions is as follows:

1. Find the stationary points of the temperature function, where the fluxes are zero. This can be done by solving the flux equation in the domain of interest.
2. Compute the second derivatives of the temperature function: $\frac{d^2 f}{dx^2}$ and $\frac{d^2 f}{dy^2}$ at these stationary points.

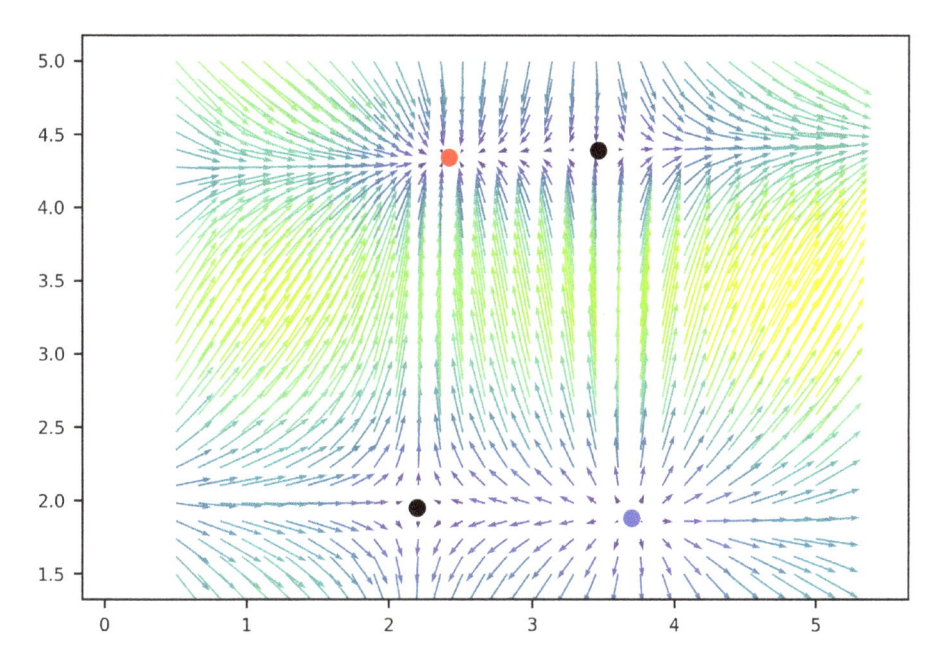

Figure 3.9. Gradient vector field of the temperature function with stationary points marked. Note that the local heat flux at a point is proportional to the negative gradient of the temperature function (Fourier's law).

3. If both $\frac{d^2 f}{dx^2}$ and $\frac{d^2 f}{dy^2}$ are negative, the temperature function is likely at a maximum. Hence, there may be a heat source keeping the temperature locally high there.

4. If both $\frac{d^2 f}{dx^2}$ and $\frac{d^2 f}{dy^2}$ are positive, the temperature function is likely at a minimum. Hence, there may be a heat sink keeping the temperature locally low there.

5. If $\frac{d^2 f}{dx^2}$ and $\frac{d^2 f}{dy^2}$ have opposite signs, the temperature function is at a maximum along one axis and a minimum along another. This point is known as a **saddle point**. The heat flows in and out from different directions.

The codes for computing the second partial derivatives are as follows:

```
1  # Compute the 2nd partitial derivatives of the temperature function:
2  for sln in slns:
3      d2fdx2 = Tf.subs(dic).diff(x, 2).subs({x:sln[0],y:sln[1]})
4      d2fdy2 = Tf.subs(dic).diff(y, 2).subs({x:sln[0],y:sln[1]})
5      print(f"At {sln},\n    d2fdx2 = {d2fdx2}, \n    d2fdy2 = {d2fdy2}")
```

```
At Matrix([[2.1996268], [1.9510884]]),
    d2fdx2 = -1.18972752569825,
    d2fdy2 = 2.19821603433783
At Matrix([[3.6966488], [1.8820845]]),
    d2fdx2 = 1.19405019740662,
    d2fdy2 = 2.13686300919912
At Matrix([[3.4649158], [4.3902847]]),
    d2fdx2 = 0.945826450489281,
    d2fdy2 = -2.14801114028946
At Matrix([[2.4205012], [4.3421618]]),
    d2fdx2 = -0.952790358994990,
    d2fdy2 = -2.19071994212435
```

It is found that the second point is a heat sink (blue dot in Fig. 3.9) because the second partial derivatives along both coordinates are positive. The fourth point is a heat source (red dot) because the second partial derivatives along both coordinates are negative. The first and third points are saddle points of the temperature function because the second partial derivatives along the two coordinates have opposite signs.

The analysis above uses a simple analogy to the 1D case, providing only an intuitive understanding. The effects of the mixed partial derivatives are ignored. A rigorous analysis requires the Hessian matrix and the concept of convexity of nD functions.

3.7 Hessian matrix of nD functions

3.7.1 *Definition*

As seen in the discussion above, the gradient operator for nD functions is analogous to the derivative of a 1D function. Due to the multi-variable nature, the gradient is not a single value but a vector, with each component representing a partial derivative of the function.

The next question would be: What is the analogy of the second-order derivative of a 1D function in nD functions? Naturally, one can expect a matrix operator to be involved. Indeed, this is the case. First, we should expect the gradient to operate twice for a similar effect to the second-order derivative. Since the gradient is a column vector, the possible way to perform such an operation twice is by letting the gradient operator (column vector) operate again on its transpose (which is the derivatives, a row vector). Then, the double operations act on an nD function. This is essentially the outer product of the gradient operator vector [11]. It leads to the so-called **Hessian matrix** of the function.

Assume that $f(\mathbf{x})$ is a continuously differentiable function, its Hessian matrix is often denoted as $\mathbf{H}_f(\mathbf{x})$ and is defined as

$$
\underbrace{\nabla \nabla^\top f(\mathbf{x})}_{\mathbf{H}_f(\mathbf{x})} = \begin{bmatrix} \frac{\partial}{\partial x_1} \\ \frac{\partial}{\partial x_2} \\ \vdots \\ \frac{\partial}{\partial x_n} \end{bmatrix} \begin{bmatrix} \frac{\partial}{\partial x_1} & \frac{\partial}{\partial x_2} & \cdots & \frac{\partial}{\partial x_n} \end{bmatrix} f(\mathbf{x})
$$

$$
= \begin{bmatrix} \frac{\partial^2}{\partial x_1^2} & \frac{\partial^2}{\partial x_1 \partial x_2} & \cdots & \frac{\partial^2}{\partial x_1 \partial x_n} \\ \frac{\partial^2}{\partial x_2 \partial x_1} & \frac{\partial^2}{\partial x_2^2} & \cdots & \frac{\partial^2}{\partial x_2 \partial x_n} \\ \vdots & \vdots & \ddots & \vdots \\ \frac{\partial^2}{\partial x_n \partial x_1} & \frac{\partial^2}{\partial x_n \partial x_2} & \cdots & \frac{\partial^2}{\partial x_n^2} \end{bmatrix} f(\mathbf{x})
$$

$$
= \begin{bmatrix} \frac{\partial^2 f(\mathbf{x})}{\partial x_1^2} & \frac{\partial^2 f(\mathbf{x})}{\partial x_1 \partial x_2} & \cdots & \frac{\partial^2 f(\mathbf{x})}{\partial x_1 \partial x_n} \\ \frac{\partial^2 f(\mathbf{x})}{\partial x_2 \partial x_1} & \frac{\partial^2 f(\mathbf{x})}{\partial x_2^2} & \cdots & \frac{\partial^2 f(\mathbf{x})}{\partial x_2 \partial x_n} \\ \vdots & \vdots & \ddots & \vdots \\ \frac{\partial^2 f(\mathbf{x})}{\partial x_n \partial x_1} & \frac{\partial^2 f(\mathbf{x})}{\partial x_n \partial x_2} & \cdots & \frac{\partial^2 f(\mathbf{x})}{\partial x_n^2} \end{bmatrix} \tag{3.32}
$$

It is seen that the Hessian matrix is square and symmetric due to the symmetry of the partial derivatives of continuously differentiable functions (Clairaut's theorem). Hence, at any point $\mathbf{x}$, the Hessian matrix is diagonalizable and has n real eigenvalues. The definiteness of $\mathbf{H}$ determines the **convexity** property of the function at point $\mathbf{x}$.

The Hessian matrix is fully populated in general. Thus, computing and storing it can be expensive for high-dimensional functions. There are techniques for efficiently approximating the Hessian matrix, such as BFGS [12]. Since the Hessian matrix essentially represents the rate of change in the gradient, such approximations may utilize the local variation property of the gradient vector.

3.7.2 *Example: Convexity of a function at stationary points*

Consider the same example studied in the previous section. We now have a better instrument to study it rigorously by computing the Hessian matrix of the temperature function defined in Eq. (3.31). We use the locations of the stationary points obtained earlier for this study:

```
1  x, y, a, b, c, d, e = symbols('x, y, a, b, c, d, e')        # variables
2  Tf = sp.sin(a*x)**2 + sp.cos(b*y) + c*x + d*y + e*x*y     # the function
3  dic = {a:.8, b:1.5, c:0.1, d:0.1, e:0.1}    # the dictinary used earlier
4
5  # The solutions of the stationary points obtained earlier:
6  slns = [Matrix([[2.19962681951243],[1.95108836369552]]),
7          Matrix([[3.69664881608535],[1.88208453601711]]),
8          Matrix([[3.46491582706229],[4.39028468191995]]),
9          Matrix([[2.42050116977703],[4.34216180458632]])]
10
11 display(Math(f" \\text{{The temperature function, Tf: }}{latex(Tf)}"))
12
13 Hessian = sp.hessian(Tf.subs(dic), (x, y))
14 display(Math(f" \\text{{The Hassian matrix of Tf: }}{latex(Hessian)}"))
```

The temperature function, Tf: $cx + dy + exy + \sin^2(ax) + \cos(by)$

The Hassian matrix of Tf:

$$\begin{bmatrix} -1.28\sin^2(0.8x) + 1.28\cos^2(0.8x) & 0.1 \\ 0.1 & -2.25\cos(1.5y) \end{bmatrix}$$

```
1  # Hassian matrixes at all the four stationary points:
2  Hfs = [Hessian.subs(dic).subs({x:s[0],y:s[1]}).evalf(8) for s in slns]
3  Hfs[:2]                                        # at the first two points
```

$$\begin{bmatrix} \begin{bmatrix} -1.1897275 & 0.1 \\ 0.1 & 2.198216 \end{bmatrix}, & \begin{bmatrix} 1.1940502 & 0.1 \\ 0.1 & 2.136863 \end{bmatrix} \end{bmatrix}$$

```
1  Hfs[2:]                                         # at the last two points
```

$$\begin{bmatrix} \begin{bmatrix} 0.94582645 & 0.1 \\ 0.1 & -2.1480111 \end{bmatrix}, & \begin{bmatrix} -0.95279036 & 0.1 \\ 0.1 & -2.1907199 \end{bmatrix} \end{bmatrix}$$

We can now compute the eigenvalues of the Hessian matrix at these extreme points. First, at the source point (the fourth point):

```
1  # Eigenvalues of Hesian matrix at point-4, marked red:
2  Hf  = [np.array(H, dtype=float) for H in Hfs]
3  print(f"Eigenvalues of Hessian matrix at point-4 = {lg.eig(Hf[3])[0]}")
```

```
Eigenvalues of Hessian matrix at point-4 = [-9.4476e-01 -2.1987e+00]
```

Since these two eigenvalues of the Hassian matrix are all negative, the function is said to be **concave** there.

```
1  # Eigenvalues of Hessian matrix at the 2nd point, marked blue:
2  Hf  = [np.array(H, dtype=float) for H in Hfs]
3  print(f"Eigenvalues of Hessian matrix at point-2 = {lg.eig(Hf[1])[0]}")
```

```
Eigenvalues of Hessian matrix at point-2 = [1.1836e+00 2.1474e+00]
```

Since these two eigenvalues of the Hessian matrix are all positive, the function there is said to be **convex** (locally near the sink point):

```
1  # Eigenvalues of Hessian matrix at these two saddle points:
2  Hf  = [np.array(H, dtype=float) for H in Hfs]
3  print(f"Eigenvalues of Hessian matrix at point-1 = {lg.eig(Hf[0])[0]}")
4  print(f"Eigenvalues of Hessian matrix at point-3 = {lg.eig(Hf[2])[0]}")
```

```
Eigenvalues of Hessian matrix at point-1 = [-1.1927e+00 2.2012e+00]
Eigenvalues of Hessian matrix at point-3 = [9.4906e-01 -2.1512e+00]
```

Since the eigenvalues of the Hessian matrices at these two **saddle** points have opposite signs, the function is locally **indeterminate** at these points.

When a Hessian matrix operates on a vector on both the right-hand side and the left-hand side (with its transpose), it produces a scalar. This operation represents a dot product sandwiched between a matrix, known as a matrix-modulated dot product or vMv form. We will use this operation in the Taylor expansion of an nD function.

3.8 nD Taylor series expansion

3.8.1 *Formulation*

We know from the Chapter 2 that many differentiable 1D functions have a Taylor expansion. A differentiable nD function can have the same. Using the gradient vector and the Hessian matrix, we can have Taylor's expansion of nD functions up to the quadratic term. It is expressed as follows: Together with that for the 1D counterpart for comparison,

$$f(x + \Delta x) = f(x) + \frac{df(x)}{dx}\Delta x + \frac{1}{2}\frac{d}{dx}\left(\frac{df(x)}{dx}\right)\Delta x^2 + \mathcal{O}(\Delta x^3)$$

$$f(\mathbf{x} + \Delta\mathbf{x}) = f(\mathbf{x}) + \nabla f(\mathbf{x})^\top \Delta\mathbf{x} + \frac{1}{2}\Delta\mathbf{x}^\top \underbrace{\nabla\nabla^\top f(\mathbf{x})}_{\mathbf{H}(\mathbf{x})} \Delta\mathbf{x} + \mathcal{O}(\|\Delta\mathbf{x}\|^3)$$

$$(3.33)$$

It is observed that the Taylor expansion for nD functions is a natural extension of that for 1D functions. Here, the first derivative of a 1D function is replaced with the gradient of the nD function, and the second derivative of the 1D function corresponds to the Hessian matrix in nD. Care is crucial in the expression of an nD Taylor expansion due to involving matrix operations. The order of these operations is critical and non-commutative. Each term must result in a scalar, akin to the nD function itself, necessitating the careful placement of transposes for vector and matrix operations in the correct sequence. In contrast, for 1D functions, there is no concern over the order of terms since all terms are scalars and thus commutative.

One may also use the dot-product notation. In this case, we worry only about the order, but no transposes are needed. The following expression is identical:

$$f(\mathbf{x} + \Delta\mathbf{x}) = f(\mathbf{x}) + \nabla f(\mathbf{x}) \cdot \Delta\mathbf{x} + \frac{1}{2}\Delta\mathbf{x} \cdot \mathbf{H}(\mathbf{x}) \cdot \Delta\mathbf{x} + \mathcal{O}(\|\Delta\mathbf{x}\|^3) \quad (3.34)$$

Since the Hessian matrix is symmetric (and thus diagonalizable with n real eigenvalues), the third term on the right-hand side takes a quadratic form. Note that the Taylor expansion applies only in the vicinity of $\mathbf{x}$ unless the function is quadratic throughout its domain. This mirrors the 1D case studied in detail in Chapter 2.

3.8.2 *Definiteness of the Hessian matrix*

Determining whether this stationary point is a maximum or another type depends on the definiteness of the Hessian matrix, which is akin to the 1D case but more intricate. The conditions are as follows:

1. **Negative definite** (**H** has all negative eigenvalues): The stationary point is a maximum.
2. **Positive definite** (**H** has all positive eigenvalues): The stationary point is a minimum.
3. **Indefinite** (**H** has both positive and negative eigenvalues): The stationary point is a saddle point.
4. **Zero definite along some axis** (**H** has all zero eigenvalues along that axis): The function value does not change along that axis.

This property is valuable for optimizing nD functions, inversion problems, and training ML models where objective, cost, or loss functions are involved, and methods like the Newton or quasi-Newton algorithm are applied.

In many cases, one can simply use a linear approximation or linearization by taking only the first two terms in Eq. (3.34). The formula is the same as given in Eq. (3.4) at the beginning of this chapter.

Approximations using orders higher than two are not very common for nD functions. This is because all the third-order derivatives of the nD function must be computed, and hence the complexity grows extremely fast to higher-order tensors.

3.8.3 *Approximation of 2D functions via Taylor expansion*

Let us use the 2D Taylor expansion to approximate the temperature function studied earlier. We redefine the function:

```
1  x, y, a, b, c, d, e = symbols('x, y, a, b, c, d, e')      # variables
2  Tf = sp.sin(a*x)**2 + sp.cos(b*y) + c*x + d*y + e*x*y     # the function
3  dic = {a:.8, b:1.5, c:0.1, d:0.1, e:0.1}
4  Tfc = Tf.subs(dic)
5  Tfc
```

$$0.1xy + 0.1x + 0.1y + \sin^2(0.8x) + \cos(1.5y)$$

Consider an approximation in the vicinity of $(x_0, y_0) = (2, 2)$, and use up to quadratic terms. The following code produces the approximated function:

```
1  x0 = 2; y0 = 2
2  n = 3
3  Taylor_Tf=sp.series(Tfc, x, x0, n).removeO().series(y, y0, n).removeO()
4  Taylor_Tf_ = gr.printM(Taylor_Tf.expand(), n_dgt=6) # limits the digids
5  display(Math(f" \\text{{Taylor series of Tf:}} {latex(Taylor_Tf_)}"))
```

Taylor series of Tf: $-0.638909x^2 + 0.1xy + 2.60894x + 1.11374y^2 - 4.56665y + 2.42525$

The approximated function becomes a polynomial in x and y near $(2, 2)$. Let us plot both surfaces of the original and approximated functions:

```
1  dx = 1.5
2  x_ = np.linspace(x0, x0+dx, 300)          # range in an axis for plot
3  y_ = np.linspace(y0, y0+dx, 300)
4  X, Y = np.meshgrid(x_, y_)
5
6  Tf_np = lambdify((x, y), Tfc, "numpy")    # convert to numpy functions
7  Taylor_Tf_np = lambdify((x, y), Taylor_Tf, "numpy")
8
9  plt.figure(); plt.ioff()
10 #plt.rcParams.update({'font.size': 4})
11 fig = plt.figure()
12 ax = fig.add_subplot(111, projection='3d')
13 ax.plot_surface(X,Y,Tf_np(X,Y), alpha=0.7, cmap="seismic")
14 ax.plot_surface(X,Y,Taylor_Tf_np(X,Y), alpha=0.7, cmap="jet")
15 ax.tick_params(axis='x', pad=-4)          # pad tick labels
16 ax.tick_params(axis='y', pad=-4)
17 ax.tick_params(axis='z', pad=-4)
18
19 plt.savefig('imagesDI/Taylor_Tf.png',dpi=500,bbox_inches='tight')
20 #plt.show()
```

Figure 3.10 shows that the approximation is good near the point $(2, 2)$ where the gradient and Hessian matrix are computed. Therefore, it is widely used in local approximations in the same way as 1D cases discussed in Chapter 2.

3.9 Lagrange polynomials and derivatives in multi-dimensions

The Lagrange polynomial and its use in the approximation of functions in 1D have been discussed in Chapter 5 of Ref. [1]. It is now extended to nD.

3.9.1 *Definition and formulation*

Consider a 2D domain $[a, b] \times [c, d] \in \mathbb{R}^2$ (in FEM [5], it is a 2D element). It is discretized into $m \times n$ **regular** grids or intervals with $N_n = (m+1) \times (n+1)$ nodes: (x_i, y_j), $i = 1, 2, \ldots, m$; $j = 1, 2, \ldots, n$. If the values of a scalar function $f(x, y)$ at all these nodes are denoted as $f(x_i, y_j)$, the function value at any point $\mathbf{x} = (x, y) \in [a, b] \times [c, d]$ can then be interpolated using

$$f(x, y) = \sum_{i=0}^{n} \sum_{j=0}^{m} f(x_i, y_j) \underbrace{l_i(x) l_j(y)}_{N_{ij}(x,y)}$$

$$= \sum_{i=0}^{n} \sum_{j=0}^{m} f(x_i, y_j) N_{ij}(x, y)$$

$$= \sum_{k=1}^{N_n} f(\mathbf{x}_k) N_k(\mathbf{x}) \tag{3.35}$$

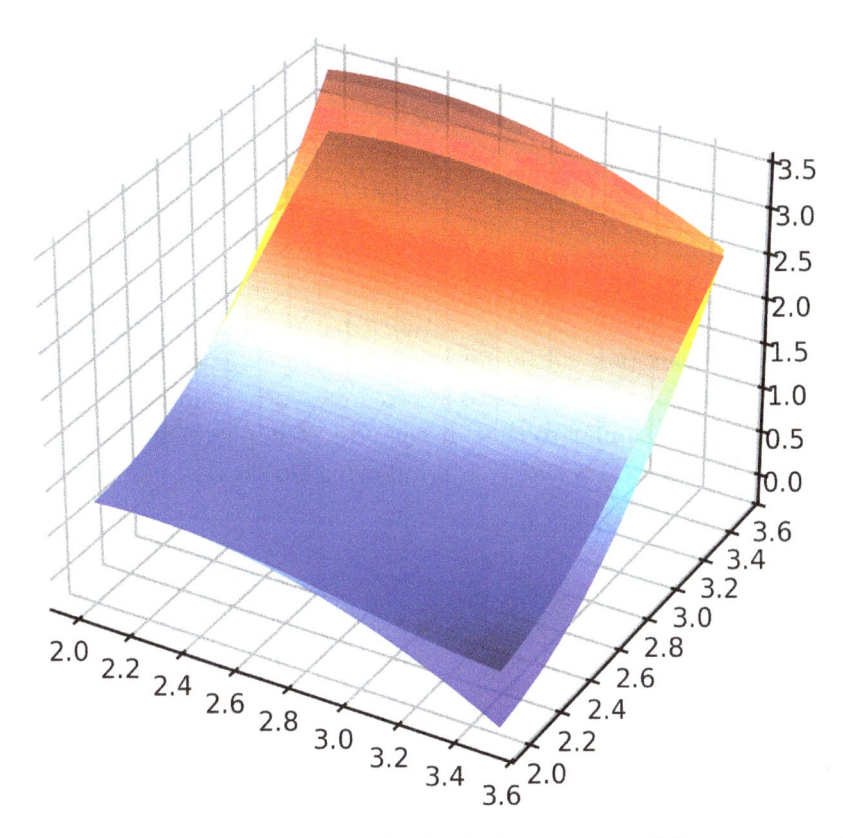

Figure 3.10. Approximation via second-order Taylor expansion of the temperature function over a 2D domain near point $(2, 2)$.

where k stands for the kth node in the element with $k = (i + 1) + j(n + 1)$, $N_n = (n + 1) + m(n + 1) = (n + 1)(m + 1)$, $\mathbf{x}_k$ is the coordinate of the kth node, $l_i(x)$ is a Lagrange interpolator corresponding to node at x_i, and $l_j(y)$ is that at y_j. They all have exactly the same form (see Section 5.4 in Ref. [1]):

$$l_i(x) = \prod_{i=0, p=1, p \neq i}^{m} \frac{x - x_p}{x_i - x_p}$$

$$= \frac{(x - x_1)(x - x_2) \cdots (x - x_{p-1})(x - x_p)(x - x_{p+1}) \cdots (x - x_n)}{(x_i - x_1)(x_i - x_2) \cdots (x_i - x_{p-1})(\quad 1 \quad)(x_i - x_{p+1}) \cdots (x_i - x_n)}$$

$$l_j(y) = \prod_{j=0,q=1,q\neq j}^{n} \frac{y - y_q}{y_j - y_q}$$

$$= \frac{(y - y_1)(y - y_2)\cdots(y - y_{q-1})(y - y_q)(y - y_{q+1})\cdots(y - y_n)}{(y_i - y_1)(y_i - y_2)\cdots(y_i - y_{q-1})(\quad 1 \quad)(y_i - y_{q+1})\cdots(y_i - y_n)}$$

$$(3.36)$$

Note that $N_{ij}(x,y) = l_i(x) \cdot l_j(y)$ can be renumbered to become N_k, which are a type of nodal **shape functions** for 2D problems used in the FEM for interpolations within an element. These shape functions become basis functions that can be used to approximate other functions. The theory and procedure are similar to that discussed in Chapter 6 of Ref. [1].

These shape functions satisfy the following properties:

1. The **delta function** property:

$$N_k(\mathbf{x}_l) = \delta_{kl} = \begin{cases} 1 & \text{when } k = l \\ 0 & \text{when } k \neq l \end{cases} \tag{3.37}$$

where δ_{kl} is the Kronecker delta. This property can be stated: *The shape function has value 1 at its home node and zero at all the remote nodes.*

2. The **partitions of unity** property:

$$\sum_{k=0}^{N_n} N_k(\mathbf{x}_k) = 1 \tag{3.38}$$

This property state that *each of the nodal shape functions is a part of the unit.*

These two properties can be proven in the same way as given in Section 5.4.2 in Ref. [1] by using the property of the Lagrange interpolators.

3.9.2　*Coordinate mapping, square to rectangle*

As discussed in Chapter 5 of Ref. [1], we often generate the nodal shape functions in the so-called natural coordinates. For nD functions, this is even more important because we may not be able to successfully compute the nodal shape functions in the physical coordinate system if the domain is not regular [5,7]. For a regular domain or element in 2D, the shape is rectangular with length $2a$ and height $2b$, as shown in Fig. 3.11(a). The corresponding domain in the natural coordinates is a dimensionless square of $[-1,1] \times [-1,1]$.

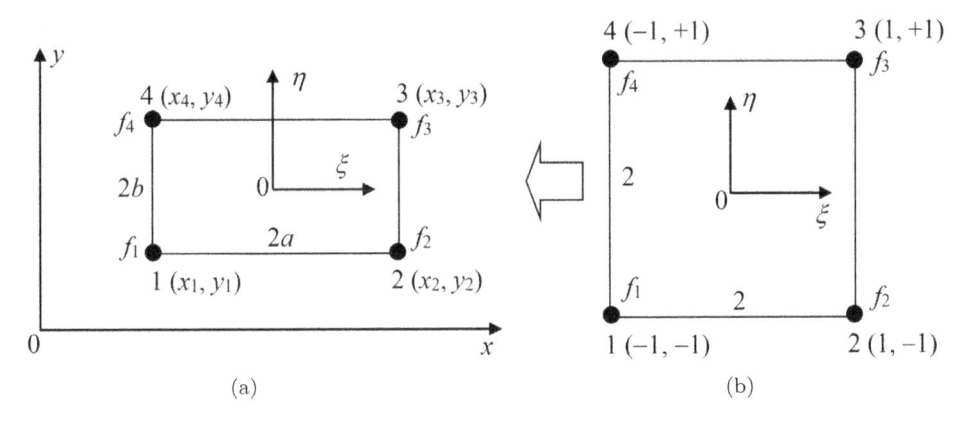

Figure 3.11. Coordinate mapping between regular and square domains: (a) a rectangular domain in the physical coordinate system and (b) a standard square domain in the natural coordinate system.

It is called standard square. Its edge length is 2, and its area is 4, as shown in Fig. 3.11(b).

The coordinate mapping between the rectangular domain in (x, y) and the square domain in (ξ, η) is given as follows:

$$\xi = \frac{1}{a}\left(x - \frac{x_1 + x_2}{2}\right) \quad \text{or} \quad x = a\xi + \frac{x_1 + x_2}{2}$$

$$\eta = \frac{1}{b}\left(y - \frac{y_1 + y_4}{2}\right) \quad \text{or} \quad y = b\eta + \frac{y_1 + y_4}{2} \tag{3.39}$$

The mapping is clearly linear. The nodal coordinate correspondence shown in Fig. 3.11 can be confirmed easily using the following code:

```
1  x, y, ξ, η =symbols('x, y, ξ, η')      # define 2D coordinates variables
2  x1, x2, y1, y4 =symbols('x1, x2, y1, y4')          # nodal coordinates
3
4  x4 = x1; x3 = x2; y2 = y1; y3 = y4           # rectangle conditions
5  a = (x2-x1)/2; b = (y4-y1)/2
6
7  ξ=(x - (x1+x2)/2)/a; η=(y - (y1+y4)/2)/b       # x,y to ξ, η mappying
8
9  ξi = [ξ.subs(x,xi).simplify() for xi in [x1, x2, x3, x4]]
10 print(f" Physical coordinates xi = {[x1, x2, x3, x4]}")       # xi ⇔ ξi
11 print(f" Natural  coordinates ξi = {ξi}\n")
12
13 ηi = [η.subs(y,yi).simplify() for yi in [y1, y2, y3, y4]]
14 print(f" Physical coordinates yi = {[y1, y2, y3, y4]}")       # yi ⇔ ηi
15 print(f" Natural  coordinates ηi = {ηi}")
```

```
Physical coordinates xi = [x1, x2, x2, x1]
Natural  coordinates ξi = [-1, 1, 1, -1]

Physical coordinates yi = [y1, y1, y4, y4]
Natural  coordinates ηi = [-1, -1, 1, 1]
```

It is thus clear we have a one-to-one nodal correspondence. Since the mapping is linear, as described by Eq. (3.39), the mapping is bijective: a point in one domain has a unique one-to-one correspondence in the other.

3.9.3 *Derivative computation for rectangular-square mapping*

The rectangular-square mapping is, in fact, a coordinate transformation, but a very simple one involving only stretching or compression. The partial derivatives of a function $f(x, y)$ defined in physical coordinate can be found using

$$\frac{\partial f}{\partial x} = \frac{\partial f}{\partial \xi}\frac{\partial \xi}{\partial x} = \frac{1}{a}\frac{\partial f}{\partial \xi}$$

$$\frac{\partial f}{\partial y} = \frac{\partial f}{\partial \eta}\frac{\partial \eta}{\partial x} = \frac{1}{b}\frac{\partial f}{\partial \eta} \tag{3.40}$$

The scaling in the horizontal direction is $\frac{1}{a}$ and that in the vertical direction is $\frac{1}{b}$. Therefore, computing the partial derivatives $\frac{\partial f}{\partial x}$ and $\frac{\partial f}{\partial y}$ is straightforward by scaling $\frac{\partial f}{\partial \xi}$ and $\frac{\partial f}{\partial \eta}$.

We need now to create nodal shape functions and compute their derivatives over a standard square domain.

3.9.4 *Codes for 2D Lagrange interpolators for square domains*

We write the following code function to compute and plot shape functions for 2D square domains using the Lagrange interpolators. These shape functions will be used to approximate functions in 2D space:

```python
 1 def N_Lagrange(nodeξi, nodeηj, *p_name):
 2     '''Compute and plot shape functions generated using lagrange
 3        interpolators.
 4        Inputs:  nodeξi, nodal coordinates in the horizontal direction.
 5                 nodeηj, nodal coordinates in the vertical direction.
 6                 example: nodeξi = [-1, 0, 1]; nodeηj = [-1, 0, 1]
 7                 p_name, filename for the plot to be saved.
 8     '''
 9     ξ, η =symbols('ξ, η')                    # define2D natrual coordinate
10     X, Y = np.mgrid[-1:1:100j, -1:1:100j]            # mesh grid for plot
11
12     lξs = [lξ for lξ in gr.LagrangeP(ξ, nodeξi)] # lagng. interpolators
13     lηs = [lη for lη in gr.LagrangeP(η, nodeηj)]
14     lambda_fx = [lambdify(ξ,f) for f in lξs]      # convert to numpy func.
15     lambda_fy = [lambdify(η,f) for f in lηs]
16     Nijs = sp.Matrix(lξs)@ sp.Matrix(lηs).T           # a 2D matrix
17
18     plt.figure(); plt.ioff()
19     plt.rcParams.update({'font.size': 4})
20     plt.rcParams["figure.autolayout"] = True
21     fig_s = plt.figure(figsize=(6,4))
22
23     ni = len(nodeξi); nj = len(nodeηj)
24     for i in range(ni):
25         for j in range(nj):
26             ij = nj*i+j+1
27             ax = fig_s.add_subplot(ni,nj,ij, projection='3d')
28             N = lambda_fx[i](X)*lambda_fy[j](Y)       # shape function
29             ax.set_xlabel('$x$', labelpad=-11)#, fontsize=6)
30             ax.set_ylabel('$y$', labelpad=-11)#, fontsize=6)
31             ax.tick_params(axis='x', pad=-5)          # pad tick labels
32             ax.tick_params(axis='y', pad=-5)
33             ax.tick_params(axis='z', pad=-5)
34             ax.set_zticks([0, 0.5, 1.])               # tick labels
35             ax.set_box_aspect((1, 1, .5))
36             ax.plot_surface(X,Y,N,cmap="seismic")        #cmap="jet")
37             plt.title('N'+str(i)+str(j))
38
39     fig_s.tight_layout()
40     plt.savefig('imagesDI/'+p_name[0]+'.png',dpi=500,bbox_inches='tight')
41     #plt.show()
42     return lξs, lηs, Nijs
```

3.9.5 *Example: Bilinear shape functions with four nodes*

Using the code provided above, we now compute and plot the simplest but most widely used shape functions called bilinear shape functions [5]. Such an element has a square shape with four straight-line edges and four nodes, called a Q4 element. It is often defined in the natural coordinates ξ and η.

The following code computes these shape functions in a square element in the natural coordinates ξ and η, forming a domain of $[-1, 1] \times [-1, 1]$. A square element may be mapped to irregular quadrilateral elements in physical coordinates x and y:

```
1  ξ, η =symbols('ξ, η')                    # define2D natrual coordinate
2  nodeξi = [-1, 1]; nodeηj = [-1, 1]
3  lξs,lηs,Nij4 = N_Lagrange(nodeξi, nodeηj, 'Nxy1_4')   # Compute&plot Ns
4  #plt.show()
```

We obtained a total of four pieces of shape functions for each node of a four-noded square domain, as shown in Fig. 3.12. These shape functions

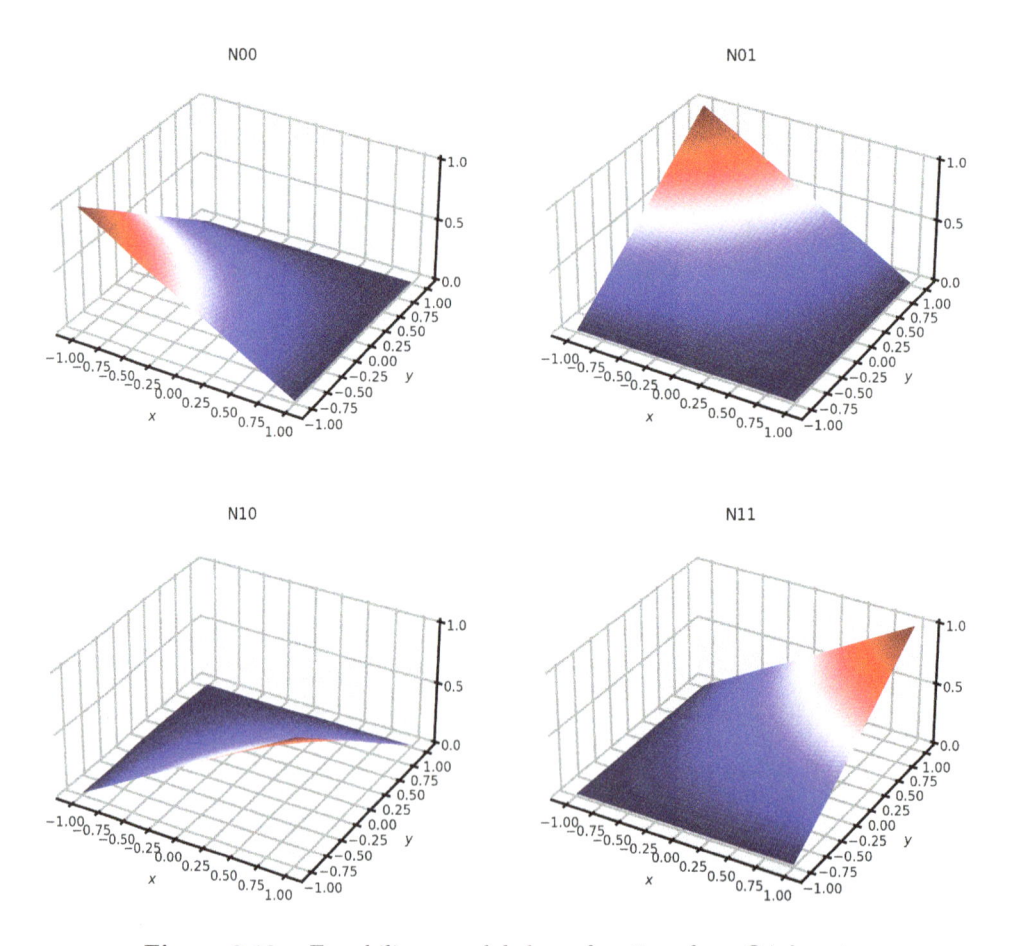

Figure 3.12.   Four bilinear nodal shape functions for a Q4 domain.

change linearly in only two directions: the ξ-direction and the η-direction, hence the name 'bilinear shape functions'. The formulas for these functions are given as follows:

```
1  print(f'Lagrange interpolators: {lξs}')
2  print(f'Lagrange interpolators: {lηs}')
3  print(f'Formed shape functions N:')
4  Nij4
```

```
Lagrange interpolators: [1/2 - ξ/2, ξ/2 + 1/2]
Lagrange interpolators: [1/2 - η/2, η/2 + 1/2]
Formed shape functions N:
```

$$\left[\left(\tfrac{1}{2}-\tfrac{\eta}{2}\right)\left(\tfrac{1}{2}-\tfrac{\xi}{2}\right)\left(\tfrac{1}{2}-\tfrac{\xi}{2}\right)\left(\tfrac{\eta}{2}+\tfrac{1}{2}\right)\right.$$
$$\left.\left(\tfrac{1}{2}-\tfrac{\eta}{2}\right)\left(\tfrac{\xi}{2}+\tfrac{1}{2}\right)\left(\tfrac{\eta}{2}+\tfrac{1}{2}\right)\left(\tfrac{\xi}{2}+\tfrac{1}{2}\right)\right]$$

In the standard FEM formulation, the shape function matrix is usually arranged as a row matrix: $\mathbf{N} = [N_1, N_2, N_3, N_4]$. Conventionally, nodes in the Q4 domain are often arranged counterclockwise (CCW). The following utility code generates nodal shape functions N_i in the natural coordinates and then arranges them CCW to match the node numbering:

```
1  def N_Lagrange_fC(nodeξi,nodeηj):
2      '''Compute 4 shape functions generated using lagrange
3      interpolators. Use natural coordinates (square domain/element).
4      Nodes arrangement, counterclockwise: 4   3
5                                           1   2
6      '''
7      ξ, η =sp.symbols('ξ, η')            # define2D natrual coordinate
8      lξs = [lξ for lξ in gr.LagrangeP(ξ, nodeξi)]
9      lηs = [lη for lη in gr.LagrangeP(η, nodeηj)]
10
11     Nij = [lξs[0]*lηs[0], lξs[1]*lηs[0], lξs[1]*lηs[1], lξs[0]*lηs[1]]
12     Nij = [Ni.simplify() for Ni in Nij]
13
14     return lξs, lηs, Nij
```

```
1  # Generate counterclockwise arranged nodal shape functions:
2  nodeξi = [-1, 1]; nodeηj = [-1, 1]
3  _, _, N = N_Lagrange_fC(nodeξi,nodeηj)
4  N                                       # N = [N1,N2,N3,N4], CCW
```

$$\left[\frac{(\eta-1)(\xi-1)}{4}, \ -\frac{(\eta-1)(\xi+1)}{4}, \ \frac{(\eta+1)(\xi+1)}{4}, \ -\frac{(\eta+1)(\xi-1)}{4}\right]$$

3.9.6 *Partial derivatives of bi-linear shape functions*

Partial derivatives of shape functions are often needed in applications. For example, in FEM formulations, we must use the partial derivatives of the displacement function to compute strains. Using SymPy, the partial derivatives of these shape functions can be computed with ease. The code is given as follows:

```
1  dNdξ = [sp.diff(Ni, ξ) for Ni in N]
2  display(Math(f" \\frac{{∂N}}{{∂ξ}} = {latex(dNdξ)}"))
3
4  dNdη = [sp.diff(Ni, η) for Ni in N]
5  display(Math(f" \\frac{{∂N}}{{∂η}} = {latex(dNdη)}"))
```

$$\frac{\partial N}{\partial \xi} = \left[\frac{\eta}{4} - \frac{1}{4}, \ \frac{1}{4} - \frac{\eta}{4}, \ \frac{\eta}{4} + \frac{1}{4}, \ -\frac{\eta}{4} - \frac{1}{4} \right]$$

$$\frac{\partial N}{\partial \eta} = \left[\frac{\xi}{4} - \frac{1}{4}, \ -\frac{\xi}{4} - \frac{1}{4}, \ \frac{\xi}{4} + \frac{1}{4}, \ \frac{1}{4} - \frac{\xi}{4} \right]$$

Since these shape functions are not purely linear, their partial derivatives are not constants but linear functions of ξ and η. To compute the partial derivatives $\frac{\partial f}{\partial x}$ and $\frac{\partial f}{\partial y}$, we simply use Eq. (3.40) and $\frac{\partial f}{\partial \xi}$, $\frac{\partial f}{\partial \eta}$. We will demonstrate this in later examples.

3.9.7 *Example: Quadratic shape functions with nine nodes*

Next, we can compute and plot the shape functions for quadratic elements with nine nodes, also in the natural coordinates ξ and η, forming a domain of $[-1, 1] \times [-1, 1]$. These nine nodes are located at the four corners, four midpoints of the edges, and one at the center of the element. Therefore, for this element, we need to place a node at its center:

```
1  nodeξi = [-1, 0, 1]; nodeηj = [-1, 0, 1]
2  lξs,lηs,Nijs = N_Lagrange(nodeξi, nodeηj, 'Nxy1_9')    # Compute&plot Ns
3  print(f'Lagrange interpolators: {lξs}')
4  print(f'Lagrange interpolators: {lηs}')
5  print(f'Formed shape functions N in (3×3) matrix form:')
6  sp.simplify(Nijs)
7  #plt.show()
```

```
Lagrange interpolators: [-ξ*(1/2 - ξ/2), (1 - ξ)*(ξ + 1), ξ*(ξ/2 + 1/2)]
Lagrange interpolators: [-η*(1/2 - η/2), (1 - η)*(η + 1), η*(η/2 + 1/2)]
Formed shape functions N in (3×3) matrix form:
```

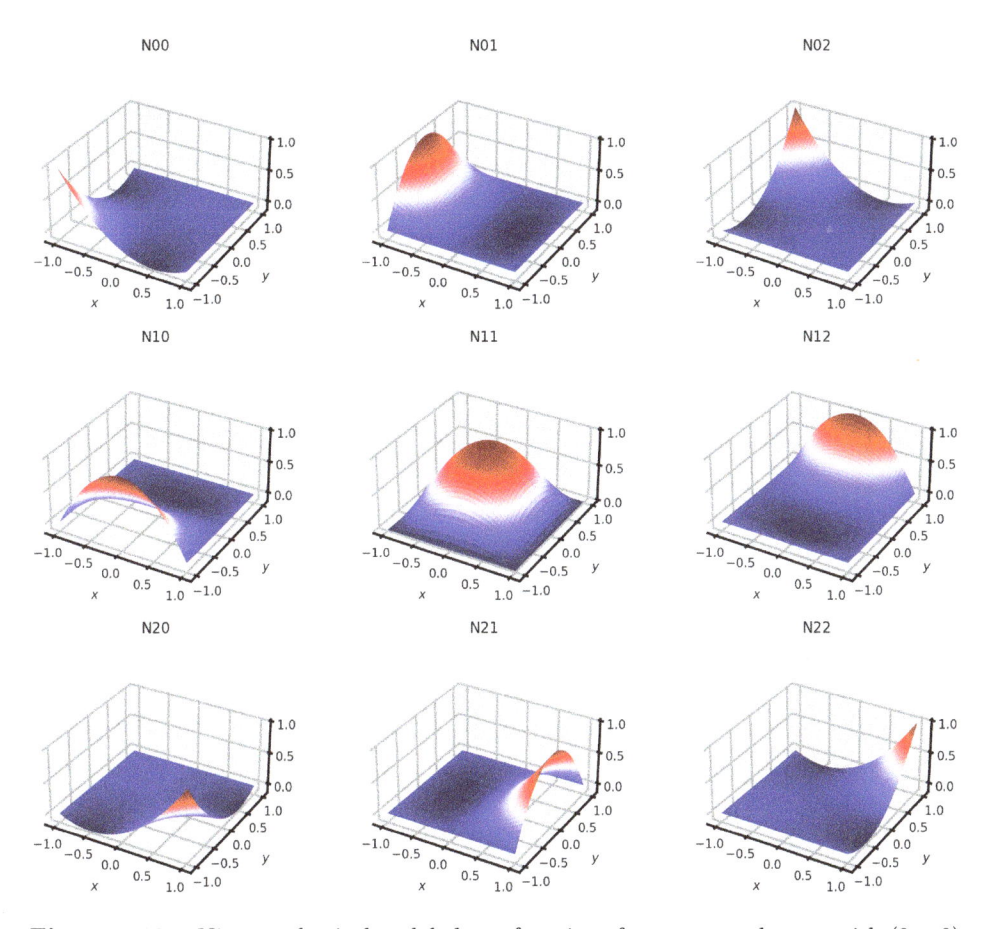

Figure 3.13. Nine quadratical nodal shape functions for a square element with (3×3) nodes (one node is at the center of the element).

$$
\begin{bmatrix}
\dfrac{\eta\xi(\eta-1)(\xi-1)}{4} & -\dfrac{\xi(\eta-1)(\eta+1)(\xi-1)}{2} & \dfrac{\eta\xi(\eta+1)(\xi-1)}{4} \\[3mm]
-\dfrac{\eta(\eta-1)(\xi-1)(\xi+1)}{2} & (\eta-1)(\eta+1)(\xi-1)(\xi+1) & -\dfrac{\eta(\eta+1)(\xi-1)(\xi+1)}{2} \\[3mm]
\dfrac{\eta\xi(\eta-1)(\xi+1)}{4} & -\dfrac{\xi(\eta-1)(\eta+1)(\xi+1)}{2} & \dfrac{\eta\xi(\eta+1)(\xi+1)}{4}
\end{bmatrix}
$$

We obtained a total of nine shape functions for each node of a nine-noded square domain, as shown in Fig. (3.13).

3.9.8 *Partial derivatives of quadratic shape functions*

The partial derivatives of these nodal shape functions can be computed using the following code:

```
1  #sp.flatten(Nijs) this converts to N = [N1,N2,...,N9] (optional)
2  print(f'Diffs of the Lagrange interpolators: {sp.diff(sp.Matrix(lξs))}')
3  print(f'Diffs of the Lagrange interpolators: {sp.diff(sp.Matrix(lηs))}')
4  print(f'Partial derivatives of N w.r.t. ξ (first 2 columes):')
5  sp.diff(sp.Matrix(sp.simplify(Nijs)), ξ)[:,:2]   # print first 2 columns
```

Diffs of the Lagrange interpolators: Matrix([[ξ - 1/2], [-2*ξ], [ξ + 1/2]])
Diffs of the Lagrange interpolators: Matrix([[η - 1/2], [-2*η], [η + 1/2]])
Partial derivatives of N w.r.t. ξ (first 2 columes):

$$
\begin{bmatrix}
\frac{\eta\xi(\eta-1)}{4} + \frac{\eta(\eta-1)(\xi-1)}{4} & -\frac{\xi(\eta-1)(\eta+1)}{2} - \frac{(\eta-1)(\eta+1)(\xi-1)}{2} \\[2mm]
-\frac{\eta(\eta-1)(\xi-1)}{2} - \frac{\eta(\eta-1)(\xi+1)}{2} & (\eta-1)(\eta+1)(\xi-1) + (\eta-1)(\eta+1)(\xi+1) \\[2mm]
\frac{\eta\xi(\eta-1)}{4} + \frac{\eta(\eta-1)(\xi+1)}{4} & -\frac{\xi(\eta-1)(\eta+1)}{2} - \frac{(\eta-1)(\eta+1)(\xi+1)}{2}
\end{bmatrix}
$$

```
1  print(f'Partial derivatives of N w.r.t. η (first 2 columns):')
2  sp.diff(sp.Matrix(sp.simplify(Nijs)), η)[:,:2]   # print last 2 columns
```

Partial derivatives of N w.r.t. η (first 2 columns):

$$
\begin{bmatrix}
\frac{\eta\xi(\xi-1)}{4} + \frac{\xi(\eta-1)(\xi-1)}{4} & -\frac{\xi(\eta-1)(\xi-1)}{2} - \frac{\xi(\eta+1)(\xi-1)}{2} \\[2mm]
-\frac{\eta(\xi-1)(\xi+1)}{2} - \frac{(\eta-1)(\xi-1)(\xi+1)}{2} & (\eta-1)(\xi-1)(\xi+1) + (\eta+1)(\xi-1)(\xi+1) \\[2mm]
\frac{\eta\xi(\xi+1)}{4} + \frac{\xi(\eta-1)(\xi+1)}{4} & -\frac{\xi(\eta-1)(\xi+1)}{2} - \frac{\xi(\eta+1)(\xi+1)}{2}
\end{bmatrix}
$$

Their partial derivatives are still quadratic in ξ and η.

3.9.9 *Example: Bilinear–quadratic shape functions with six nodes*

Using N_Lagrange(), we can conveniently compute and plot the shape functions for various types of shape functions. For example, we may create an element with six nodes: two along the ξ-direction and three along the η-direction. The following is an example:

```
1  nodeξi = [-1, 1]                      # 2 nodes in the x-direction
2  nodeηj = [-1, 0, 1]                   # 3 nodes in the y-direction
3  lξs, lηs, Nijs = N_Lagrange(nodeξi, nodeηj, 'Nxy1_6') # compute&plot Ns
4  print(f'Lagrange interpolators: {lξs}')
5  print(f'Lagrange interpolators: {lηs}')
6  print(f'Formed shape functions N in (2×3) matrix form:')
7  Nijs
```

Lagrange interpolators: [1/2 - ξ/2, ξ/2 + 1/2]
Lagrange interpolators: [-η*(1/2 - η/2), (1 - η)*(η + 1), η*(η/2 + 1/2)]
Formed shape functions N in (2×3) matrix form:

$$\left[\begin{matrix} -\eta\left(\frac{1}{2}-\frac{\eta}{2}\right)\left(\frac{1}{2}-\frac{\xi}{2}\right) & \left(\frac{1}{2}-\frac{\xi}{2}\right)(1-\eta)(\eta+1) & \eta\left(\frac{1}{2}-\frac{\xi}{2}\right)\left(\frac{\eta}{2}+\frac{1}{2}\right) \\ -\eta\left(\frac{1}{2}-\frac{\eta}{2}\right)\left(\frac{\xi}{2}+\frac{1}{2}\right) & (1-\eta)(\eta+1)\left(\frac{\xi}{2}+\frac{1}{2}\right) & \eta\left(\frac{\eta}{2}+\frac{1}{2}\right)\left(\frac{\xi}{2}+\frac{1}{2}\right) \end{matrix}\right]$$

We obtained a total of six shape functions for each node of a six-noded square domain, as shown in Fig. 3.14. These shape functions change linearly along the ξ-direction, but quadratically along the η-direction. Readers may compute the derivatives of these shape functions, as done previously.

3.10 Approximation of functions and derivatives: irregular domain

When the shape of the domain or element is not regular in the Cartesian coordinates (x, y), which is more often the case in actual applications, we cannot directly use the Lagrangian elements developed earlier. The most effective way to overcome this is to perform a **coordinate transformation** to convert the domain to a regular (often square) one. This is a more general skewed **mapping** in addition to the simple stretching and compression discussed earlier. This section presents the related formulation and techniques for coordinate transformation for functions and their derivatives in irregular domains. We discuss 2D formulas first followed by their extension to 3D.

3.10.1 *Coordinate mapping in 2D domains*

Suppose that the relationship between a Cartesian coordinates (x, y) and another coordinate system (ξ, η) are given in the form of

$$x = x(\xi, \eta); \quad y = y(\xi, \eta) \tag{3.41}$$

Consider a function $f(x, y)$ defined in the coordinate system (x, y). Its value in the coordinate system (ξ, η) becomes a composite function $f(x(\xi, \eta), y(\xi, \eta))$. Its partial derivatives $\frac{\partial f}{\partial x}$ and $\frac{\partial f}{\partial y}$ need to be computed using the chain rule of differentiation, as given in Chapter 2. There are a couple of ways to do this. For the case of (x, y) written in (ξ, η), we write $\frac{\partial f}{\partial \xi}$ and $\frac{\partial f}{\partial \eta}$ as

$$\frac{\partial f}{\partial \xi} = \frac{\partial f}{\partial x}\frac{\partial x}{\partial \xi} + \frac{\partial f}{\partial y}\frac{\partial y}{\partial \xi}$$

$$\frac{\partial f}{\partial \eta} = \frac{\partial f}{\partial x}\frac{\partial x}{\partial \eta} + \frac{\partial f}{\partial y}\frac{\partial y}{\partial \eta} \tag{3.42}$$

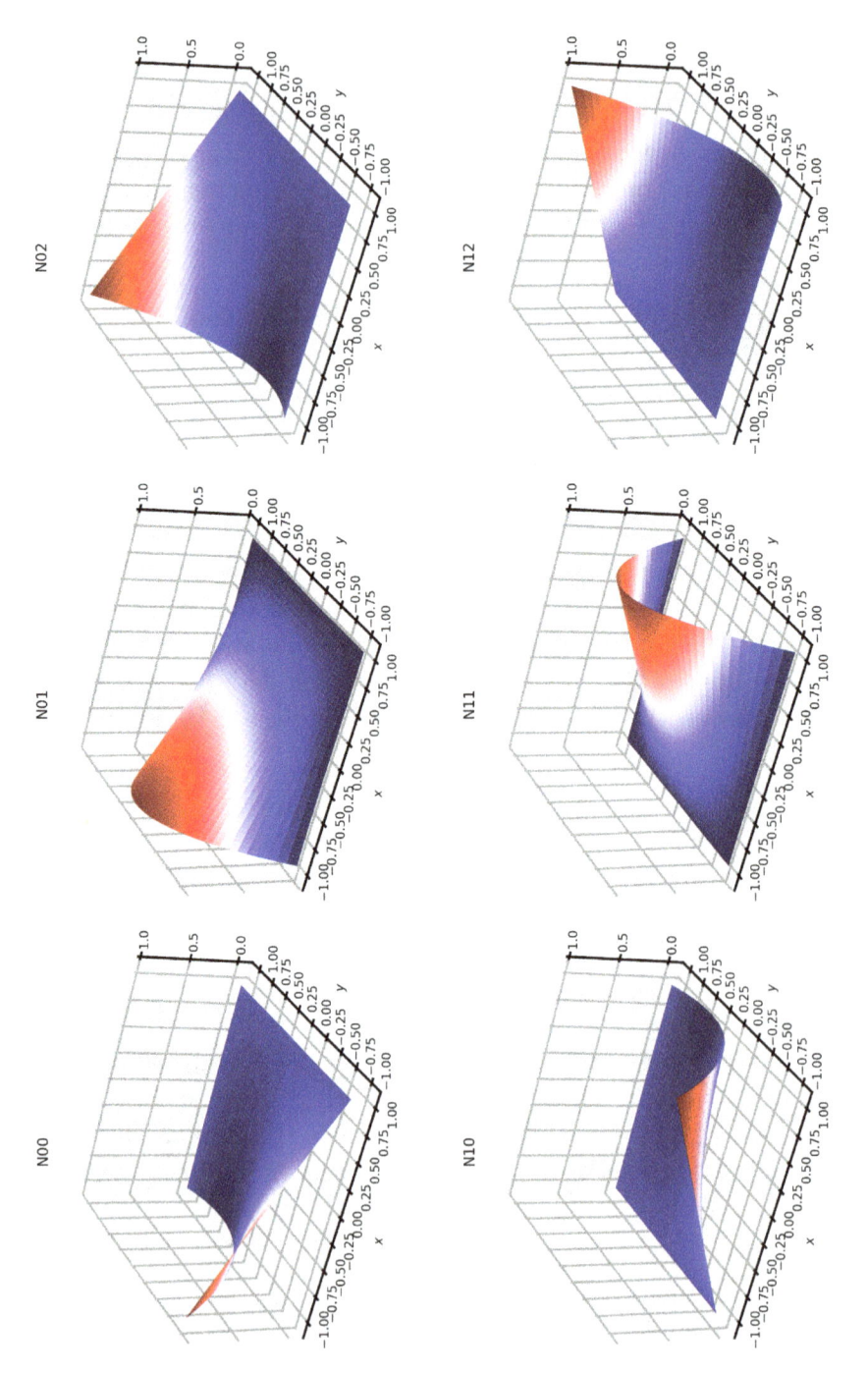

Figure 3.14. Six nodal shape functions for a square element with (3×3) nodes.

or in matrix form

$$\begin{bmatrix} \frac{\partial f}{\partial \xi} \\ \frac{\partial f}{\partial \eta} \end{bmatrix} = \underbrace{\begin{bmatrix} \frac{\partial x}{\partial \xi} & \frac{\partial y}{\partial \xi} \\ \frac{\partial x}{\partial \eta} & \frac{\partial y}{\partial \eta} \end{bmatrix}}_{\mathbf{J}^{\mathsf{T}}} \begin{bmatrix} \frac{\partial f}{\partial x} \\ \frac{\partial f}{\partial y} \end{bmatrix} \tag{3.43}$$

where J is called the **Jacobian matrix** defined in this book in the following form:

$$\mathbf{J} = \begin{bmatrix} \frac{\partial x}{\partial \xi} & \frac{\partial x}{\partial \eta} \\ \frac{\partial y}{\partial \xi} & \frac{\partial y}{\partial \eta} \end{bmatrix} \tag{3.44}$$

3.10.2 *Mapping based on a square domain*

The Jacobian matrix gives the differential relations between these two coordinate systems, (x, y) and (ξ, η). It is often seen in the literature whenever a transformation of two coordinate systems is involved (may be in a different form). Using the Lagrangian nodal shape functions, Eq. (3.41) can be written as

$$x = \sum_{i=0}^{n}\sum_{j=0}^{m} x_{ij} N_{ij}(\xi, \eta) \underset{\text{or}}{=} \sum_{k=1}^{N_n} x_k N_k(\xi, \eta)$$

$$y = \sum_{i=0}^{n}\sum_{j=0}^{m} y_{ij} N_{ij}(\xi, \eta) \underset{\text{or}}{=} \sum_{k=1}^{N_n} y_k N_k(\xi, \eta) \tag{3.45}$$

If the bilinear shape functions are used for a quadrilateral domain with four nodes (Q4), the mapping is schematically shown in Fig. 3.15.

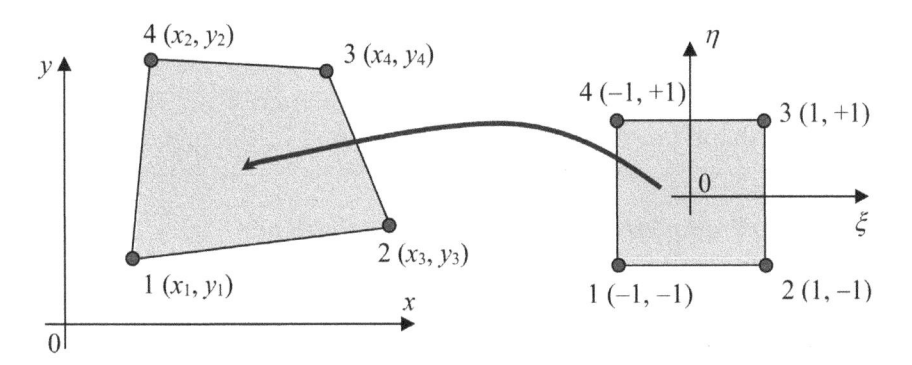

Figure 3.15. Coordinate mapping using bilinear shape functions for a Q4 domain. Left: a general quadrilateral in the physical coordinate (x, y); right: a standard square in the natural coordinate (ξ, η). Nodes are numbered CCW.

Clearly, such a mapping involves skewing, not just stretching and compression. Using Eq. (3.45) and the delta function property of the shape function, it is easy to see one-to-one nodal correspondence:

$$(x_k, y_k) \iff (\xi_k, \eta_k) \tag{3.46}$$

Now, because these shape functions are bilinear, all the edges of the domain in both the physical and natural coordinates are straight lines. Therefore, edge 1–2 in the physical coordinates corresponds to edge 1–2 in the natural coordinates. The same goes for all the other edges:

$$\text{edge}_{i-j,\text{Q4}} \iff \text{edge}_{i-j,\text{standard square}} \tag{3.47}$$

This means that the domain of the Q4 domain in the physical coordinate corresponds precisely to the standard square domain in the natural coordinates:

$$\text{Area of Q4} \iff \text{Area of standard square} \tag{3.48}$$

The mapping in the interior of these two domains is bilinear following Eq. (3.45). Hence, the mapping is bijective: A point in one domain has a unique one-to-one correspondence in the other.

3.10.3 *Evaluation of the Jacobian matrix*

Using the shape function, the Jacobian matrix is given as

$$\mathbf{J} = \begin{bmatrix} \sum_{k=1}^{N_n} x_k \frac{\partial N_k(\xi,\eta)}{\partial \xi} & \sum_{k=1}^{N_n} x_k \frac{\partial N_k(\xi,\eta)}{\partial \eta} \\ \sum_{k=1}^{N_n} y_k \frac{\partial N_k(\xi,\eta)}{\partial \xi} & \sum_{k=1}^{N_n} y_k \frac{\partial N_k(\xi,\eta)}{\partial \eta} \end{bmatrix} \tag{3.49}$$

For a Q4, $N_n = 4$. Once the coordinates (x_k, y_k) of the vertices of the domain is known, the Jacobian matrix can computed using the derivatives of the nodal shape functions, which are computed in the previous section.

3.10.4 *Approximation of partial derivatives of functions*

Using Eq. (3.43), the partial derivatives $\frac{\partial f}{\partial x}$ and $\frac{\partial f}{\partial y}$ can now be computed using

$$\begin{bmatrix} \frac{\partial f}{\partial x} \\ \frac{\partial f}{\partial y} \end{bmatrix} = [\mathbf{J}^{-1}]^\top \begin{bmatrix} \frac{\partial f}{\partial \xi} \\ \frac{\partial f}{\partial \eta} \end{bmatrix} \tag{3.50}$$

When $f(x, y)$ is approximated using Eq. (3.35), its partial derivatives can be approximated using

$$\begin{bmatrix} \frac{\partial f}{\partial x} \\ \frac{\partial f}{\partial y} \end{bmatrix} = [\mathbf{J}^{-1}]^\top \begin{bmatrix} \sum_{k=1}^{N_n} f(\mathbf{x}_k) \frac{\partial N_k(\xi, \eta)}{\partial \xi} \\ \sum_{k=1}^{N_n} f(\mathbf{x}_k) \frac{\partial N_k(\xi, \eta)}{\partial \eta} \end{bmatrix} \tag{3.51}$$

The partial derivatives of the nodal shape functions with respect to the natural coordinates ξ and η can now be used to compute $\frac{\partial f}{\partial x}$ and $\frac{\partial f}{\partial y}$. Note that we need to compute the Jacobian matrix and its inverse. The existence of the inverse of the Jacobian matrix depends on the shape of the domain. In general, the shape must be convex [5].

3.10.5 *Coordinate mapping in 3D domains*

Following the same procedure, formulas for mapping between a 3D physical coordinates (x, y, z) and a natural coordinate system (ξ, η, ζ) can be written out directly:

$$x = x(\xi, \eta, \zeta); \quad y = y(\xi, \eta, \zeta); \quad z = y(\xi, \eta, \zeta) \tag{3.52}$$

$$\begin{bmatrix} \frac{\partial f}{\partial \xi} \\ \frac{\partial f}{\partial \eta} \\ \frac{\partial f}{\partial \zeta} \end{bmatrix} = \underbrace{\begin{bmatrix} \frac{\partial x}{\partial \xi} & \frac{\partial y}{\partial \xi} & \frac{\partial z}{\partial \xi} \\ \frac{\partial x}{\partial \eta} & \frac{\partial y}{\partial \eta} & \frac{\partial z}{\partial \eta} \\ \frac{\partial x}{\partial \zeta} & \frac{\partial y}{\partial \zeta} & \frac{\partial z}{\partial \zeta} \end{bmatrix}}_{\mathbf{J}^\top} \begin{bmatrix} \frac{\partial f}{\partial x} \\ \frac{\partial f}{\partial y} \\ \frac{\partial f}{\partial z} \end{bmatrix} \tag{3.53}$$

The Jacobian matrix is given as

$$\mathbf{J} = \begin{bmatrix} \frac{\partial x}{\partial \xi} & \frac{\partial x}{\partial \eta} & \frac{\partial x}{\partial \zeta} \\ \frac{\partial y}{\partial \xi} & \frac{\partial y}{\partial \eta} & \frac{\partial y}{\partial \zeta} \\ \frac{\partial z}{\partial \xi} & \frac{\partial z}{\partial \eta} & \frac{\partial z}{\partial \zeta} \end{bmatrix} \tag{3.54}$$

The other formulas for 3D is the same as those for 2D with an addition of one dimension. The 3D formulas will be used in Chapter 5 when computing integrals over 3D irregular domains.

3.10.6 *Case study: Function approximation via shape functions*

We will write the code to approximate a given function using Lagrange shape functions and plot the contours for the approximated function with estimated root-mean-squared-error (RMSE). First, let us write the following utility code for later use:

```
1  def N_Lagrange_fξη(nodeξi,nodeηj):
2      '''Compute the 4 shape functions using lagrange interpolators.
3         Use natural coordinate for square elements. 2D array form.
4      '''
5      ξ, η =sp.symbols('ξ, η')                   # define2D natrual coordinate
6      lξs = [lξ for lξ in gr.LagrangeP(ξ, nodeξi)]
7      lηs = [lη for lη in gr.LagrangeP(η, nodeηj)]
8      Nijs = sp.Matrix(lξs)@sp.Matrix(lηs).T              # a 2D array
9
10     return lξs, lηs, Nijs
```

```
1  def intrpl_Q4(vertices, ξ, η):
2      '''Generate coordinates in a quadrilateral (Q4) in physical
3         coordinates (x, y) corresponding to that of (ξ,η) for a sqaure,
4         using bilinear interpolation and shape functions in a natural
5         coordinate system, N(ξ,η).
6         Inputes: ξ,η -- given coordinates in a natural coordinate system.
7                  vertices -- 4 vertices of the Q4 domain, CCW order
8         Reture: x, y -- coordinates in a physical coordinates system.
9         This code wis written in reference to one suggestion by ChatGPT.
10     '''
11     x0, y0 = vertices[0]     # Bottom-left
12     x1, y1 = vertices[1]     # Bottom-right
13     x2, y2 = vertices[2]     # Top-right
14     x3, y3 = vertices[3]     # Top-left
15
16     # for ξ, η in [-1,1]×[-1,1] standard square
17     x =((η-1)*(ξ-1)*x0+(1-η)*(ξ+1)*x1+(η+1)*(ξ+1)*x2+(η+1)*(1-ξ)*x3)/4
18     y =((η-1)*(ξ-1)*y0+(1-η)*(ξ+1)*y1+(η+1)*(ξ+1)*y2+(η+1)*(1-ξ)*y3)/4
19
20     # for ξ, η in [0,1]×[0,1],  unit square, alternative
21     #x = (1-ξ)*(1-η)*x0 + ξ*(1-η)*x1 + ξ*η*x2 + (1-ξ)*η*x3
22     #y = (1-ξ)*(1-η)*y0 + ξ*(1-η)*y1 + ξ*η*y2 + (1-ξ)*η*y3
23
24     return x, y
```

```
1  def nodes_Q4(vertices, nodes_per_side):
2      '''Generate nodes in a quadrilateral (Q4) using bilinear
3         interpolation.
4         Inputes: nodes_per_side -- number of node on each side of Q4.
5                  vertices -- 4 vertices of the Q4 domain, CCW order
6         Reture: x, y -- coordinates in a physical coordinates system.
7         This code wis written in reference to one suggestion by ChatGPT.
8      '''
9      nodes = []
10     for i in range(nodes_per_side):
11         for j in range(nodes_per_side):
12             ξ = 2*i/(nodes_per_side-1) - 1
13             η = 2*j/(nodes_per_side-1) - 1
14             # Bilinear interpolation
15             node =((η-1)*(ξ-1)*vertices[0]-(η-1)*(ξ+1)*vertices[1]\
16                  + (η+1)*(ξ+1)*vertices[2]-(η+1)*(ξ-1)*vertices[3])/4
17             nodes.append(node)
18     return np.array(nodes)                     # a 2D array (nodes, 2)
```

```
1   def appox_fxy_Q4_Lag(np_f_true, vertices, nodes_IJ):
2       '''Compute the approximated function of a given function over Q4
3          domain, using Lagrangian shape functions.
4          Inpute: np_f_true -- given function (numpy) to be approximated.
5                  vertices -- cordinates of the vertices of Q4 in (x,y)
6                  nodes_IJ -- nodes number in 2 directions in Q4, eg.(5,5)
7                              for the approximation.
8          Reture: nodes -- used in the Lagrangian shape functions.
9                  f_lag_approx -- approximated function in ξ, η.
10      '''
11      nodeξi = np.linspace(-1., 1., nodes_IJ[0])          # nodes along ξ
12      nodeηj = np.linspace(-1., 1., nodes_IJ[1])              # along η
13      _, _, Nijs = N_Lagrange_fξη(nodeξi, nodeηj)       # shape functions
14
15      # compute the physical nodal coordinates, xi and yj:
16      nodes = nodes_Q4(vertices, nodes_IJ[0])
17      nodes_xy = nodes.reshape(nodes_IJ[0], nodes_IJ[0],2)
18
19      f_lag_approx = 0                   # Generate the Lagrange function
20      for i in range(len(nodeξi)):             # using Nijs to approximate
21          for j in range(len(nodeηj)):
22              xy = nodes_xy[i,j,:]                          # xy: (x,y)
23              f_lag_approx += Nijs[i,j]*np_f_true(xy[0], xy[1])
24
25      return nodes, f_lag_approx          # approximated function in ξ, η
```

```
1   def plot_Lgrng_interp2d(f_true, vertices, nodes_cases, *p_name):
2       '''Approximate a 2D true function using a set of Lagrange shape
3          functions. Plot the Approximated and the true function.
4          Inpute: np_f_true -- given function (numpy) to be approximated.
5                  vertices -- cordinates of the vertices of Q4 in (x,y)
6                  nodes_cases -- cases of number nodes in 2 directions in
7                      Q4 for the approximation. eg. 2 cases: [(3,3), (5,5)]
8                  p_name -- filename for the plot to be saved.
9       '''
10      plt.figure(); plt.ioff()
11      ξ, η =symbols('ξ, η')                      # 2D natural coordinates
12      plt.rcParams.update({'font.size': 5})
13      fig_s = plt.figure(figsize=(6,6))
14
15      Φ, Ψ = np.mgrid[-1:1:100j, -1:1:100j]       # regular grids in ξ, η
16      X, Y = intrpl_Q4(vertices, Φ, Ψ)             # grids X, Y for plots
17
18      # Plot the contour of the given true function
19      levels = 51
20      ax = fig_s.add_subplot(2,2,1)
21      ax.contour(X, Y, f_true(X,Y),cmap='jet',linewidths=.5,levels=levels)
22      plt.plot(*np.vstack([vertices, vertices[0]]).T, 'k-')       # plot Q4
23      ax.set_aspect(1.0)
24      plt.title(f'True function')
25
26      # Plot different cases of the approximated functions using Lagrange
27      for k in range(len(nodes_cases)):
28          ax = fig_s.add_subplot(2,2,k+2)
29
```

```
30      nodes,f_lag_a = appox_fxy_Q4_Lag(f_true,vertices,nodes_cases[k])
31      np_f = lambdify([ξ, η],f_lag_a)      # convert to numpy function
32
33      plt.contour(X,Y,np_f(Φ,Ψ),cmap='jet',linewidths=.5,levels=levels)
34      ax.set_aspect(1.0)
35      rmse = sqrt(np.mean((np_f(Φ,Ψ)-f_true(X,Y))**2))   # rmse error
36      plt.title(f'Approximated with {nodes_cases[k]}, rmse={rmse:.4e}')
37      plt.plot(*np.vstack([vertices, vertices[0]]).T, 'k-') # plot Q4
38      plt.scatter(nodes[:, 0], nodes[:, 1], color='blue', s=10)
39
40      plt.savefig('imagesDI/'+p_name[0]+'.png',dpi=500,bbox_inches='tight')
41      #plt.show()
```

We now define the true function and the quadrilateral domain with four vertices (Q4) in the physical coordinates. The function will be approximated over the Q4 domain, which can have a regular or irregular shape. The following code approximates the function over a unit square:

```
1  # Define a true function, bi-fourth-order polynomial:
2
3  def f_true(X, Y):
4      return .2 + .5*X - .5*X*Y + X**2 + X**4 -.8*Y**4 -.3*X**3*Y**2
5                          # Its highest combined order is 5
6
7  # Define a unit square with 4 vertices, regular counterclockwise (CCW):
8  vertexReg = np.array([[0.0, 0.0],        # Vertex Bottom-left
9                        [1.0, 0.0],        # Vertex Bottom-right
10                       [1.0, 1.0],        # Vertex Top-right
11                       [0.0, 1.0]])       # Vertex Top-left
12
13 # Define a starndar square with 4 vertices, regular (CCW):
14 #vertex22 = np.array([[-1,-1], [1, -1], [1, 1], [-1, 1]])
15
16 nodes_cases = [(4, 4),(5, 5),(6, 6)] # node-densities for interpolation
17 plot_Lgrng_interp2d(f_true,vertexReg,nodes_cases,'Lgrng_approxR4Poly')
18 #plt.show()
```

We found that there are errors when the order of the Lagrange interpolators is lower than 4, for example, using (4, 4) nodes, as shown in Fig. 3.16. When we use (5, 5) nodes, we can generate complete fourth-order shape functions, and the error becomes zero (machine accuracy). The same is found when using (6, 6) nodes. This means that we can use more nodes, but not less, to produce exact results. This finding is the same as we observed in Ref. [1].

Next, we approximate the same function over an irregular domain. All the other settings remain the same.

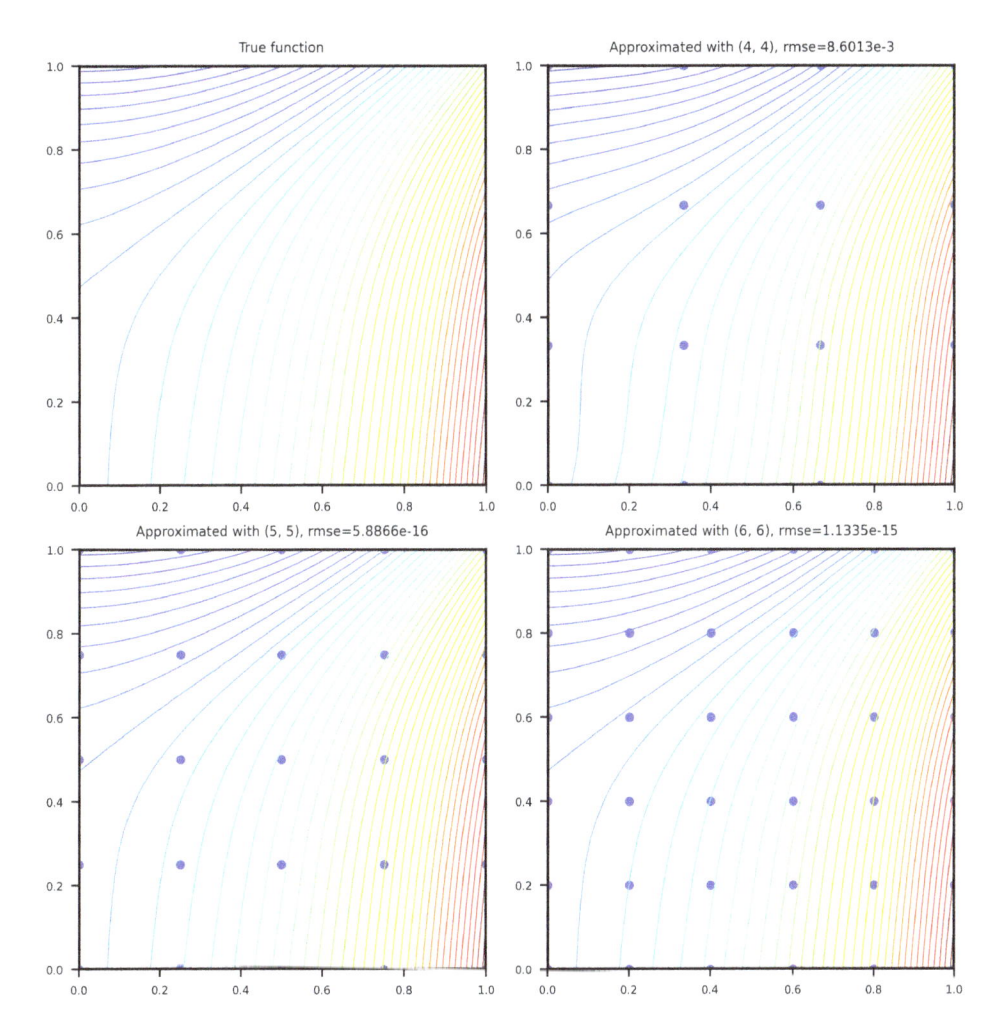

Figure 3.16. Approximation of a bi-fourth-order polynomial using Lagrange interpolators over a regular domain defined by four vertices.

```
1  # Define an irregular Q4, 4 vertices, irregular counterclockwise (CCW):
2  vertexIrr = np.array([[0.0, 0.0],        # Vertex Bottom-left
3                         [1.0, 0.1],        # Vertex Bottom-right
4                         [0.8, 1.1],        # Vertex Top-right
5                         [0.1, 0.8]])       # Vertex Top-left
6
7  plot_Lgrng_interp2d(f_true,vertexIrr,nodes_cases,'Lgrng_approxQ4Poly')
8  #plt.show()
```

This time, we found that there are errors when the order of the Lagrange interpolators is lower than 5, for example, using (5, 5) nodes, as shown in Fig. 3.17. When we use (6, 6) nodes, we can generate complete fifth-order

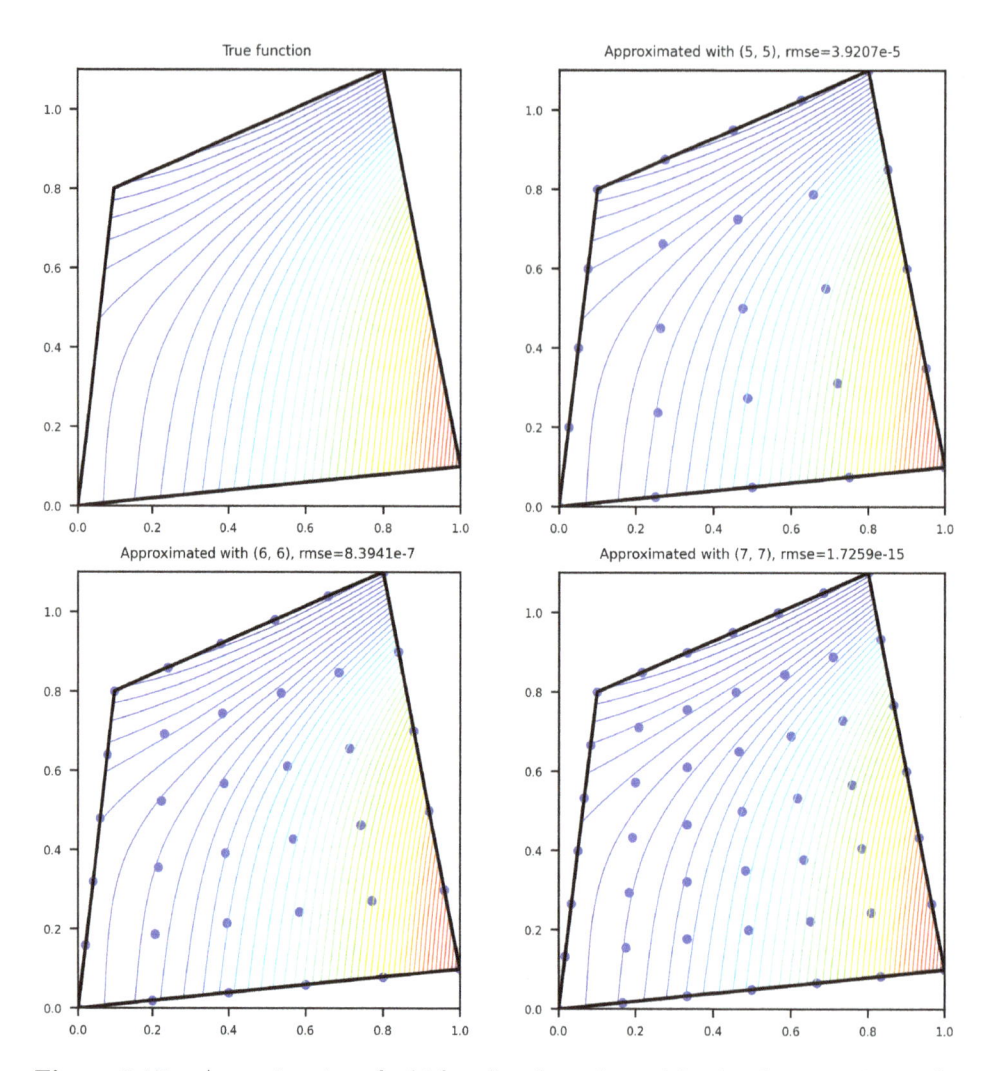

Figure 3.17. Approximation of a bi-fourth-order polynomial using Lagrange interpolators over an irregular domain defined by four vertices.

shape functions, and the error becomes zero (machine accuracy). This is because when the domain becomes irregular, we must perform mapping that involves skewing. The bi-fourth-order polynomial (its last term has a combined order of 5) now becomes a complete fifth-order polynomial. We thus note the following:

> When approximation is over an irregular domain, the order of the Lagrange shape functions must at least equal the highest combined order of the true function to reproduce it. The mapping makes the combined order in the true function become the complete order.

If we increase the combined order by 1 in the true function, the function will have a complete order of 6 after the mapping, and hence we must use a (7, 7) grid to reproduce the true function. The following code confirms this prediction:

```
1  def f_true(X, Y):                   # a polynomial function, bi-4th-order
2      return .2 + .5*X - .5*X*Y + X**2 + X**4 -.8*Y**4 -.3*X**3*Y**3
3                          # with the highest combined order of 6
4
5  nodes_cases = [(5, 5),(6, 6),(7, 7)] # node-densities for interpolation
6  plot_Lgrng_interp2d(f_true,vertexIrr,nodes_cases,'Lgrng_approxR4Poly6')
7  #plt.show()
```

As expected, we found that there are errors when the order of the Lagrange interpolators is lower than 6, for example, using (6, 6) nodes, as shown in Fig. 3.18. When we use (7, 7) nodes, we can generate complete sixth-order shape functions, and the error becomes zero (machine accuracy). This is because of the mapping that also involves skewing. The bi-fourth-order polynomial (its last term has a combined order of six) becomes a complete sixth-order polynomial.

Next, we create a true function that is non-polynomial, which is the deflection of a square plate centered at (0,0):

```
1  def f_true(X, Y):                   # a polynomial function, bi-4th-order
2      return np.sin(np.pi*(X+1)/2)*np.sin(np.pi*(Y+1)/2)
3
4  nodes_cases = [(5, 5),(6, 6),(7, 7)] # node-densities for interpolation
5  plot_Lgrng_interp2d(f_true,vertexReg,nodes_cases,'Lgrng_approxR4plate')
6  #plt.show()
```

In this case, we found that using higher-order shape functions reduces the error but does not reduce it to zero, as shown in Fig. 3.19. This is because a trigonometric function is not in the polynomial space and cannot be reproduced, as discussed in Chapter 6 of Ref. [1]. We see again that the theories studied for 1D can largely be extended to nD functions.

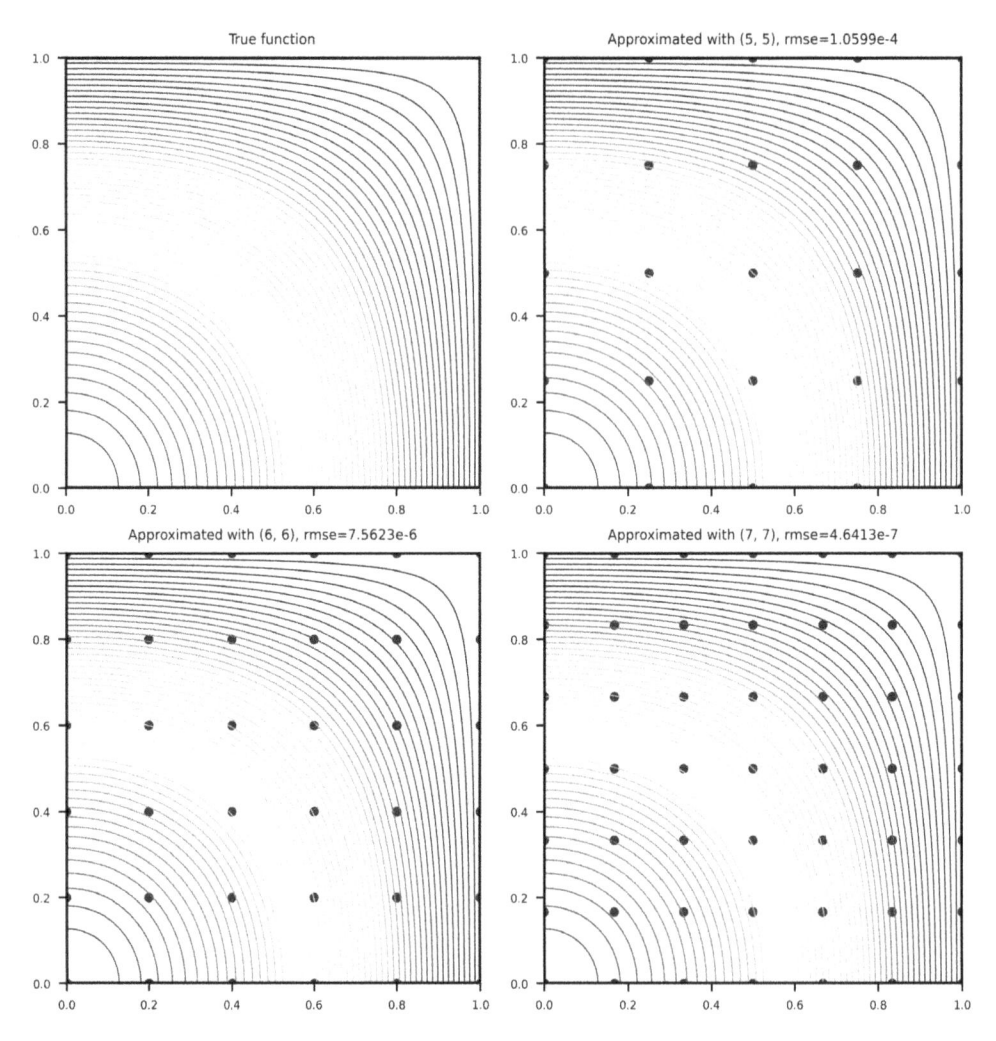

Figure 3.18. Approximation of a bi-fourth-order polynomial with a combined order of 6 using Lagrange interpolators over an irregular domain defined by four vertices.

The approximation can also be done over an irregular domain:

```
1  nodes_cases = [(5, 5),(6, 6),(7, 7)] # node-densities for interpolation
2  plot_Lgrng_interp2d(f_true,vertexIrr,nodes_cases,'Lgrng_approxQ4plate')
3  #plt.show()
```

Similar to the case using a regular domain, using higher-order shape functions reduces the error, but will not reduce it to zero, as shown in Fig. 3.20.

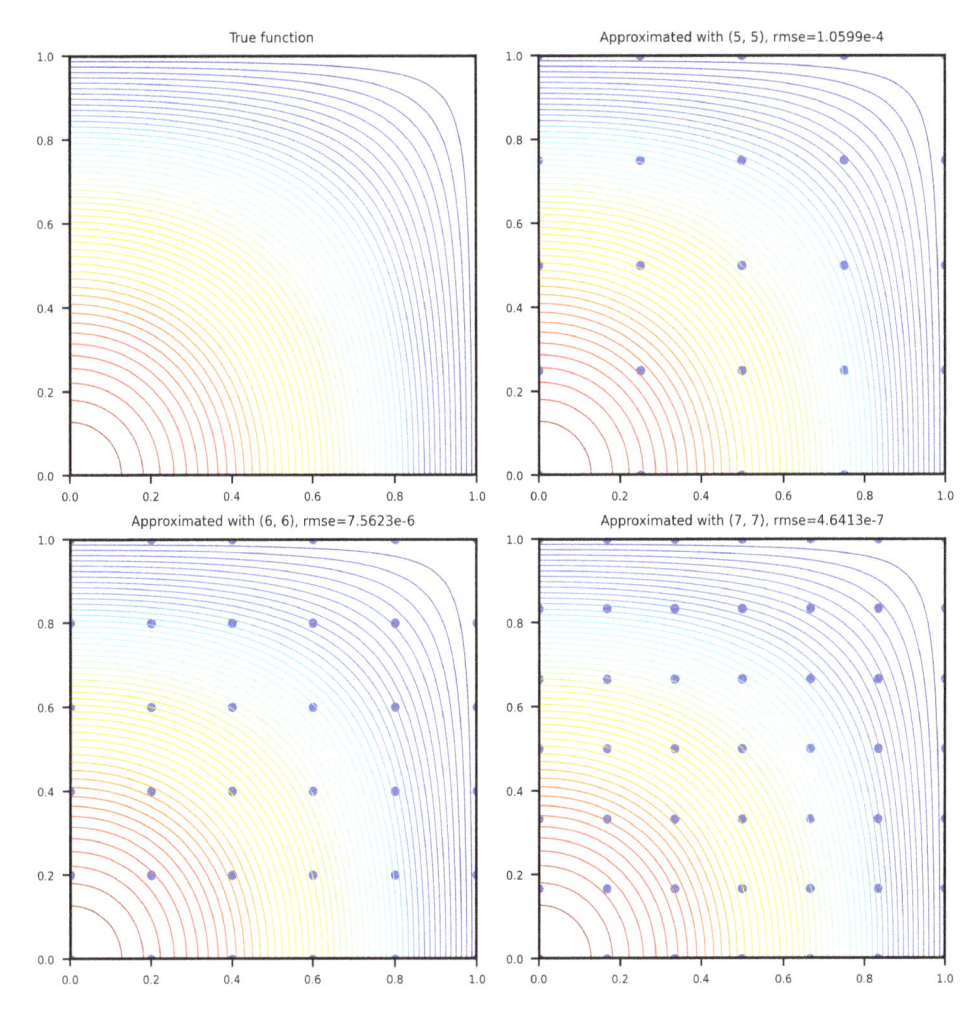

Figure 3.19. Approximation of a non-polynomial function using Lagrange interpolators over a regular domain defined by four vertices.

3.10.7 *Example: Partial derivative approximation*

We write code for approximating the derivatives of a given function using the derivatives of the Lagrange shape functions and plot the contour plots of the approximations with the estimated RMSE. The major equations to use are Eqs. (3.51) and (3.49). First, let us write the following utility code for later use:

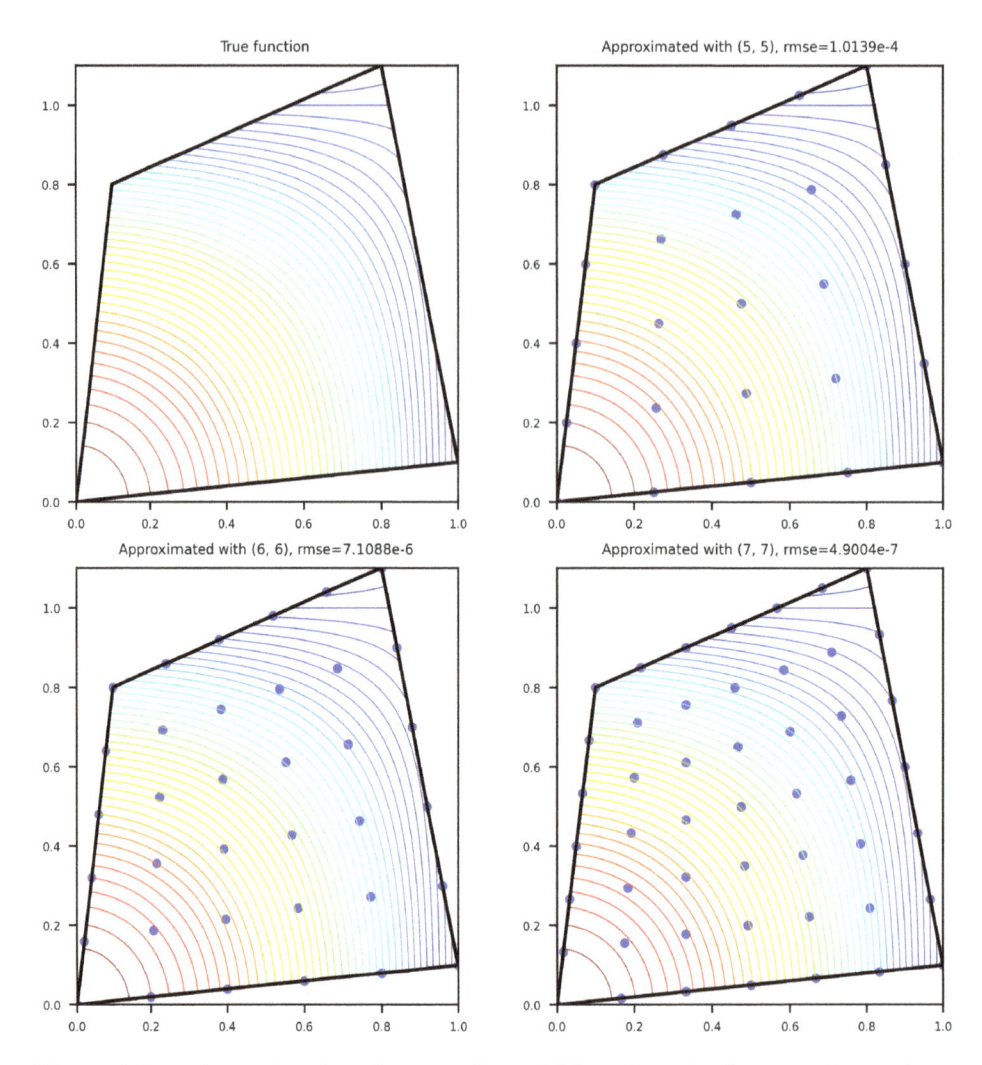

Figure 3.20. Approximation of a non-polynomial function using Lagrange interpolators over an irregular domain defined by four vertices.

```
1  def appox_dfxy_Q4_Lag(np_f_true, vertices, nodes_IJ):
2      '''Compute the approximated derivatives of a given function using
3          Lagrangian shape functions and their derivatives for a Q4.
4          Inpute: np_f_true -- given function to be approximated.
5                  vertices -- cordinates of the vertices of the Q4 in (x,y)
6                  nodes_IJ -- node number used in the Q4 for approximation.
7          Return: nodes -- used in the Lagrangian shape functions of Q4
8                  np_dfxy_a -- list of two approximated partial derivatives
9      '''
```

```
10      ξ, η =symbols('ξ, η')                          # 2D natural coordinates
11      nodeξi = np.linspace(-1., 1., nodes_IJ[0])          # nodes along ξ
12      nodeηj = np.linspace(-1., 1., nodes_IJ[1])                  # along η
13      _, _, Nijs = N_Lagrange_fξη(nodeξi, nodeηj)  # shape functions (2,2)
14      dNijsdξ = Nijs.diff(ξ)
15      dNijsdη = Nijs.diff(η)
16
17      # compute the physical nodal coordinates in the domain, xi and yj:
18      nodes = nodes_Q4(vertices, nodes_IJ[0])
19      nodes_xy = nodes.reshape(nodes_IJ[0], nodes_IJ[0],2)
20
21      dfξ_a, dfη_a = 0, 0      # df/dξ & df/dη of Lagrangian approximation
22      j11, j12, j21, j22 = 0, 0, 0, 0                  # for Jacobian matrix
23      for i in range(len(nodeξi)):            # using Nijs to approximate
24          for j in range(len(nodeηj)):
25              xy = nodes_xy[i,j,:]                        # xy: (x,y)
26              dfξ_a += dNijsdξ[i,j]*np_f_true(xy[0], xy[1])
27              dfη_a += dNijsdη[i,j]*np_f_true(xy[0], xy[1])
28
29              j11 += dNijsdξ[i,j]*xy[0]                # for Jacobian matrix
30              j12 += dNijsdη[i,j]*xy[0]
31              j21 += dNijsdξ[i,j]*xy[1]
32              j22 += dNijsdη[i,j]*xy[1]
33
34      Ji = (sp.Matrix([[j11, j12],      # inverse of the Jacobian matrix
35                       [j21, j22]])).inv()
36
37      # Approximated derivative: df/dx,df/dy:
38      dfxy_a=[Ji[0,0]*dfξ_a+Ji[1,0]*dfη_a,Ji[0,1]*dfξ_a+Ji[1,1]*dfη_a]
39      np_dfxy_a = [lambdify([ξ, η],df) for df in dfxy_a]     # to numpy f
40
41      return nodes, np_dfxy_a #list of 2 approximated partial derivatives
```

```
1  def plot_Lgrng_df(f_true, vertices, nodes_IJ, *p_name):
2       '''Approximate a 2D true function using a set of Lagrange shape
3          functions. Plot the Approximated and the true functions.
4       '''
5       plt.figure(); plt.ioff()
6       x, y, ξ, η =symbols('x, y, ξ, η')              # 2D natural coordinates
7       plt.rcParams.update({'font.size': 5})
8       fig_s = plt.figure(figsize=(6,6))
9
10      np_f_t = lambdify([x, y],f_true)       # convert to numpy function
11      dfxy_true = [f_true.diff(x), f_true.diff(y)]     # derivatives of f
12      np_dfxy_t = [lambdify([x, y], df) for df in dfxy_true]
13
14      Φ, Ψ = np.mgrid[-1:1:100j, -1:1:100j]  # regular mesh-grids in ξ, η
15      X, Y = intrpl_Q4(vertices, Φ, Ψ)          # mesh-grids X, Y for plots
16
```

```
17      # Plot the contour of the derivatives of the given true function
18      nls = 50                              # set levels for contour plot
19      titles = [f'True df/dx', f'True df/dy']
20      for k in range(2):                    # plot 2 figures in the first raw
21          ax = fig_s.add_subplot(2,2,k+1)
22          ax.contour(X,Y, np_dfxy_t[k](X,Y),cmap='jet',linewidths=.5,
23                                          levels=nls)
24          plt.plot(*np.vstack([vertices, vertices[0]]).T, 'k-') # plot Q4
25          ax.set_aspect(1.0); plt.title(titles[k])
26
27      # Compute the approximated derivatives of the function:
28      nodes, np_dfxy_a = appox_dfxy_Q4_Lag(np_f_t, vertices, nodes_IJ)
29
30      # Plot the approximated derivatives of the function:
31      for k in range(2):                    # plot 2 figures in the 2nd raw
32          ax = fig_s.add_subplot(2,2,3+k)
33          plt.contour(X,Y,np_dfxy_a[k](Φ,Ψ),cmap='jet',linewidths=.5
34                                          ,levels=nls)
35          ax.set_aspect(1.0)
36          rmse = sqrt(np.mean((np_dfxy_a[k](Φ,Ψ)-np_dfxy_t[k](X,Y))**2))
37          plt.title(f'Approximated with {nodes_IJ},rmse error={rmse:.4e}')
38          plt.plot(*np.vstack([vertices, vertices[0]]).T, 'k-') # plot Q4
39          plt.scatter(nodes[:, 0], nodes[:, 1], color='blue', s=10)
40
41      plt.savefig('imagesDI/'+p_name[0]+'.png',dpi=500,bbox_inches='tight')
42      #plt.show()
```

```
1  x, y, ξ, η =symbols('x, y, ξ, η')                # 2D natural coordinates
2  f_true = .2 + .5*x - .5*x*y + x**2 + x**4 -.8*y**4 -.3*x**3*y**2
3                              # Its highest combined order is 5
4  nodes_IJ = (4, 4)                       # nodes density for interpolation
5  plot_Lgrng_df(f_true, vertexReg, nodes_IJ, 'Lgrng_approxR4dPoly')
6  #plt.show()
```

We found that there are errors in the derivative approximation when the order of the Lagrange interpolators is lower than 4, for example, using (4, 4) nodes, as shown in Fig. 3.21. If we use (5, 5) nodes, we can generate complete fourth-order shape functions and their derivatives, and the error becomes zero (machine accuracy). The same is found when using (6, 6) nodes. We can use more nodes, but not fewer, to produce exact results.

Next, we approximate the same function but over an irregular domain. All the other settings are the same:

```
1  x, y, ξ, η =symbols('x, y, ξ, η')                # 2D natural coordinates
2  f_true = .2 + .5*x - .5*x*y + x**2 + x**4 -.8*y**4 -.3*x**3*y**2
3                              # Its highest combined order is 5
4  nodes_IJ = (5, 5)                       #nodes density for interpolation
5  plot_Lgrng_df(f_true, vertexIrr, nodes_IJ, 'Lgrng_approxQ4dPoly')
6  #plt.show()
```

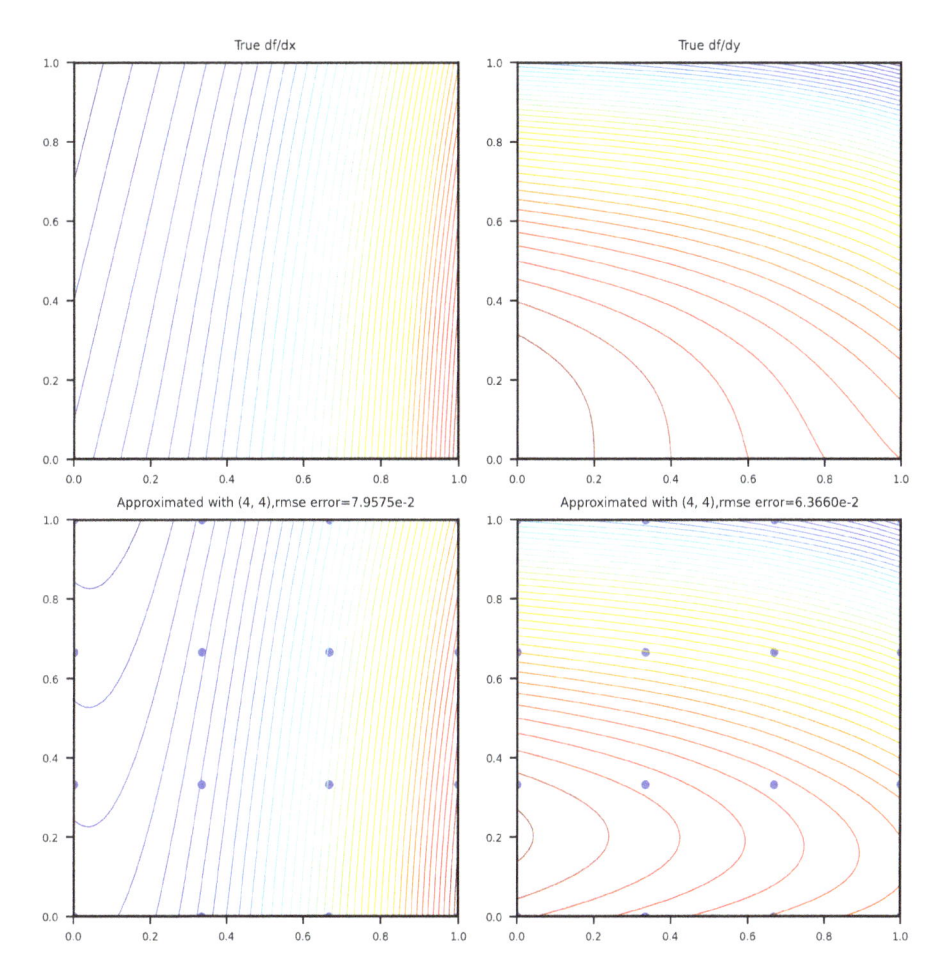

Figure 3.21. Approximation of the derivative of a bi-fourth-order polynomial using Lagrange interpolators over a regular domain defined by four vertices.

We found that there are errors in the derivative approximation when the order of the Lagrange interpolators is lower than 5, for example, using (5, 5) nodes, as shown in Fig. 3.22. If we use (6, 6) nodes, we can generate complete fifth-order shape functions, and the error in the derivative becomes zero (machine accuracy). The reason is the same as for the approximation of the function itself.

3.11 Polynomial in high dimensions

3.11.1 *General formulation and generation in SymPy*

In Chapter 2, we discussed 1D polynomials and their derivatives in detail. This section extends the discussion to nD functions using SymPy tools.

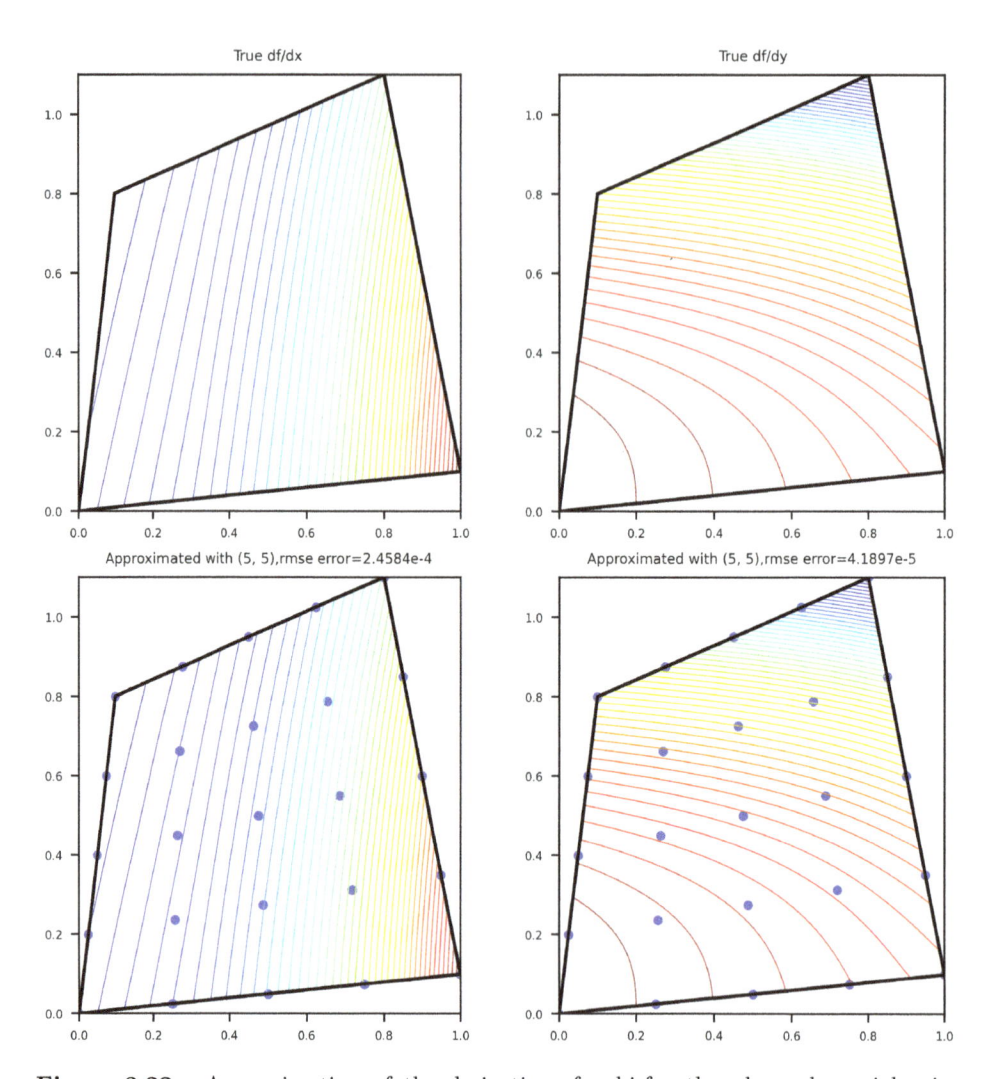

Figure 3.22. Approximation of the derivative of a bi-fourth-order polynomial using Lagrange interpolators over an irregular domain defined by four vertices.

Consider an nD function with independent variables $x_1, x_2, \ldots, x_n$, each in $\mathbb{R}$, which defines a Cartesian coordinate space $\mathbb{R}^n$. The general form of the polynomial function can be expressed as

$$f(\mathbf{x}) = \sum_{\nu_1 \nu_2 \ldots \nu_n} c_{\nu_1 \nu_2 \ldots \nu_n} x_1^{\nu_1} x_2^{\nu_2} \ldots x_n^{\nu_n} \tag{3.55}$$

where $\nu_1, \nu_2, \ldots, \nu_n$ are integers and $c_{\nu_1 \nu_2 \ldots \nu_n}$ are scalar constants. Due to the higher dimension, we see the complexity in indexing the coefficients and the order.

Let us write the following SymPy code to generate nD polynomials:

```
1  def poly_gen(n_var, p_order):
2      '''Create a multivariable polynomials.
3      input:  n_var, the number of variables, or dimension of the space;
4              p_order, the order of the polynomial
5      return: poly_f, sympy object, polynomial of p_order in n_var space.
6              coeffs, disctionary, the coefficients of the polynomial.
7      '''
8      variables=symbols('x1:'+str(n_var+1)) # name of variables:x1,x2,...
9
10     # Create indexes for the coefficients:
11     indexcs=[i for i in itertools.product(range(n_var+1),repeat=n_var)
12                         if sum(i)<p_order+1]
13     print(f'Number of terms in the polynomial = {len(indexcs)}')
14
15     c = sp.IndexedBase('c')                # base-name of the coefficients
16     coeffs = {i: c[i] for i in indexcs}    # dict holds the coefficients
17     poly_f = sp.Poly(coeffs, *variables)
18
19     return poly_f, coeffs
```

In the following example, we set the number of variables to 4 and the order of the polynomial to 2. Using **poly_gen()**, a second-order polynomial in a 4D real space can be easily generated:

```
1  n_var = 4; p_order = 2
2  poly_f, c42 = poly_gen(n_var, p_order)
3  poly_f
```

Number of terms in the polynomial = 15

$$
\begin{aligned}
\mathrm{Poly}\,\big(&c_{2,0,0,0}x_1^2 + c_{1,1,0,0}x_1x_2 + c_{1,0,1,0}x_1x_3 + c_{1,0,0,1}x_1x_4 + c_{1,0,0,0}x_1 \\
&+ c_{0,2,0,0}x_2^2 + c_{0,1,1,0}x_2x_3 + c_{0,1,0,1}x_2x_4 + c_{0,1,0,0}x_2 + c_{0,0,2,0}x_3^2 + c_{0,0,1,1}x_3x_4 \\
&+ c_{0,0,1,0}x_3 + c_{0,0,0,2}x_4^2 + c_{0,0,0,1}x_4 + c_{0,0,0,0},\, x_1, x_2, x_3, x_4, domain = \mathrm{EX}\big)
\end{aligned}
$$

There are four indices because the polynomial is in 4D space. Note that SymPy has an 'EX domain' which allows coefficients to be any SymPy expression. The numbers in each of the indices follow Eq. (3.55). To disply all the coefficients, use the following:

```
1  c42                                      # display all the coefficients
```

$$
\begin{aligned}
\{&(0,\ 0,\ 0,\ 0): c_{0,0,0,0},\ (0,\ 0,\ 0,\ 1): c_{0,0,0,1},\ (0,\ 0,\ 0,\ 2): \\
&c_{0,0,0,2},\ (0,\ 0,\ 1,\ 0): c_{0,0,1,0},\ (0,\ 0,\ 1,\ 1): c_{0,0,1,1},\ (0,\ 0,\ 2,\ 0): \\
&c_{0,0,2,0},\ (0,\ 1,\ 0,\ 0): c_{0,1,0,0},\ (0,\ 1,\ 0,\ 1): c_{0,1,0,1},\ (0,\ 1,\ 1,\ 0): \\
&c_{0,1,1,0},\ (0,\ 2,\ 0,\ 0): c_{0,2,0,0},\ (1,\ 0,\ 0,\ 0): c_{1,0,0,0},\ (1,\ 0,\ 0,\ 1): \\
&c_{1,0,0,1},\ (1,\ 0,\ 1,\ 0): c_{1,0,1,0},\ (1,\ 1,\ 0,\ 0): c_{1,1,0,0},\ (2,\ 0,\ 0,\ 0): c_{2,0,0,0}\}
\end{aligned}
$$

It is a dictionary. Hence, to access a particular coefficient value, use the following:

```
1  c42[0,0,1,0]
```

$c_{0,0,1,0}$

3.11.2 *Example: Linear function and derivatives in 3D*

As seen, the number of terms in the polynomial grows very fast with both the number of variables and the order of the polynomial. For problems with a large number of variables (such as ML problems), lower-order polynomials are more often used for better scalability. When the number of variables is small, we may use higher-order polynomials.

Considering 3D space, the independent variables are more often denoted as x, y, and z. In particular, a linear polynomial function is most often used:

$$f(x, y, z) = ax + by + cz + d \qquad (3.56)$$

where a, b, c, d are constants. This is the type of function we used in creating the shape functions for T4 elements/domains.

Equation (3.56) is familiar to many readers. If using `poly_gen()`, the linear polynomial is represented as follows:

```
1  n_var = 3; p_order = 1
2  poly_f, c31 = poly_gen(n_var, p_order)
3  poly_f                    # SymPy polynomial object with all atributes
```

Number of terms in the polynomial = 4

Poly $(c_{1,0,0}x_1 + c_{0,1,0}x_2 + c_{0,0,1}x_3 + c_{0,0,0}, x_1, x_2, x_3, domain = \text{EX})$

The partial derivatives of this function becomes the following:

```
1  x1, x2, x3 = symbols('x1, x2, x3')
2  dfdx = poly_f.diff(x1).as_expr()
3  dfdy = poly_f.diff(x2).as_expr()
4  dfdz = poly_f.diff(x3).as_expr()
5  derivatives = [dfdx,dfdy,dfdy]
6  derivatives
```

$[c_{1,0,0}, c_{0,1,0}, c_{0,1,0}]$

As expected, all the first derivatives are constants. Its Hessian matrix must be zero. Let us confirm this using the following:

```
1 sp.hessian(poly_f.as_expr(), [x1,x2,x3]) # as_expr() gets only the poly
```

$$\begin{bmatrix} 0 & 0 & 0 \\ 0 & 0 & 0 \\ 0 & 0 & 0 \end{bmatrix}$$

3.11.3 Example: Quadratic polynomial in 3D

Let us form a quadratic function in a 3D space:

```
1 n_vars = 3; p_order = 2
2 poly_f, c32 = poly_gen(n_vars, p_order)
3 poly_f
```

Number of terms in the polynomial = 10

$$\text{Poly}\left(c_{2,0,0}x_1^2 + c_{1,1,0}x_1x_2 + c_{1,0,1}x_1x_3 + c_{1,0,0}x_1 + c_{0,2,0}x_2^2 + c_{0,1,1}x_2x_3 \right.$$
$$\left. + c_{0,1,0}x_2 + c_{0,0,2}x_3^2 + c_{0,0,1}x_3 + c_{0,0,0}, x_1, x_2, x_3, domain = \text{EX}\right)$$

This expression has 10 terms. Let us compute its derivatives and use these to form its gradient:

```
1 dfdx = poly_f.diff(x1).as_expr()
2 dfdy = poly_f.diff(x2).as_expr()
3 dfdz = poly_f.diff(x3).as_expr()
4 sp.Matrix([dfdx, dfdy, dfdz])
```

$$\begin{bmatrix} 2x_1c_{2,0,0} + x_2c_{1,1,0} + x_3c_{1,0,1} + c_{1,0,0} \\ x_1c_{1,1,0} + 2x_2c_{0,2,0} + x_3c_{0,1,1} + c_{0,1,0} \\ x_1c_{1,0,1} + x_2c_{0,1,1} + 2x_3c_{0,0,2} + c_{0,0,1} \end{bmatrix}$$

Or simply use gr.grad_f() to compute the gradient:

```
1 X = sp.Matrix([x1, x2, x3])
2 gr.grad_f(poly_f.as_expr(), X)                        # Gradient ∇f
```

$$\begin{bmatrix} 2x_1c_{2,0,0} + x_2c_{1,1,0} + x_3c_{1,0,1} + c_{1,0,0} \\ x_1c_{1,1,0} + 2x_2c_{0,2,0} + x_3c_{0,1,1} + c_{0,1,0} \\ x_1c_{1,0,1} + x_2c_{0,1,1} + 2x_3c_{0,0,2} + c_{0,0,1} \end{bmatrix}$$

As expected, all the first derivatives are linear functions. Its Hessian matrix must contain constants. Let us confirm it using the following:

```
1  sp.hessian(poly_f.as_expr(), [x1, x2, x3])
```

$$\begin{bmatrix} 2c_{2,0,0} & c_{1,1,0} & c_{1,0,1} \\ c_{1,1,0} & 2c_{0,2,0} & c_{0,1,1} \\ c_{1,0,1} & c_{0,1,1} & 2c_{0,0,2} \end{bmatrix}$$

We obtained the expected results.

3.11.4 *Case study: Hessian matrix and extreme values*

The tasks of this example include the following:

1. creating a quadratic function in a 2D space;
2. computing its gradient;
3. computing its Hessian matrix;
4. computing the eigenvalues of the Hessian matrix;
5. examining the relationship between these eigenvalues and the extrema of the function;
6. generating plots to show various cases.

First, let us form a quadratic function in a 2D space:

```
1  n_vars = 2; p_order = 2
2  fq2d, c = poly_gen(n_vars, p_order)
3  fq2d
```

Number of terms in the polynomial = 6

$$\text{Poly}\left(c_{2,0}x_1^2 + c_{1,1}x_1x_2 + c_{1,0}x_1 + c_{0,2}x_2^2 + c_{0,1}x_2 + c_{0,0}, x_1, x_2, domain = \text{EX}\right)$$

The gradient vector becomes the following:

```
1  X = sp.Matrix([x1, x2])
2  gr.grad_f(fq2d.as_expr(), X)                                    # Gradient ∇f
```

$$\begin{bmatrix} 2x_1c_{2,0} + x_2c_{1,1} + c_{1,0} \\ x_1c_{1,1} + 2x_2c_{0,2} + c_{0,1} \end{bmatrix}$$

As expected, all the first derivatives are linear functions. Its Hessian matrix must contain constants. Let us confirm the following:

```
1  Hf = sp.hessian(fq2d.as_expr(), [x1, x2])
2  Hf
```

$$\begin{bmatrix} 2c_{2,0} & c_{1,1} \\ c_{1,1} & 2c_{0,2} \end{bmatrix}$$

In examining a 1D quadratic function, we found that it has a unique stationary point. This is also true for nD quadratic functions. Let us write the code to find the stationary point for quadratic nD functions.

For the quadratic function created, let us set its coefficients with specific given values. In this example, we will design four cases in a list, each with a different set of values for the coefficients:

```
1  cases = [{c[2,0]:-5, c[1,1]:1, c[1,0]:1, c[0,2]:-5, c[0,1]:1, c[0,0]:1},
2           {c[2,0]:+5, c[1,1]:1, c[1,0]:1, c[0,2]:-5, c[0,1]:1, c[0,0]:1},
3           {c[2,0]:+5, c[1,1]:1, c[1,0]:1, c[0,2]:+5, c[0,1]:1, c[0,0]:1},
4            c[2,0]:+5, c[1,1]:1, c[1,0]:1, c[0,2]: 0, c[0,1]:1, c[0,0]:1}]
```

The following code finds the stationary point, the Hessian matrix at that point, and its eigenvalues for all these four cases:

```
1  stationary_points = []              # place holder for stationary points
2
3  for i in range(len(cases)):
4      print(f'Case {i+1}: {cases[i]}')
5      Hfc = Hf.subs(cases[i])
6      print(f'Eigenvalues of Hessian matrix: {Hfc.eigenvals().keys()}')
7
8      fq2d_c = fq2d.subs(cases[i])
9
10     # find the stationary point:
11     sc = sp.solve([fq2d_c.diff(x1),fq2d_c.diff(x2)],[x1,x2] )
12     stationary_points.append([sc[x1], sc[x2]])
13
14     print(f'The stationary point:{sc[x1], sc[x2]}\n')
15
16 print(stationary_points)
```

```
Case 1: {c[2, 0]: -5, c[1, 1]: 1, c[1, 0]: 1, c[0, 2]: -5, c[0, 1]: 1,
c[0, 0]: 1}
Eigenvalues of Hessian matrix: dict_keys([-9, -11])
The stationary point:(1/9, 1/9)

Case 2: {c[2, 0]: 5, c[1, 1]: 1, c[1, 0]: 1, c[0, 2]: -5, c[0, 1]: 1, c[0, 0]: 1}

Eigenvalues of Hessian matrix: dict_keys([-sqrt(101), sqrt(101)])
The stationary point:(-11/101, 9/101)

Case 3: {c[2, 0]: 5, c[1, 1]: 1, c[1, 0]: 1, c[0, 2]: 5, c[0, 1]: 1, c[0, 0]: 1}

Eigenvalues of Hessian matrix: dict_keys([11, 9])
The stationary point:(-1/11, -1/11)

Case 4: {c[2, 0]: 5, c[1, 1]: 1, c[1, 0]: 1, c[0, 2]: 0, c[0, 1]: 1, c[0, 0]: 1}

Eigenvalues of Hessian matrix: dict_keys([5 - sqrt(26), 5 + sqrt(26)])
The stationary point:(-1, 9)

[[1/9, 1/9], [-11/101, 9/101], [-1/11, -1/11], [-1, 9]]
```

From the eigenvalues of the Hessian matrix, we found the following:

1. **H** for Case 1 is negative definite, and hence the stationary point is at a maximum.
2. The matrix **H** for Case 3 is positive definite, and hence the stationary point is at a minimum.
3. For Case 2, **H** is indefinite. Therefore, the stationary point is at a saddle point. This means that along one direction it is a maximum, while along another direction it is a minimum.
4. For Case 4, the dominant eigenvalue is positive, indicating that it is mostly at a minimum along the x-axis.

Let us confirm this by plotting these functions with the stationary points on them. The results are plotted in Fig. 3.23.

```
1  xi = np.linspace(-20., 20., 100)
2  yi = np.linspace(-20., 20., 100)
3  lxi, lyi = len(xi), len(yi)
4  print('length of xi=',lxi,'length of yi=',lyi)
5
6  plt.figure(); plt.ioff()
7  plt.rcParams['grid.linewidth'] = 0.2
8  X1, X2 = np.meshgrid(xi, yi)
9  fig_s = plt.figure(figsize=(8,5))
10 for i in range(len(cases)):
11     np_f = lambdify([x1,x2], fq2d.subs(cases[i]), "numpy")
12     Z = np_f(X1, X2)
13     ax = fig_s.add_subplot(2,2,i+1,projection='3d')
```

```
14      #ax.view_init(35, -10)
15      ax.set_xlabel('$x$', fontsize=4, labelpad=-14) #, rotation=80)
16      ax.set_ylabel('$y$', fontsize=4, labelpad=-12)
17      #ax.set_zlabel('Function Value', fontsize=9, rotation=0)
18      ax.yaxis._axinfo['label']['space_factor'] = 1.0
19      #ax.xaxis._axinfo["grid"]['linewidth'] = 0.2
20      ax.tick_params(axis='z', which='major', pad=-2)
21      ax.tick_params(axis='x', which='major', pad=-5)
22      ax.tick_params(axis='y', which='major', pad=-5)
23      ax.plot_surface(X1,X2,Z, color='b',rstride=1,cstride=1,
24                      shade=False,cmap="seismic", linewidth=0, alpha=.7)
25                      #PiYG_r, seismic, cool
26      ax.scatter(stationary_points[i][0], stationary_points[i][1], \
27              color='r', s=20, zorder=9)
28
29  fig_s.tight_layout()          # otherwise right y-label is clipped
30  plt.savefig('imagesDI/Quadratic3D.png',dpi=500,bbox_inches='tight')
31  #plt.show()
```

```
length of xi= 100 length of yi= 100
```

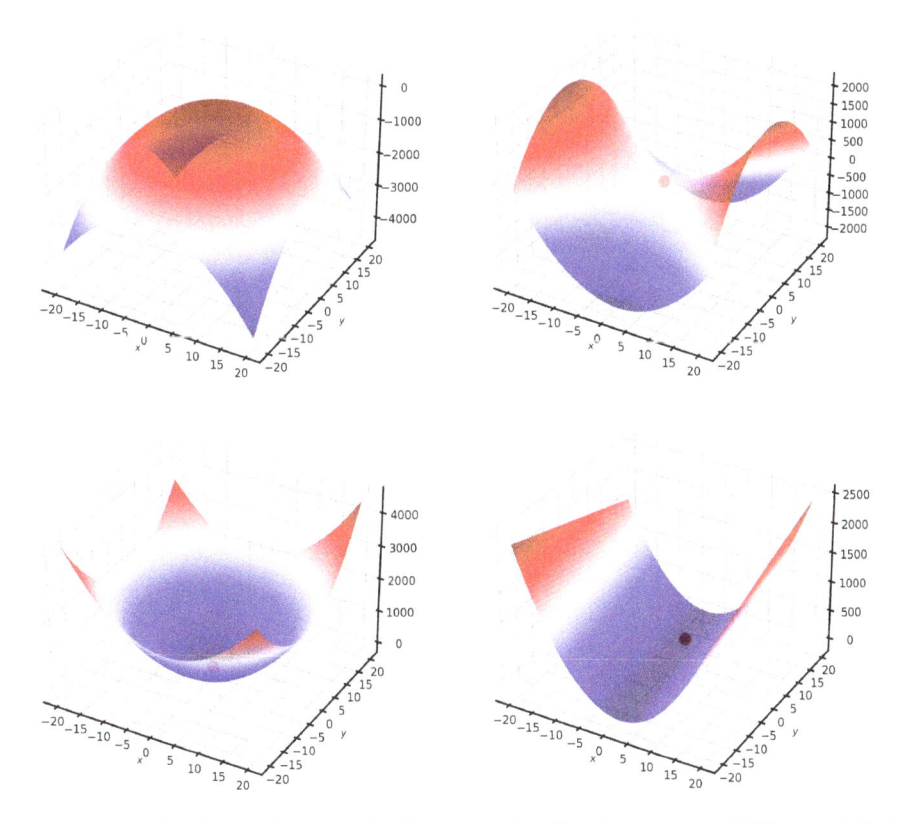

Figure 3.23. Quadratic functions in 2D space with Hessian matrix of different definiteness: top-left: with a maximum point; bottom-left: with a minimum point; top-right: with a saddle point; bottom-right: with a minimum point along x-axis.

3.12 Remarks

This chapter discussed theories, rules, techniques, and applications of derivatives of nD functions. We conclude this chapter with the following remarks:

1. An nD function has multiple independent variables and partial derivatives if it is differentiable. Each partial derivative can essentially be treated as the derivative of a 1D function. Therefore, all the discussions given in Chapter 2 hold for the partial derivatives of an nD function.
2. The derivative of an nD function is a row vector, and the gradient is the transpose of the derivative, hence it is a column vector. It contains all the slopes of the function in all its directions. These are used in gradient descent algorithms to find minimizers/optimizers.
3. The second-order derivative of an nD function is a Hessian matrix, whose eigenvalues reveal the local convexity of the function.
4. Taylor's expansion of an nD function has a similar form to that of a 1D function, but the first derivative becomes the gradient vector, and the second derivative becomes the Hessian matrix. This provides an effective way to approximate a complicated function locally.
5. Lagrange polynomials are useful in approximating a function and its partial derivatives over an interval where the Lagrange interpolators are constructed.
6. For irregular domains, coordinate mapping is performed using shape functions, resulting in the Jacobian matrix. Computations can be done using formulas similar to those for regular square domains, with the aid of the Jacobian matrix. For more complicated domains, one may need to divide the domain into elements and perform the mapping for each element. This is typically used in FEM methods.
7. A Python code is presented for systematically generating polynomials with multiple variables in higher dimensions, allowing for easy computation of their gradients and Hessian matrices.

The following chapter discusses another important operation for functions: integration.

References

[1] G.R. Liu, *Numbers and Functions: Theory, Formulation, and Python Codes*, World Scientific, 2024.
[2] G.R. Liu, *Machine Learning with Python: Theory and Applications*, World Scientific, 2023.

[3]   Y.C. Fan, *Advanced Mathematics*, 1979.

[4]   Gilbert Strang, *Calculus*, 1991. Available online (http://ocw.mit.edu/OcwWeb/resources/RES-18-001Spring-2005/ResourceHome/index.htm).

[5]   G.R. Liu and Siu Sin Quek, *The Finite Element Method: A Practical Course*, Butterworth-Heinemann, 2013.

[6]   G.R. Liu and T.T. Nguyen, *Smoothed Finite Element Methods*, Taylor and Francis Group, New York, 2010.

[7]   G.R. Liu, *Mesh Free Methods: Moving beyond the Finite Element Method*, Taylor and Francis Group, New York, 2010.

[8]   A.P. Boresi and R.J. Schmidt, *Advanced Mechanics of Materials*, John Wiley & Sons, New York, 2002.

[9]   Stephen Timoshenko and James N. Goodier, *Theory of Elasticity*, 1970. Available online (http://books.google.com/books?id=yFISAAAAIAAJ&dq=theory+of+elasticity&ei=ICiKSsr3G4jwkQSbxMyPCg).

[10] Z.L. Xu, *Elasticity*, Vol. 1 & 2, People's Publisher, China, 1979.

[11] G.R. Liu, *Mechanics of Materials: Formulations and Solutions with Python*, World Scientific, 2024.

[12] Jorge Nocedal and J. Stephen, *Numerical Optimization*, 2006.

[13] GR Liu, *Vector Calculus: Formulation, Applications, and Python Codes*, World Scientific, New Jersey, 2025.

Chapter 4

Integration of One-Dimensional Functions

```
 1  # Place cursor in this cell, and press Ctrl+Enter to import dependences.
 2  import sys                          # for accessing the computer system
 3  sys.path.append('../grbin/') # Add in the path to your system
 4
 5  from commonImports import *        # Import dependences from '../grbin/'
 6  import grcodes as gr                 # Import the module of the author
 7  importlib.reload(gr)                # When grcodes is modified, reload it
 8
 9  init_printing(use_unicode=True)     # For latex-like quality printing
10  np.set_printoptions(precision=4,suppress=True,
11                      formatter={'float_kind': '{:.4e}'.format})
```

Integration of functions is a frequently performed, fundamental, and useful operation in computational methods. It is often referred to as an **antiderivative**. We have discussed the derivatives of functions in detail in Chapter 3. This chapter is devoted to topics on antiderivatives of functions.

The focus will be on one-dimensional (1D) functions so that fundamental concepts, theories, formulation, techniques, and properties of both indefinite and definite integrals can be studied in detail. Handling the complexity of higher dimensionality will be discussed in Chapter 5.

This chapter is written with reference to textbooks in Refs. [1–4]. Many of the proofs can be found in the first three text books. Wikipedia pages, particularly those on integrals (https://en.wikipedia.org/wiki/Integral), serve as valuable additional references.

Both NumPy and SymPy are used in the development of code for the demonstration examples. "Discussions" with ChatGPT, Gemini, and Bing have also greatly helped in coding and in the preparation of this chapter.

4.1 Definition

There are two types of integration: definite integration and indefinite integration. The function to be integrated is often called **integrand** denoted generically as $f(x)$. The definite integration integrates the function over a finite domain called interval, which produces a value often denoted as I. The indefinite integration produces the general form of antiderivative function denoted generically as $F(x)$. The differentiation of an antiderivative gets back its corresponding integrand $f(x)$.

4.1.1 *Area and integration*

In contrast to derivatives that reveal the local feature of a function, an integration reveals some important **global** feature of the integrand function. For 1D non-negative functions, the definite integration gives the area between the function curve and the x-axis in the closed **interval** often denoted as $[a, b]$. The result will always be a positive value. For a general function, the integral gives the **signed area** of the region bounded by the function and the x-axis, as shown in Fig. 4.1.

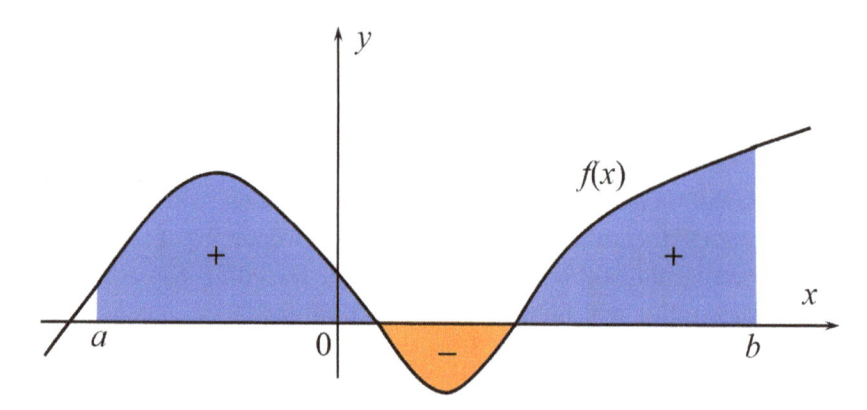

Figure 4.1. A definite integral of a function gives the signed area of the region bounded by the curve of the function and the x-axis. In the above graph, the integral is the blue $(+)$ area subtracted by the orange $(-)$ area.

Source: Image made in reference to that at https://commons.wikimedia.org/ by KSmrq under CC BY-SA 3.0 license.

The integrated value is the blue (+) area subtracted by the orange (−) area. Therefore, the integral result can be negative. In either case, the integral gives just one scalar value for a given function: the signed area. In general, it is an operation to obtain the accumulated effects of the function, which can carry different physical meaning.

4.1.2 *Example: Area of speed, distance*

Consider a function $f(t)$ that is the speed of a traveling car. The integration of $f(t)$ over time t is the area under $f(t)$ from 0 to time t, which gives the distance that the car traveled. It is denoted as the antiderivative $F(t)$. If the speed of the car is constant, say $f(t) = 2m/s$, as shown in Fig. 4.2(a), the distance traveled is the shaded rectangular area from 0 to time t. It is presented by $F(t)$ which is a function of t, as shown in Fig. 4.2(b). When $t = 0s$, the area is zero and $F(0) = 0$. When $t = 3s$, the area is $A = 2m/s \times 3s = 6m$ and $F(3s) = A = 6m$.

As another example, if a distribution of force is defined as a function of space over a solid, its integral will be the net total force exerted on the solid.

There are numerous such examples in science, engineering, and nature, and the physical meaning of the integral can differ. The theory, formulation, and techniques discussed in this chapter are universal and applicable to all disciplines.

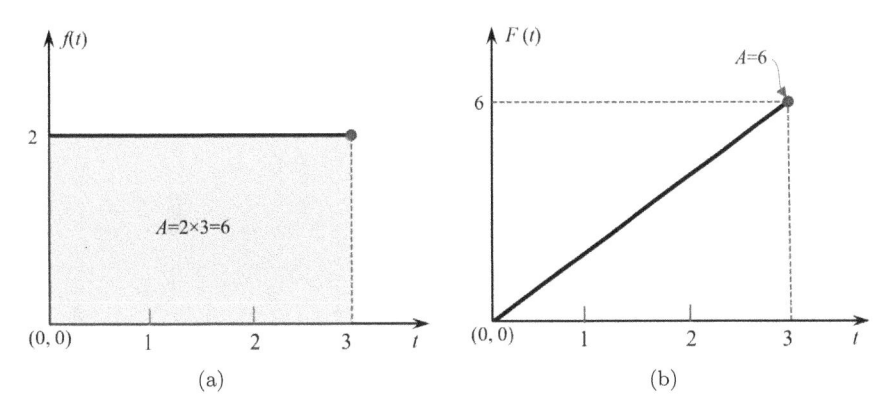

Figure 4.2. When speed of a car is the integrand $f(t)$, its antiderivative $F(t)$ becomes a function that measures the distance the car traveled.

4.1.3 *Integration via Riemann sum*

The definite integration can be expressed as

$$I = \int_a^b f(x)\,dx = \lim_{\substack{m\to\infty \\ x_{i-1}\leq x_i^* \leq x_i \\ \text{all } (x_i - x_{i-1})\to 0}} \left(\sum_{i=1}^{m} f(x_i^*)(x_i - x_{i-1}) \right) \qquad (4.1)$$

where a and b are the lower and upper limits for the integration interval, which are real numbers that define a finite range for the integration, and the sampling of the function is at x^* within the subinterval: $x_{i-1} \leq x_i^* \leq x_i$, as shown in Fig. 4.3. The summation in Eq. (4.1) is called the **Riemann sum**. If the limit of the Riemann sum converges to the signed area under the curve of the function $f(x)$, as all these subintervals approach zero, it is said to be Riemann integrable.

A definite integration of a scalar function over a specific interval produces a scalar I. If the integrand is a vector or matrix of functions, the result will be a vector or matrix containing scalars. This means that the integration gets into each of the functions in the vector or matrix.

The integral symbol $\int$ is a stretched S because it was rooted from summation. Examination of integrability of a function uses summation

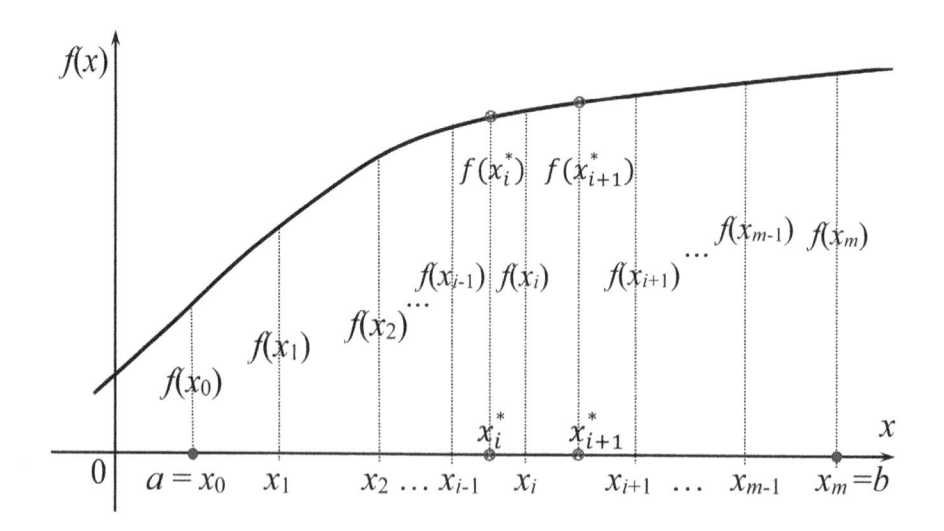

Figure 4.3.   Definite integration of a function over interval $[a, b]$.

(the Riemann sum, for example). In a numerical integration, the $\int$ also becomes a summation.

4.1.4 *Convergence of the Riemann sum*

Several different theories of integration exist, including the Riemann integral, Darboux integral, and the Lebesgue integral. A list of these and other integration theories can be found on the wikipage (https://en.wikipedia.org/wiki/Integral). Our discussion will focus on the Riemann integral, which is widely used in practical applications due to its straightforward concept.

The Riemann integral divides the integration interval into subintervals, in each of which the function is sampled. For each subinterval, a rectangle can be formed: The length of the subinterval is the base, and the function value sampled at **a point** (such as the left endpoint, right endpoint, or another point within the subinterval) is the height, as shown in an example given in Fig. 4.4.

If a refined division of the interval can be found such that the Riemann sum of the areas of all these rectangles is bounded and approaches the signed area formed by the curve of the function and the x-axis within a predefined level of accuracy, we obtain the integral value.

The integrability of a function is strongly influenced by the behavior of the function, including its boundedness, continuity, and the integration interval:

In most cases, a bounded and continuous function with a finite number of jumping points in a closed interval is Riemann integrable. A convergent Riemann sum can be found for such functions. Many of the functions from real-life problems belongs to this category.

If the integrand is unbounded at a point, such as a singular point in the interval, its integrability may become questionable. Unbounded continuous functions with growth rate slower than linear might still be Riemann integrable, depending on their specific behavior near the unbounded point. However, unbounded functions with faster growth are generally not integrable. Counterexamples to these generalities can also be found.

Let us see some examples.

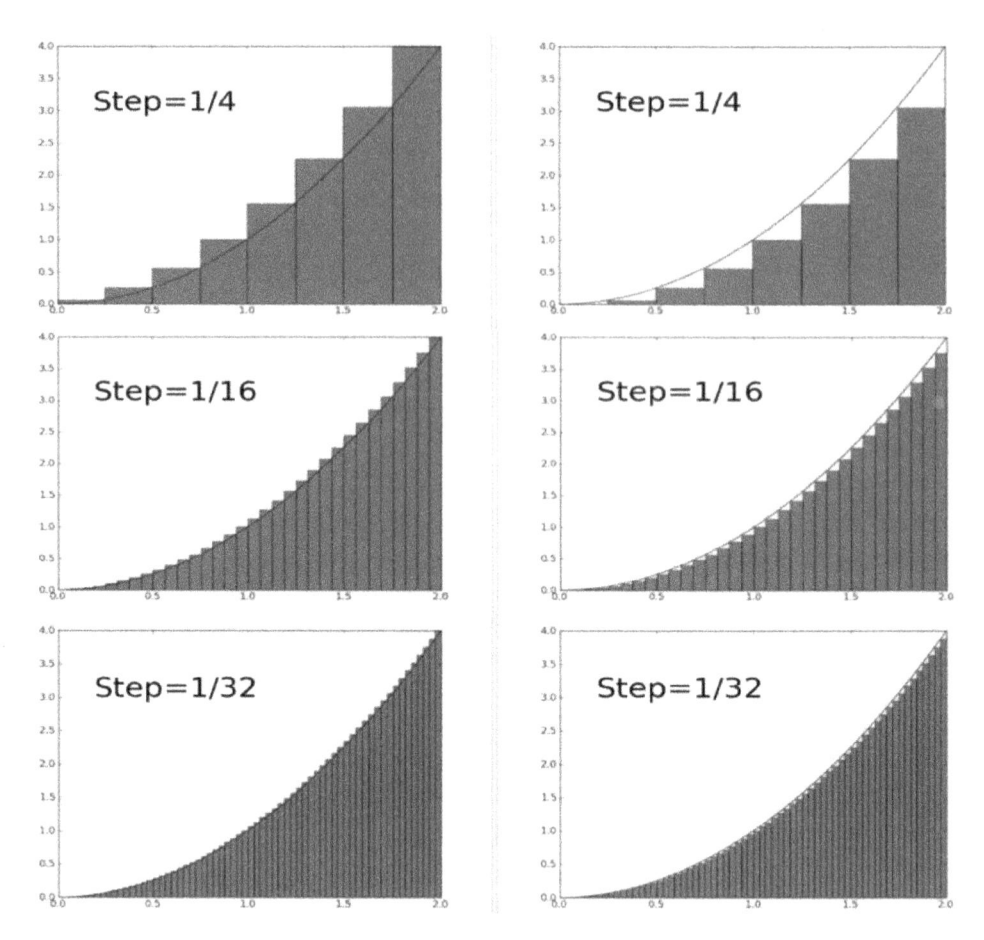

Figure 4.4. Division of the interval in the Riemann integral using different step sizes for function $f(x) = x^2$. Left: maximum heights in each subinterval are taken, which gives the upper Darboux sum. Right: minimum heights are taken, which gives the lower Darboux sum.
Source: Image created using the animation available at https://commons.wikimedia.org/ by IkamusumeFan under CC BY-SA 3.0 license.

4.1.4.1 *Example: Integrability of simple power functions*

A simple example is the power function: $\frac{1}{x^p}$, which is essentially the same as the monomials studied in Ref. [4]. This function is not defined for $x = 0$ for $p > 0$; it is singular there. As $x \to 0$, $\frac{1}{x^p} \to \infty$. When the integration interval is closed with 0, $[0, 1]$, $\frac{1}{x^p}$ may or may not be integrable, depending on the value of p that controls how fast the function approaches ∞ as $x \to 0$.

To have a better view on this issue, let us plot the distribution of the power function over $(0, 1]$ with different p using the following code:

```
1  # To plot the distribution of power functions in a given interval:
2
3  colors = ['b', 'r', 'g', 'c', 'm']
4  xL = 0.001; xR = 1.0; N = 200
5  ps = [-0.5, 0, 0.5, 1, 1.5]
6  x  = np.linspace(xL, xR, N)                            # on x-axis
7
8  plt.figure(); plt.ioff()
9  fig, ax = plt.subplots(1,1,figsize=(4,2.8))
10 ax.grid(c='r', linestyle=':', linewidth=0.3)
11
12 for k, p in enumerate(ps):
13     ax.plot(x, [1/xi**p for xi in x],c=colors[k],label='p='+str(p),lw=1)
14     plt.fill_between(x,[1/xi**p for xi in x],[0]*len(x),\
15                      where=(x>=xL)&(x<=xR), color=colors[k], alpha=0.1)
16
17 ax.scatter(1, 1, c='r' )
18 ax.set_xlabel('x')
19 ax.set_ylabel(r"$1/x^p$"); ax.set_ylim(0, 6)
20 ax.axvline(x=0, c="k", lw=0.6); ax.axhline(y=0, c="k", lw=0.6)
21 ax.legend()
22
23 plt.savefig('imagesDI/powerf0_1.png', dpi=500)
24 #plt.show();
```

It is seen from Fig. 4.5 that when $(p \leq 0)$, the function is bounded and continuous in $[0, 1]$, so we simply compute the area:

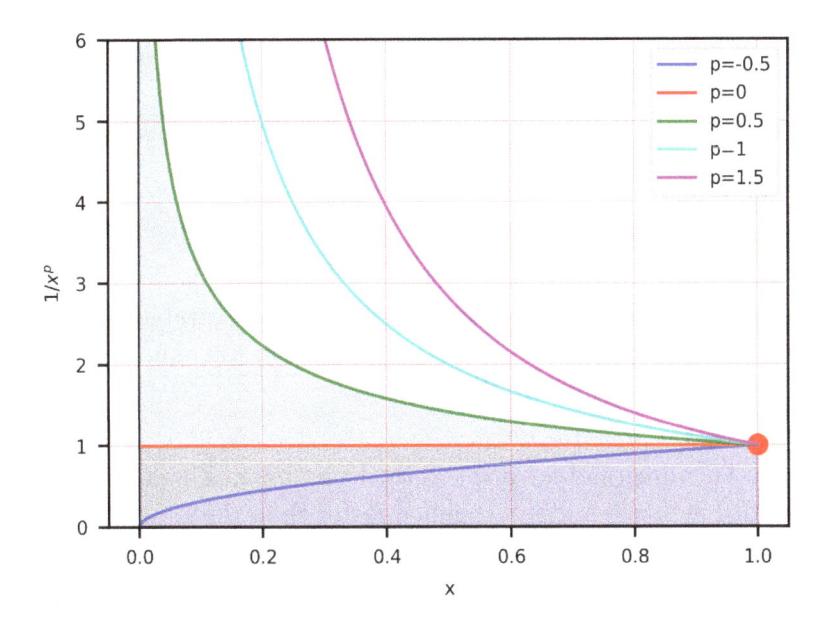

Figure 4.5. Distribution of the power functions with different p vlaues and the area under these functions.

```
1  # Example: integration of f(x)=1/x^p over [0,1] for different p:
2  x = sp.symbols("x")
3  a = 0; b = 1
4  ps = [-2, -1, -0.5, 0.]                              # p ≤ 0
5
6  print(f"For p ≤ 0, integratble over [{a}, {b}], I is finite:")
7  for p in ps:
8      print(sp.integrate(1/x**p,  (x, a, b)))
```

```
For p ≤ 0, integratble over [0, 1], I is finite:
1/3
1/2
0.666666666666667
1
```

For $p > 0$, the function is continuous, but **unbounded** at 0. The unboundedness does not necessarily make the integrand non-integrable. It depends on how fast the integrand grows to ∞ as $x \to 0$. Let us examine it case by case.

Case I: For $(0 < p < 1)$, the function $\frac{1}{x^p}$ growth rate is less than 1, and the definite integral over $[0, 1]$ is Riemann integrable. It can be integrated using the following code:

```
1  # Example: integration of f(x)=1/x^p over [0,1]:
2  ps = [0.0001, 0.5, 0.99999]                          # p < 1
3
4  print(f"For 0 < p < 1, integratble over [{a}, {b}], I is finite:")
5  for p in ps:
6      print(f" p = {p:.6f}, I = {sp.integrate(1/x**p, (x,a,b)).evalf(9)}")
```

```
For 0 < p < 1, integratble over [0, 1], I is finite:
p = 0.000100, I = 1.00010001
p = 0.500000, I = 2.00000000
p = 0.999990, I = 100000.000
```

It is found that all the integrals are Riemann integrable and give finite value. The integrand for $p = 0.5$ is the green line in Fig. 4.5, it is unbounded at $x = 0$, but the area over $[0, 1]$ is bounded.

For $p \geq 1$, the definite integral over the domain $[0, 1]$ is not Riemann integrable. The integrand for $p = 1$ is the cyan line in Fig. 4.5. The function is unbounded at $x = 0$ and grows much faster than the green line ($p = 0.5$). The area over $[0, 1]$ is also unbounded. The unboundedness is even more severe for $p = 1.5$, denoted by the magenta line in Fig. 4.5. The following code confirms all these:

```
1  x = sp.symbols("x")
2  ps = [1, 1.5, 2,]                                          # p ≥ 1
3
4  print("For p ≥ 1, integratble over [0, 1], I is infinite:")
5  for p in ps:
6      print(f" p = {p:.6f}, I = {sp.integrate(1/x**p, (x,a,b)).evalf(9)}")
```

```
For p ≥ 1, integratble over [0, 1], I is infinite:
 p = 1.000000, I = oo
 p = 1.500000, I = oo
 p = 2.000000, I = oo
```

In all these cases, $\frac{1}{x^p}$ grows very fast to ∞ as $x \to 0$, which is faster than linear. The Riemann sum diverges.

If our integration interval becomes $(0, 1]$, which does not include 0, the power function does not have a chance to grow to infinity. In this case, the function is bounded, the interval is finite, and the integral is Riemann integrable, regardless of the p value:

```
1  x = sp.symbols("x")
2  ps = [0.1, 0.5, 1, 1.5]                                    # any p
3  a = 0.00001; b =1                                          # a > 0, b > a
4  print("For p < 1, integratble over (0, 1], I is finite:")
5  for p in ps:
6      print(f" p = {p:.6f}, I = {sp.integrate(1/x**p, (x,a,b)).evalf(9)}")
```

```
For p < 1, integratble over (0, 1], I is finite:
 p = 0.100000, I = 1.11107597
 p = 0.500000, I = 1.99367544
 p = 1.000000, I = 11.5129255
 p = 1.500000, I = 630.455532
```

4.1.4.2 *Example: Integrability of general rational functions*

Consider more general rational functions $f(x) = \frac{1}{(x-d)^p}$. The same conclusions made for the previous example hold. The only difference is the location of the singularity point. In the previous example, the singularity is at $x = 0$, but for $f(x) = \frac{1}{(x-d)^p}$, the singularity is at $x = d$. The integrability of this function will be in question if the integration interval includes d:

- When $p \geq 1$, the function is not Riemann integrable over $[d, b]$.
- When $p < 1$, the function is Riemann integrable over $[d, b]$.
- The function is Riemann integrable over $(d, b]$ regardless of the value of p.

```
1  x = sp.symbols('x', real=True)
2  d = 2; b = 3
3  ps = [0.5, 0.99, 1, 1.1]
4  for p in ps:
5      print(f" p = {p:.6f}, I = {integrate(1/(x-2)**p,(x,d,b)).evalf(9)}")
```

```
p = 0.500000, I = 2.00000000
p = 0.990000, I = 100.000000
p = 1.000000, I = oo
p = 1.100000, I = oo
```

The integrated values are finite when $p < 1$ and infinite when $p \geq 1$.

When the singularity point at $x = d$ is excluded, the rational function is integrable for all real finite p:

```
1  ps = np.array([0.5, 1, 1.5, 2])
2  ε = 0.0001
3  for p in ps:
4      print(f" p={p:.6f}, I={integrate(1/(x-d)**p, (x,d+ε,b)).evalf(9)}")
```

```
p=0.500000, I=1.98000000
p=1.000000, I=9.21034037
p=1.500000, I=198.000000
p=2.000000, I=9999.00000
```

4.1.5 *Counterexample: Growth slower than linear, not integrable*

We mentioned that unbounded continuous functions with a growth rate slower than linear might still be Riemann integrable, depending on their specific behavior near the unbounded point. We showed that the power function is integrable over $[0, 1]$ if $p < 1$. A counterexample is the following function:

$$f(x) = \frac{1}{x \ln x} \tag{4.2}$$

Its growth rate is slower than linear as $x \to 0$, but it is not integrable over $[0, 1]$:

```
1  integrate(1/(x*sp.log(x)), (x,0,1))
```

$-\infty$

4.1.6 *Counterexample: Linear growth rate, integrable*

We also mentioned that unbounded functions with a growth rate faster than or equal to linear are generally not integrable. However, counterexamples can also be found. We showed that the power function is not integrable over $[0, 1]$ if $p \geq 1$. A counterexample is the following function:

$$f(x) = \frac{\sin(x)}{x^2} \tag{4.3}$$

Its growth rate is linear as $x \to 0$, but it is integrable over $[0, 1]$:

```
1  f = sin(x)/(x**2)
2  I = integrate(f,(x,0,1))
3  display(Math(f"  ∫_0^1{latex(f)}dx = {latex(I.evalf(9))}"))
```

$$\int_0^1 \frac{\sin(x)}{x^2} dx = -0.0812827268$$

In the foregoing examples, we looked at singularities within the finite interval $[a, b]$. If the interval is infinite, such as $[a, \infty)$, the power function also encounters issues with unboundedness. Integrals with singular or unbounded integrands are called improper integrals. We will discuss integrability and techniques to deal with these in more detail in the section on improper integrals.

In this book, an integrable function refers to a Riemann integrable function by default.

4.1.7 *Indefinite integration*

An indefinite integration can be expressed as

$$\int f(x)\, dx \tag{4.4}$$

if $f(x)$ is **integrable for any interval in the domain of the function**. The purpose of the indefinite integration is to find a new function denoted as $F(x)$ called **antiderivative**, such that

$$\frac{dF(x)}{dx} = f(x) \tag{4.5}$$

Based on the discussion in Chapter 3, we know that for any constant function C, we have $\frac{dC}{dx} = 0$. Therefore, there can be multiple antiderivative functions that satisfy Eq. (4.5) because

$$\frac{d(F(x) + C)}{dx} = f(x) \tag{4.6}$$

This means that any antiderivative function added with an arbitrary constant is also an antiderivative function. Therefore, a general expression for indefinite integral becomes

$$\int f(x)\, dx = F(x) + C \tag{4.7}$$

Indefinite integration is the process of finding all possible antiderivatives of an integrand $f(x)$. Equation (4.7) defines the operator $\int$, which is the inverse of differentiation. Thus, $F(x) + C$ represents a family of antiderivatives, and $F(x) + C$ is also called an **antiderivative**. The task of indefinite integration is to find such an $F(x)$ called the **base antiderivative** for convenience in this book. When using Sympy for an infinite integration, the output is the base antiderivative by default. The family of antiderivatives can then be found by adding an arbitrary constant to the base antiderivative $F(x)$.

In the general definition, Eq. (4.7), C is called the **integral constant** because it is added to $F(x)$ whenever we find $F(x)$ through integration.

4.1.8 *Challenges in finding antiderivatives*

Due to the inverse nature, finding the base antiderivative of a given function $f(x)$ is usually much more difficult than finding the derivative of a given function. In Chapter 2, we learned that finding the derivative of a function at a point x can follow the definition of differentiation, which only requires finding the limit of an expression defined in the differentiation. It is a local operation around x, and any elementary function can have an explicit formula for its derivative.

To find the antiderivative for a function $f(x)$, however, we do not have such a local expression to work with. Essentially, we need to **guess** what kind of $F(x)$ has a derivative that becomes $f(x)$. It is well known that finding an inverse solution is often full of challenges [5].

In addition, as one might expect, $F(x)$ may not have a closed formula for many given closed-form functions $f(x)$, even if it is Riemann integrable. The Gaussian error function $\exp(-x^2)$ is an example. It looks simple and is Riemann integrable over any interval:

```
1  x = sp.symbols('x', real=True)
2  f = exp(-x**2)
3  I = sp.integrate(f, (x, -oo, oo))
4  display(Math(f" \int_{-\infty}^{\infty}{sp.latex(f)}dx = {sp.latex(I)}"))
```

$$\int_{-\infty}^{\infty} e^{-x^2} dx = \sqrt{\pi}$$

However, the explicit expression for its $F(x)$ cannot be found. We will see other examples in this chapter. For this preceding definite integral with limits of integration from $-\infty$ to ∞, the value of $\sqrt{\pi}$ is obtained through a series of techniques. The integral is first converted to a double integral (simply by squaring it) and then the integration is carried out in polar coordinates, followed by a u-substitution. In general, for such a function $f(x)$, we may need to resort to numerical techniques to find the result for definite integrals.

The difficulty is rooted in the **global** nature of the integral operation. In the case of derivatives, we are given the entire function and we find its **local** features following a given rule. Hence, this can always be done for a given function, provided the derivative exists. In contrast, for integrals, $F(x)$ represents local information about $f(x)$. This means that we need to find $F(x)$ from given local information. It's akin to being given only a part of an elephant's leg and needing to visualize the entire elephant! Let's note the following:

Finding an antiderivative of a given function is much more challenging than finding the derivative of a given function. Antiderivatives for many functions have been found, many have not yet, and many may not ever be found. It's essentially an inverse problem.

The derivative of a function is unique, but the antiderivative of a function is unique up to an additive constant.

4.1.9 *The first fundamental theorem of calculus*

Assume that the antiderivative $F(x)$ is found for an integrand $f(x)$. When evaluating the integral over a given closed interval Eq. (4.7), we have

$$\int_a^b f(x)\, dx = \big[F(x) + C\big]_a^b = F(b) - F(a) \tag{4.8}$$

In this case, the integral constant C is canceled in the process; it is always the same constant regardless at $x = a$ or $x = b$. When the integration interval is fixed, the arbitrariness is no longer there, and the integral outcome is a fixed value that is the signed area bounded by the curve of $f(x)$ and the x-axis over the interval. It becomes the difference of the antiderivative function evaluated at these two ends of the interval. The integration of $f(x)$ over $[a, b]$ is the difference of $F(x)$ at the two boundaries of the interval. It implies that if the antiderivative is found, the computation of the definite integral over any interval is straightforward. It may seem like a surprising result, but can be proven [3].

Proof. The proof of Eq. (4.8) uses the mean value theorem (MVT) given in Chapter 3. Consider a division of the interval $[a, b]$ into $a = x_0 < x_1 < x_2, \ldots, x_{n-1}, x_n = b$. Using the MVT for each of the subintervals, we have

$$\frac{dF(\xi_1)}{dx}(x_1 - a) = F(x_1) - F(a), \quad \text{where } x_1 \leq \xi_1 \leq a$$

$$\frac{dF(\xi_2)}{dx}(x_2 - x_1) = F(x_2) - F(x_1), \quad \text{where } x_2 \leq \xi_2 \leq x_1$$

$$\cdots = \cdots$$

$$\frac{dF(\xi_{n-1})}{dx}(x_{n-1} - x_{n-2}) = F(x_{n-1}) - F(x_{n-2}), \quad \text{where } x_{n-2} \leq \xi_{n-1} \leq x_{n-1}$$

$$\frac{dF(\xi_n)}{dx}(b - x_{n-1}) = F(b) - F(x_{n-1}), \quad \text{where } x_{n-1} \leq \xi_n \leq b \qquad (4.9)$$

When all $(x_{i+1} - x_i) \to 0$, the sum of all terms on the left gives $\int_a^b \frac{dF(x)}{dx} dx = \int_a^b f(x) dx$ and the sum of all terms on the right gives $F(b) - F(a)$, which is Eq. (4.8). The summation cancels all the function values $F(x_i)$ within the domain, leaving only the function values at the two **endpoints** of the interval. This completes the proof.

Note that in the proof given above, we require $F(x)$ being differentiable in $[a, b]$. This also implies that $F(x)$ is continuous in $[a, b]$. Further, $\frac{dF(x)}{dx}$ or $f(x)$ must be integrable in $[a, b]$. Otherwise, the sum on the left may not exist and be bounded.

The proof given above also provides an interesting observation. Each line in Eq. (4.9) is an approximated integration or average over its local subinterval: function value of $f(x_{i+1})$ multiplied by its local width $(x_{i+1} - x_i)$. It is a local version of Eq. (4.8). When $F(x)$ is continuous and $f(x)$ is integrable in $[a, b]$, the local expression becomes Eq. (4.8) at the limit of $(x_{i+1} - x_i) \to 0$. The summation cancellation of all the function values $F(x_i)$ within the domain is essential for Eq. (4.8) to hold.

Equation (4.8) is the **first fundamental theorem of calculus**.

Integration acts as a **smoothing** operator because it sums (averages) a set of values. This is in contrast to differentiation, which is a differencing operator that highlights changes between function values. It's crucial to pay attention to this distinction, especially when dealing with inverse problems that involve noisy data [5].

4.2 Examples: Computation of areas

4.2.1 *Area below a constant function*

To visualize a function to be integrated, we've included a reusable code snippet that can plot any function and shade the region under its curve within a user-defined interval. This code will be used throughout the chapter, and readers can skip the code details and directly apply it to the example problems:

```python
 1 def plot_f_shaded(f, g, A, B, a, b, p_name='tmp', shade=True):
 2     '''Plot a given function f over [A, B] with a shaded area between
 3        f and f0 in interval of [a, b].
 4        p_name: filename for the plot to be saved.
 5        Modified based on a suggestion by ChatGPT.
 6     '''
 7     x = np.linspace(A, B, 800)      # Generate x values for the function
 8     y = f(x)                        # sample the function values
 9     y0= g(x)
10
11     plt.figure(); plt.ioff()
12     plt.rcParams.update({'font.size': 8})
13     plt.plot(x, y, 'b-', linewidth=2)          # Plot the function f(x)
14     plt.plot(x,y0, 'r-', linewidth=1)          # Plot the function f(x)
15
16     if shade:
17         # Shade the area in [a, b] bounded by the f(x) and f0:
18         plt.fill_between(x, y, y0, where=(x >= a) & (x <= b), \
19                          color='skyblue', alpha=0.5)
20
21     plt.xlabel('x');      plt.ylabel('f(x)')
22     plt.axhline(0, c='k', linewidth=0.8)                    # X-axis
23     plt.axvline(0, c='k', linewidth=0.8)                    # Y-axis
24     plt.grid(color='red', linestyle='--', linewidth=0.3)
25
26     # Calculate y-limits for the plot
27     y_min = min(np.min(y), np.min(y0))    # Minimum y (f-g shade range)
28     y_max = max(np.max(y), np.max(y0))    # Maximum y
29     plt.ylim(y_min, y_max*1.1)               # Set y-limits for plot
30
31     # Add text annotation for the shaded area (optional)
32     #cx = (a+b+b)/4; cy = (f0+f(a)+f0+f(b))/4    # compute the centroid
33     #plt.text(cx, cy, 'Area A', horizontalalignment='center',
34     #            verticalalignment='center', fontsize=9, color='black')
35     plt.savefig('imagesDI/'+str(p_name)+'.png', dpi=500)    #plt.show()
```

Let us examine the simplest function:

$$f(x) = c \tag{4.10}$$

where c is a given constant. For this simple case, the base antiderivative is $F(x) = cx$ because $\frac{d(cx)}{dx} = c$. We can find the area under this function through definite integration. Over $[0, 2]$, the area is easily found using Eq. (4.8):

$$\int_0^2 f(x)dx = \int_0^2 cdx = [cx + C]_0^2 = 2c \tag{4.11}$$

We added an integral constant C, which is an arbitrary constant (not to be confused with c that is the given constant of function $f(x)$). Set $c = 1$, the function and the shaded region is plotted using the following code:

```
1  k_ = 0; c_ = 1.0          # use _ for specific value, k_=0 for constnat f
2  A = 0.9; B = 3.1                    # interval [A, B] for f(x)
3  a = 1.0; b = 3.0                    # interval [a, b] for shading
4  np_f = lambda x: k_*x + c_
5  f0= lambda x: 0.0*x                 # Start of y for shading the area
6
7  plot_f_shaded(np_f, f0, A, B, a, b, p_name='Intc') # Plot f,shaded area
8  #plt.show()
```

The area of the shaded region under the constant function is plotted in Fig. 4.6.

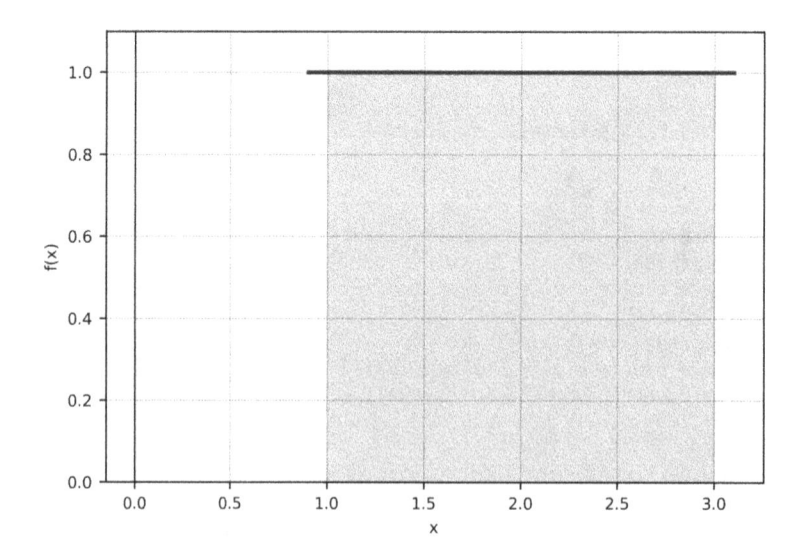

Figure 4.6. Integration of a constant function, which gives the area of the shaded region in the integration interval and under the function.

The area obtained via this definite integration is $A = 2$, which is exactly the same computed using the geometry of the rectangle shown in Fig. 4.6:

```
1  x, k, c = symbols('x, k, c')
2  f = k*x + c                          # define the SymPy function
3  np_a = 0.0; np_b = 2.0          # interval [a, b] for the shaded region
4  A = sp.integrate(f,(x, np_a, np_b))              # Area in [a, b]
5  display(Math(f" \\text{{Area formula: }} {sp.latex(A)}"))
6  print(f"Area between [{np_a},{np_b}] = {A.subs({k:k_,c:c_})}")
```

Area formula: $2.0c + 2.0k$

```
Area between [0.0,2.0] = 2.00000000000000
```

This example is simple, but the code works for any given function. Let us use it for a more complicated integrand.

4.2.2 *Example: Area below a square-root function*

Let us take a look at a square-root function:

$$f(x) = k\sqrt{x} \tag{4.12}$$

where k is a given constant. For this case, an antiderivative becomes $F(x) = \frac{2k}{3}x^{\frac{3}{2}}$ because $\frac{d}{dx}\left(\frac{2k}{3}x^{\frac{3}{2}}\right) = k\sqrt{x}$.

We first obtain the area under the square-root function over a fixed interval $[a, b]$ and then the area over interval $[a, x_b]$ with x_b as an variable. This second calculation provides a formula to compute the area under the function starting from a to any arbitrary location x_b.

Let us plot the figure for easier viewing:

```
1  f = k*sp.sqrt(x)                        # Define sympy function
2  k_ = 1.0                      # k_ coefficient value of the function
3  np_f = lambdify(x, f.subs(k,k_), 'numpy')          # convert to numpy
4
5  A = 0.0; B = 2.1                   # interval [A, B] for the function
6  np_a = 0.5; np_b = 2.0         # interval [a, b] for the shaded region
7  plot_f_shaded(np_f, f0,  A, B, np_a, np_b,p_name='Intx_sqrt')
8  #plt.show()
```

```
1  xb = symbols('x_b')                      # xb is an variable
2  A_formula = sp.integrate(f,(x,np_a,xb))
3  display(Math(f" \\text{{Area formula}}: {sp.latex(A_formula)}"))
4  print(f"Area between [{np_a},{np_b}]={A_formula.subs({k:k_,xb:np_b})}")
```

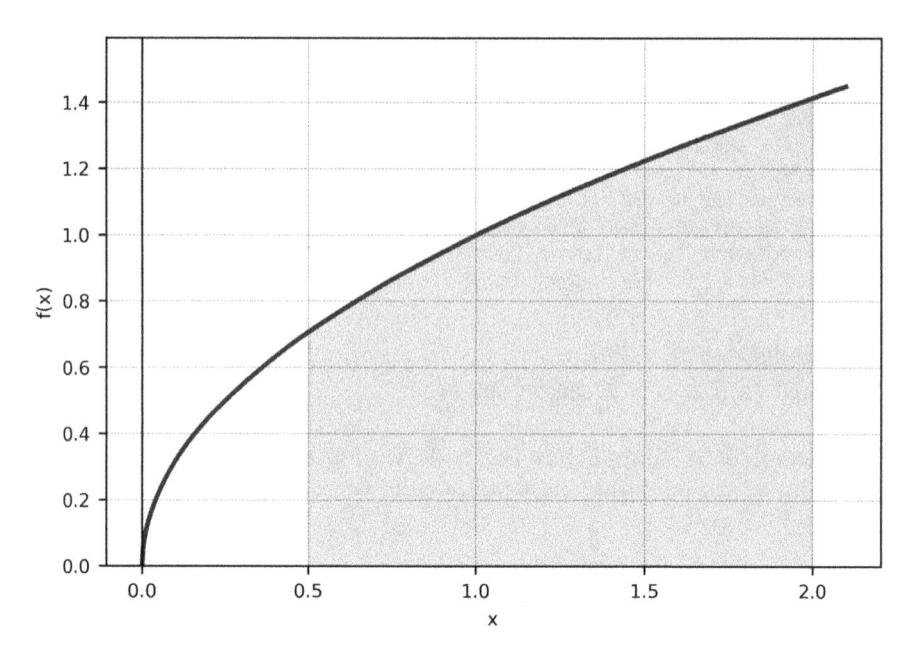

Figure 4.7. A root square function and the shaded region under the function for a definite integration.

Area formula : $\dfrac{2kx_b^{\frac{3}{2}}}{3} - 0.235702260395516k$

```
Area between [0.5,2.0]=1.64991582276861
```

This is the area value of the shaded region shown in Fig. 4.7.

4.2.3 *Example: Area below a trigonometric function*

Let us take a look at the sine function:

$$f(x) = \sin(x) \tag{4.13}$$

For this case, the base antiderivative becomes $F(x) = -\cos(x)$ because $\frac{d}{dx}(-\cos(x)) = \sin(x)$.

In this example, $f(x)$ value can be negative or positive. Let us compute the area under the function over a given interval $[a, b]$ and over the interval $[a, x_b]$ using definite integration. The function is plotted using the following code with shaded are a over $[a, b]$:

```
1  f = k*sp.sin(x)                            # Define sympy function
2  k_ = 1.0                          # k_ coefficient value of the function
3  np_f = lambdify(x, f.subs(k,k_), 'numpy')          # convert to numpy
4
5  A = 0.0; B = 2.5*np.pi              # interval [A, B] for the function
6  np_a = 0.; np_b = 2.*np.pi      # interval [a, b] for the shaded region
7
8  plot_f_shaded(np_f, f0, A, B, np_a, np_b, p_name='Int_sinx')
9  #plt.show()
```

The distribution of the sine function and the shaded region are shown in Fig. 4.8.

```
1  # Compute the area:
2  A_formula = sp.integrate(f,(x, np_a, xb))
3  display(Math(f" \\text{{ Area formula}}: {sp.latex(A_formula)}"))
4  print(f" Area between [{np_a},{np_b}]={A_formula.subs({k:k_,xb:np_b})}")
```

Area formula $: -k\cos\left(x_b\right) + k$
```
Area between [0.0,6.283185307179586]=0
```

This time, the signed area becomes zero for the sine function over interval $[0, 2\pi]$. This is because the positive and negative areas under the curve cancel each other out.

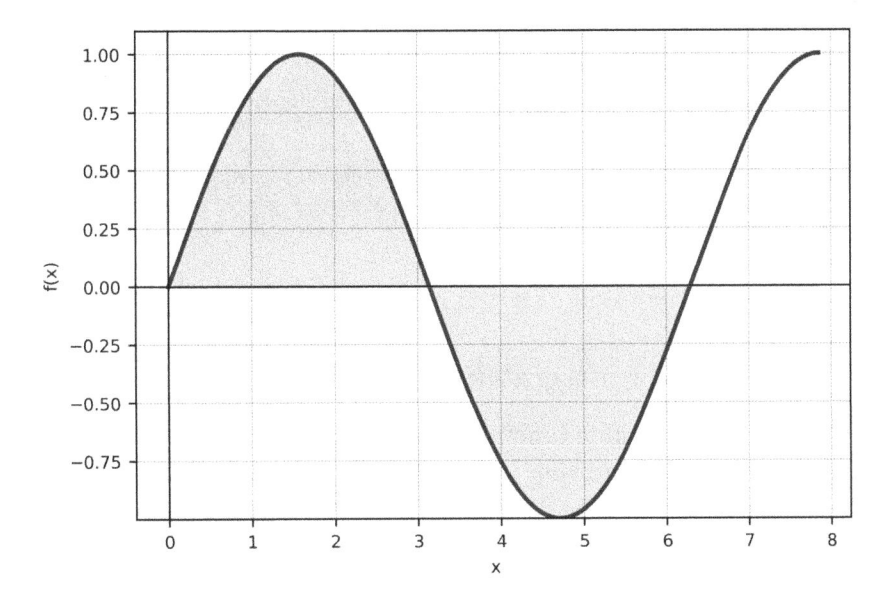

Figure 4.8. The sine function and the shaded region under the function for a definite integration.

4.2.4 *Example: Area under an absolute function*

To compute the absolute area between a function and x-axis, we use the following formula:

$$A_{\text{absolute}} = \int_a^b |f(x)|\, dx \qquad (4.14)$$

In this case, we require $f(x)$ to be **absolute integrable**. This means that the Riemann sum for $|f(x)|$ must converge and be equal to the absolute area.

For the sine function, for example, the absolute area between a function and x-axis is found using the following:

```
1  # Integration of an absolute function:
2  x = symbols('x')
3  f = sp.sin(x)                          # define the function in SymPy
4  np_a = 0.0; np_b = 2.0*np.pi
5  A = sp.integrate(f,(x, np_a, np_b))                # over [a, b]
6  print(f"Singed area between   [{np_a},{np_b}] = {A}")
7  A = sp.integrate(sp.Abs(f),(x, np_a, np_b))
8  print(f"Absolute area between [{np_a},{np_b}] = {A}")
```

```
Singed area between   [0.0,6.283185307179586] = 0
Absolute area between [0.0,6.283185307179586] = 4.00000000000000
```

The signed area enclosed between the function and the x-axis over the interval $[0, 2\pi]$ is zero. However, the same area by the absolute value of the function is positive because there is no cancellation of areas. The shaded regions are shown in Fig. 4.9.

```
1  def np_f(x): return np.abs(np.sin(x))    # define the function in numpy
2
3  A = 0.0; B = 2.5*np.pi               # interval [A, B] for the function
4  f0= lambda x: 0.0*x                  # Start of y for shading the area
5  plot_f_shaded(np_f, f0, A, B, np_a, np_b, p_name='Int_abssinx')
6  #plt.show()
```

4.2.5 *On absolute integrability*

Note that not all integrable functions are absolute integrable. Absolute integrability is a stronger condition. *If a function is absolutely integrable over an interval, the function will be integrable.* The reverse is not necessarily true.

For example, the function

$$f(x) = \frac{\sin(x)}{x} \qquad (4.15)$$

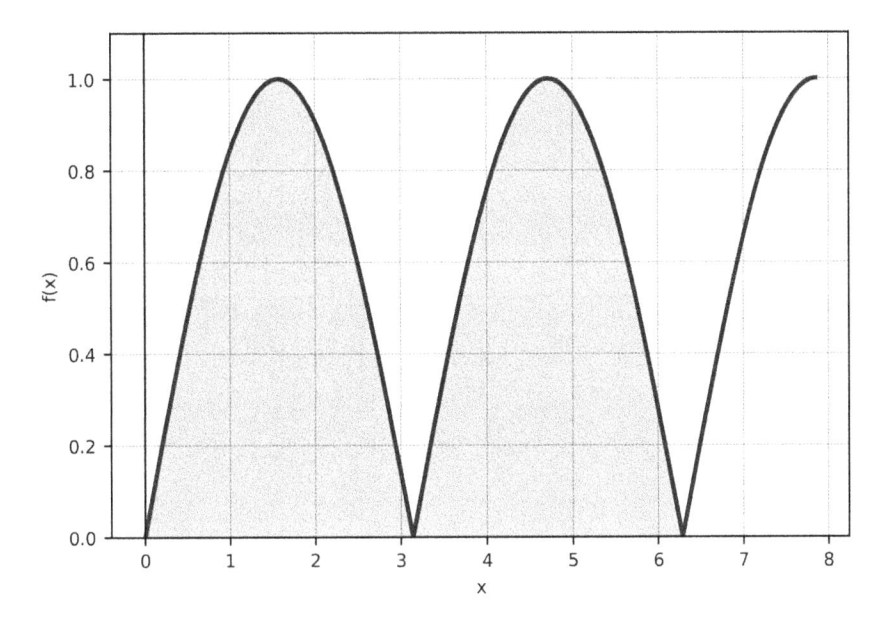

Figure 4.9. Function $|\sin(x)|$ and the shaded region.

looks simple, but its base antiderivative cannot be found. Its definite integral over $[0, \infty)$ can be found by the famous Feynman's integration tricks (https://zackyzz.github.io/feynman.html), and the value is $\pi/2$. Let us get the solution using SymPy:

```
1  x = symbols('x', real=True)
2  f = sp.sin(x)/x
3  I = sp.integrate(f, (x, 0, oo))
4  display(Math(f" ∫_0^∞ {sp.latex(f)}dx = {sp.latex(I)}"))
```

$$\int_0^\infty \frac{\sin(x)}{x}\,dx = \frac{\pi}{2}$$

However, it is not absolutely integrable over the same interval:

```
1  sp.integrate(sp.Abs(sp.sin(x)/x), (x, 0, oo))
```

$$\int_0^\infty \left|\frac{\sin(x)}{x}\right|\,dx$$

SymPy fails to integrate $|\sin(x)|$ due to the following reasons:

- **Without abs():** The integral of $\sin(x)$ oscillates between -1 and 1, causing area cancellation, resulting in a finite signed area (shown in Fig. 4.10).

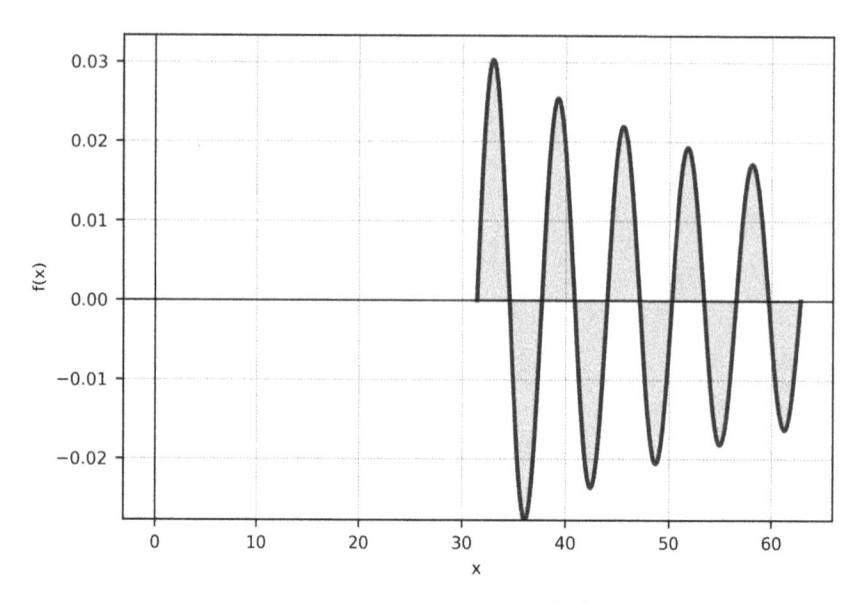

Figure 4.10. The distribution of the special function $\frac{\sin(x)}{x}$ and the shaded area for large x values.

- **With abs():** $|\sin(x)|$ is always positive. No cancellation occurs, and the integral keeps accumulating positive areas, resulting in divergence (shown in Fig. 4.11).

```
1  # Plot the the following function:
2  def np_f(x): return np.sin(x)/x              # define the function in numpy
3
4  A = 10*np.pi; B = 20*np.pi                    # interval [A, B] for the function
5  np_a = 1.*A; np_b = 1.*B          # interval [a, b] for the shaded region
6  plot_f_shaded(np_f, f0, A, B, np_a, np_b, p_name='Int_sinx_x')
7  #plt.show()
```

The distribution of $\frac{\sin(x)}{x}$ in a region with large x values is plotted in Fig. 4.10. The Riemann sum will converge as $x \to \infty$, due to the cancellations of positive and negative areas.

```
1  # Plot the the following function:
2  def np_f(x): return np.abs(np.sin(x)/x)
3  plot_f_shaded(np_f, f0, A, B, np_a, np_b, p_name='Int_abssinx_x')
4  #plt.show()
```

Distribution of $|\frac{\sin(x)}{x}|$ in a region with large x values is plotted in Fig. 4.11. The Riemann sum will not converge as $x \to \infty$.

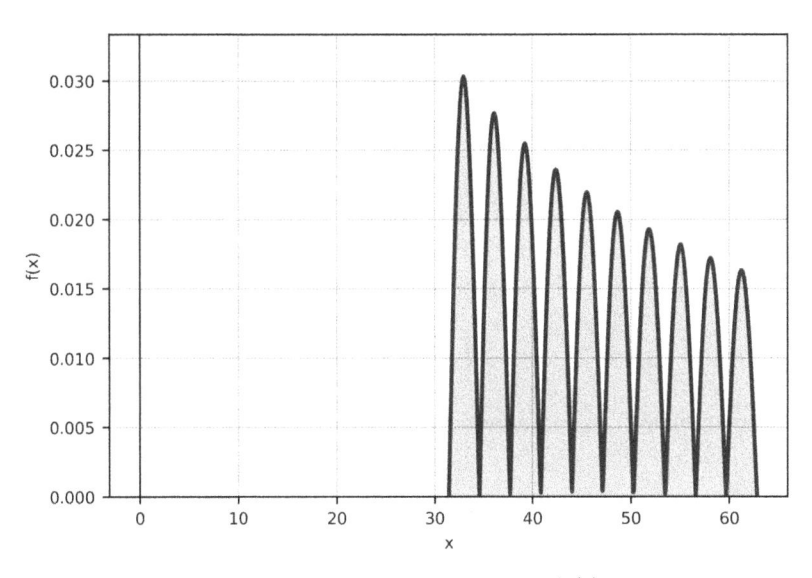

Figure 4.11. The distribution of the special function $\left|\frac{\sin(x)}{x}\right|$ and the shaded area for large x values.

4.2.6 *Example: Area under a squared function*

The integral for computing the area between function $[f(x)]^2$ and x-axis is given as

$$A_{f^2} = \int_a^b f(x)^2 dx \tag{4.16}$$

In this case, we require $f(x)$ to be **squarely integrable**. This means that the Riemann sum for $f(x)^2$ must converge and be equal to the area under $f(x)^2$.

For the squared sine function, for example, the integration is as follows:

```
1  A = sp.integrate(sp.sin(x)**2, (x, np_a, np_b))      # Area in [a, b]
2  print(f"Area between [{np_a},{np_b}] = {A}")
```

```
Area between [31.41592653589793,62.83185307179586] = 15.7079632679490
```

It is seen that the area is smaller than the absolute area, as expected because $\sin(x)$ is bounded by -1 and 1.

4.2.7 *On square integrability*

Not all integrable functions are square-integrable. Similar to absolute integrability, square integrability is a stronger condition than simple integrability. *If a function is square-integrable over an interval, the function will be integrable. The reverse is not necessarily true.* There are at least two major reasons for this:

- **Area cancellation:** Squaring a function makes its values all positive, preventing area cancellation, even if the original function oscillates with respect to the x-axis. The integral keeps accumulating positive values, leading to divergence. For example,

$$f(x) = \frac{\sin(x)}{\sqrt{x}} \tag{4.17}$$

Its integral over $[1, \infty]$ converges. However, when it is squared, the integral diverges:

```
1  f = sp.sin(x)/sqrt(x)
2  I = sp.integrate(f, (x, 1, oo))
3  display(Math(f"\\;\\;\\;\\;\\;\\;∫_1^∞{latex(f)}dx={latex(I.evalf())}"))
4  I2 = sp.integrate(f**2, (x, 1, oo))
5  display(Math(f"\\;\\;\\;\\;\\;\\;∫_1^∞{latex(f**2)}dx = {latex(I2)}"))
```

$$\int_1^\infty \frac{\sin(x)}{\sqrt{x}} dx = 0.632777533868738$$

$$\int_1^\infty \frac{\sin^2(x)}{x} dx = \infty$$

- **Doubling the growth rate:** If the integrand is unbounded and integrable over an interval, squaring it doubles its growth rate, which can lead to divergence. For example,

$$f(x) = \frac{1}{\sqrt{x}} \tag{4.18}$$

Its growth rate is $1/2$ as $x \to 0$. When it is squared, the growth rate becomes 1, and hence the integral over $[0, 1]$ diverges:

```
1  x = symbols('x', real=True)
2  f = 1/sp.sqrt(x)
3  I = sp.integrate(f, (x, 0, 1))
4  I2= sp.integrate(f**2, (x, 0, 1))
5  display(Math(f"\\;\\;\\;\\;\\;∫_0^1{sp.latex(f)}dx={sp.latex(I)}"))
6  display(Math(f"\\;\\;\\;\\;\\;∫_0^1{sp.latex(f**2)}dx={sp.latex(I2)}"))
```

$$\int_0^1 \frac{1}{\sqrt{x}} dx = 2$$

$$\int_0^1 \frac{1}{x} dx = \infty$$

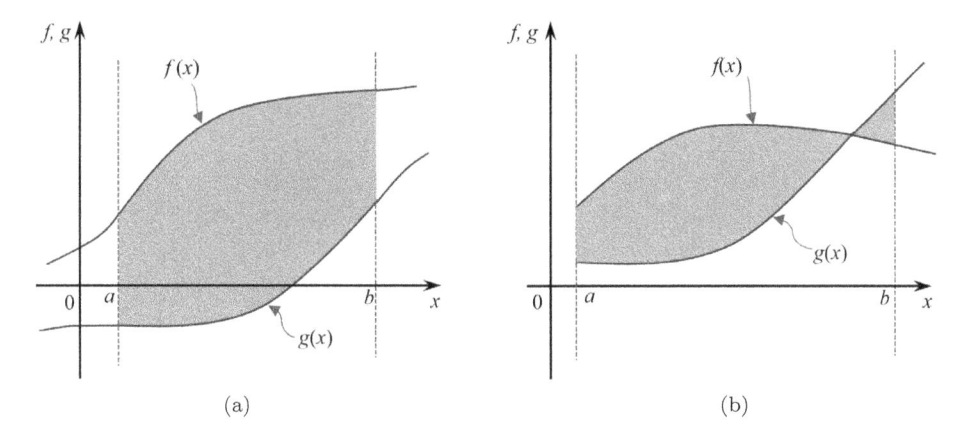

(a)   (b)

Figure 4.12.   Area between two curves in a 2D plane.

4.2.8 *Area between two curves*

Integration can be easily used to compute the area between two curves defined by two functions, as shown in Fig. 4.12.

Assuming that $f(x) \geq g(x)$ as shown in Fig. 4.12(a), the formula is as follows:

$$A_{f-g} = \int_a^b f(x) - g(x)dx \tag{4.19}$$

Otherwise, if these two curves intertwine as shown in Fig. 4.12(b), we use the following formula:

$$A_{f-g} = \int_a^b \left|[f(x) - g(x)]\right| dx \tag{4.20}$$

This makes the area between two curves positive. Let us look at an example for two intertwined curves:

```
1  # Plot the curves for functions:
2  def f(x): return np.sin(x)                  # define the function in numpy
3
4  A = 0.0; B = 2.5*np.pi                       # interval [A, B] for the function
5  np_a = 0.5; np_b = 2.3*np.pi                 # interval [a, b] for the shade
6  g = lambda x: -np.sin(x) +0.5               # Start from g for shading the area
7
8  plt.rcParams.update({'font.size': 8})
9  plot_f_shaded(f, g, A, B, np_a, np_b, p_name='Int_2sinx')
10 #plt.show()
```

The shaded region between two sine functions is shown in Fig. 4.13.

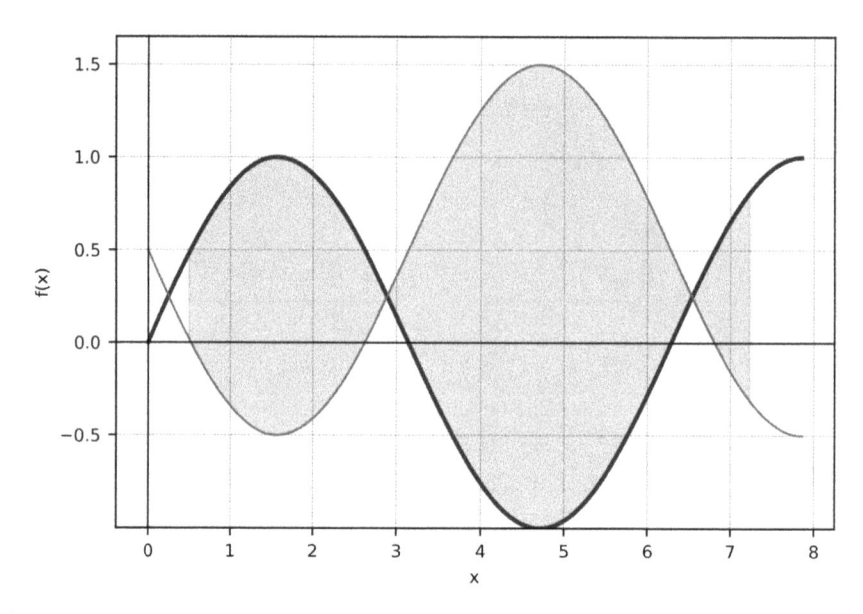

Figure 4.13. Integration of the difference of two sine functions gives the area of the shaded region between the functions.

    The area between these two intertwined curves is found using the following code:

```
1  # Area between these two intertwined curves via integration:
2  x = symbols('x')
3  f = sp.sin(x)                        # define the function in SymPy
4  g =-sp.sin(x)+0.5
5
6  A_fg = sp.integrate(sp.Abs(f-g),(x, np_a, np_b)).evalf(4)
7  print(f"Area between [{np_a},{np_b:.4f}] = {A_fg}")
```

```
Area between [0.5,7.2257] = 7.778
```

4.3 Some basic properties

We list some of the quite obvious properties of integrations:

Trivial case: Based on Eq. (4.8), when $b = a$, the integration will be zero. Therefore, in all our discussions, we exclude this trivial case and always assume $b \neq a$.

Interval splitting: Since integration is essentially a summation operation, the integration interval can be split into multiple subintervals, as long as these subintervals do not overlap and there are no gaps between them:

$$\int_a^b f(x)\,dx = \int_a^d f(x)\,dx + \int_d^b f(x)\,dx \tag{4.21}$$

where $a \le d \le b$.

The splitting is often used in the following scenarios: 1. The integrand is **piecewise continuous**. In this case, the splitting points are taken at the points of discontinuity, so that the integration can be done in pieces. 2. The integrand has points of non-differentiability (such as **kink points**). In this case, the splitting points are taken at the kink points, so that the integration can be done in pieces, each of which can be integrated smoothly.

Identical case: Equation (4.8) also gives:

$$\int_b^a f(x)\,dx = F(a) - F(b) = -\int_a^b f(x)\,dx \tag{4.22}$$

This implies that definite integration is **direction-dependent**.

Integral of an odd integrand over a symmetric interval vanishes: Considering an integrand that is an odd or antisymmetric function, $f(x) = -f(-x)$, we have

$$\int_{-a}^a f(x)\,dx = 0 \tag{4.23}$$

This is because, by definition, an integral is the signed area under the integrand, when the integrand is antisymmetric and the interval is symmetric, the signed area must be zero. The following is an example:

```
1  x = symbols('x', real=True)                      # variables
2  a, b, c = symbols('a, b, c', positive=True)       # given constants
3
4  f = a*x**3*cos(b) + c*sin(x)    # the integrand, an odd function of x
5  xL, xU = -1, 1                           # integration interval
6
7  I = integrate(f, (x, xL, xU))        # integral along x over [xL, xU]
8  display(Math(f"\\;\\;\\;\\;\\;∫_{{-1}}^1{sp.latex(f)}dx={sp.latex(I)}"))
```

$$\int_{-1}^1 ax^3 \cos\,(b) + c\sin\,(x)\,dx = 0$$

The antiderivative of an odd integrand is an even function: This is the inverse property discussed in Section 2.6. The following is an example:

```
1  C = symbols('C', real=True)                              # integral constant
2  f = a*x**3*cos(b) + c*sin(x)        # the integrand, an odd function of x
3  F = integrate(f, x) + C             # antiderivative becomes an even function
4  display(Math(f" \\;\\;\\;\\;\\;∫{sp.latex(f)}dx = {sp.latex(F)}"))
```

$$\int ax^3 \cos(b) + c\sin(x)dx = C + \frac{ax^4 \cos(b)}{4} - c\cos(x)$$

- The integral of an even function over a symmetric interval is twice the integral over half of the interval:

Considering an integrand that is an even or symmetric function, $f(x) = f(-x)$, we have

$$\int_{-a}^{a} f(x)\,dx = 2\int_{0}^{a} f(x)\,dx \tag{4.24}$$

When the integrand is symmetric and the interval is also symmetric, the signed areas of the left-half and right-half intervals are the same. This symmetry can simplify integrations both analytically and numerically. The following is an example:

```
1  a, b, c = symbols('a, b, c', positive=True)              # given constants
2
3  f = a*x**2*cos(b) + c*cos(x)       # the integrand, an even function of x
4  xL, xU = -1, 1                                         # integration interval
5
6  I_full = integrate(f, (x, xL, xU))     # integral along x over [xL, xU]
7  I_half = 2*integrate(f,(x,  0, xU))    # integral along x over [0, xU]
8  print(f"    Is I_full = 2*I_half? {I_full==I_half}")
9  display(Math(f" \\;\\;\\;\\;\\;∫_{{-1}}^1{latex(f)}dx={latex(I_half)}"))
```

```
Is I_full = 2*I_half? True
```

$$\int_{-1}^{1} ax^2 \cos(b) + c\cos(x)dx = \frac{2a\cos(b)}{3} + 2c\sin(1)$$

The antiderivative of an even integrand is an odd function: This is the inverse property discussed in Section 2.6. The following is an example:

```
1  f = a*x**2*cos(b) + c*cos(x)       # the integrand, an even function of x
2  F = integrate(f, x)                # base antiderivative becomes an odd function
3  display(Math(f" ∫{sp.latex(f)}dx = {sp.latex(F)}"))
```

$$\int ax^2 \cos(b) + c\cos(x)dx = \frac{ax^3 \cos(b)}{3} + c\sin(x)$$

4.4 Differentiation neutralizes integration

4.4.1 *The second fundamental theorem of calculus*

Using Eq. (4.7), and performing differentiation, we have

$$\frac{d}{dx}F(x) = \frac{d}{dx}\left(\int f(x)dx - C\right) = f(x) \tag{4.25}$$

This means that differentiating the antiderivative of an integrand returns the original integrand. In other words, differentiation neutralizes integration. This is known as the **second fundamental theorem of calculus**. It can be proven using an alternative expression for antiderivatives.

4.4.2 *An alternative expression of antiderivative*

Using the interval splitting property, Eq. (4.7) can be expressed alternatively as

$$F(x) = \int_a^x f(t)dt = \underbrace{\int_b^x f(t)dt}_{\text{function of } x} + \underbrace{\int_a^b f(t)dt}_{\text{constant } C} \tag{4.26}$$

in which we assume $f(x)$ is integrable over (a,b) and (b,x). In the first integral on the right-hand side of Eq. (4.26), the independent variable x is used as the upper limit of the integral. The variable inside a definite integral is called a **dummy variable**. It is used only for carrying out the definite integration and is also called the **integration variable**. After the integration is done, the integration variable disappears and is replaced by the integral limits. Therefore, the integration variable can always be changed to another symbol. In Eq. (4.26), t is used as the integration variable to avoid confusion with the independent variable x used for the upper integral limit. Equation (4.26) is pictorially shown in Fig. 4.14.

The lower integral limit a for the first integral in Eq. (4.26) can be replaced to any scalar in the domain of $f(x)$ because it changes only the constant C, and Eq. (4.26) shall still hold. Also, the arbitrary choice of a makes the constant $\int_a^b f(t)dt$ arbitrary. Hence, Eq. (4.26) is equivalent to Eq. (4.7). Equation (4.26) can be used to prove important theories.

4.4.3 *A proof of the second fundamental theorem*

Using Eq. (4.26), we can now prove the second fundamental theorem of calculus Eq. (4.25), differentiation neutralizing integration. This can be done in the following steps:

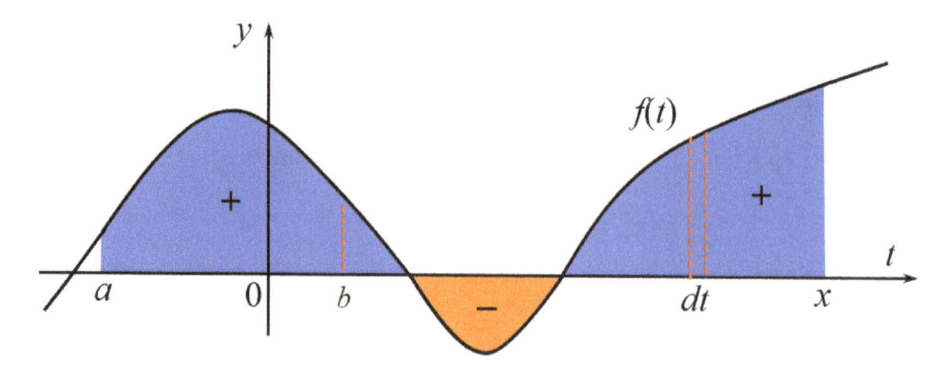

Figure 4.14. Function $F(x)$ resulted from a definite integral when the upper limit of the integral is set as the independent variable x.

Step 1: Assume that the integrand $f(x)$ is continuous and integrable and the antiderivative $F(x)$ is differentiable. Using the definitions of derivative of functions (see Chapter 3) and the Riemann integrable condition, we have

$$\frac{dF(x)}{dx} \equiv F'(x) = \lim_{\Delta x \to 0} \frac{F(x + \Delta x) - F(x)}{\Delta x}$$

$$= \lim_{\Delta x \to 0} \frac{\int_a^{x+\Delta x} f(t)dt - \int_a^x f(t)dt}{\Delta x}$$

$$= \lim_{\Delta x \to 0} \frac{\int_b^{x+\Delta x} f(t)dt + \int_b^a f(t)dt - \left[\int_b^x f(t)dt + \int_b^a f(t)dt\right]}{\Delta x}$$

$$= \lim_{\Delta x \to 0} \frac{\int_x^{x+\Delta x} f(t)dt}{\Delta x} = \lim_{\Delta x \to 0} \frac{f(\tau)\Delta x}{\Delta x}$$

$$= \lim_{\Delta x \to 0} f(\tau) = f(x) \tag{4.27}$$

where $x \le \tau \le (x + \Delta x)$. In the second-to-last step, we used the Riemann integrable condition. In the last line, we used the continuity of $f(x)$. When $\Delta x \to 0$, $f(\tau) \to f(x)$. This completes the proof of Eq. (4.25) for any continuous and bounded function $f(x)$.

Step 2: If $f(x)$ has a finite number of bounded jump discontinuities in the interval, we simply ignore these points in the proof given above. The sum of these points will be infinitesimally small when the interval is refined. Their presence does not affect the boundedness and convergence because the number of such points is finite.

Step 3: Consider a function $f(x)$ that is unbounded at a finite number of points but integrable over the entire interval. First, we exclude these singular points via interval splitting. This results in several smaller intervals, each forming a closed interval with boundaries controlled by a small ϵ away from and on both sides of the singular points. Next, for each of these closed intervals, we apply the proof given in Step 1. Finally, we take the limits as ϵ approaches zero to include these singular points. This leads to converged results for each of the smaller intervals because the function is integrable over the entire interval as a given condition.

The following is clear:

Integration neutralizes differentiation, but up to an additive constant. The statement can also be written as

$$\int \left(\frac{df(x)}{dx} \right) dx = f(x) + C \qquad (4.28)$$

4.5 Antiderivatives of often used elementary functions

In Chapter 3, we learned that all elementary functions are closed under differentiation, implying that the derivative of an elementary function can be found and is still an elementary function.

However, this is **not true** for integrals, as many elementary functions do not have closed forms of elementary functions. We have already seen several examples earlier. Nevertheless, antiderivatives of many elementary functions have been found. Following are some of the most frequently used antiderivatives of elementary functions:

$f(x)$	$F(x)$	
x^n	$\dfrac{x^{n+1}}{n+1} + C$	for $n \neq 1$
$\dfrac{1}{x}$	$\begin{cases} \ln x + C & \text{for } x \in (0, \infty) \\ \ln(-x) + C & \text{for } x \in (-\infty, 0) \end{cases}$	
$\ln x$	$x \ln x - x + C$	for $x \in (0, \infty)$
$\ln(-x)$	$x \ln(-x) - x + C$	for $x \in (-\infty, 0)$

$$\exp(ax) \quad \frac{1}{a}\exp(ax) + C \qquad \text{for } a \neq 0$$

$$a^x \quad \begin{cases} \dfrac{a^x}{\ln a} + C & \text{for } a > 0 \text{ and } a \neq 1 \\[2ex] \dfrac{(-a)^x}{\ln(-a)} + C & \text{for } a < 0 \text{ and } a \neq -1 \end{cases} \qquad (4.29)$$

$$\cos(x) \quad \sin(x) + C$$

$$\sin(x) \quad -\cos(x) + C$$

$$\sinh(x) \quad \cosh(x) + C$$

$$\cosh(x) \quad \sinh(x) + C$$

Many more antiderivatives of functions can be found using SymPy. Readers can simply use `sp.integrate()` to find them. Following is a typical example. It's important to note that $F(x)$ can only be defined in the domain of $f(x)$, so the singular points must be excluded. Otherwise, using Eq. (4.8) may lead to incorrect results:

```
1  x = symbols('x', real=True)                          # symbolic variables
2  a = symbols('a', negative=True)
3  k = symbols('k', integer=True)
4
5  C = symbols('C', real=True)                           # integral constant
6  f = sp.Abs(a)**x
7  F = integrate(f, x) + C                               # antiderivative
8  display(Math(f" ∫{sp.latex(f)}dx = {sp.latex(F)}"))
```

$$\int (-a)^x \, dx = C + \begin{cases} \dfrac{(-a)^x}{\log(-a)} & \text{for } \log(-a) \neq 0 \\[2ex] x & \text{otherwise} \end{cases}$$

The author recalls spending a significant amount of time in university practicing manual antiderivative derivation for various functions. While this honed his skills, real-world applications often involve referring to integral tables or using symbolic computation tools like SymPy for antiderivatives.

4.6 Integration by parts

Integration by parts is a technique frequently used in computational methods. The formula is given in the following:

$$\int f(x) \frac{dg(x)}{dx} \, dx = f(x)g(x) - \int g(x) \frac{df(x)}{dx} \, dx \qquad (4.30)$$

It is also often written simply as

$$\int f \, dg = fg - \int g \, df \tag{4.31}$$

Integration by parts is a powerful technique. It is essentially an inverse procedure of the product rule of differentiation introduced in Chapter 2. This connection allows for a straightforward proof of the integration by parts formula. Using the product rule of differentiation, we have

$$\frac{d}{dx}\left(f(x)g(x)\right) = f(x)\frac{dg(x)}{dx} + g(x)\frac{df(x)}{dx} \tag{4.32}$$

Integrate both sides gives Eq. (4.30).

As a simple example, let $f(x) = \ln x$ and $g(x) = x$, we have

$$\int \ln x \, dx = x \ln x - \int x \, d(\ln x) = x \ln x - \int x \frac{1}{x} dx$$

$$= x \ln x - x + C \tag{4.33}$$

This is a formula given in Eq. (4.29).

Integration by parts may be the most versatile tool in the integration toolbox. It leverages known antiderivatives of elementary functions to tackle a wider range of integrals, making it a go-to technique for many integration applications.

4.7 Integration with special functions

4.7.1 *Dirac delta and Heaviside functions*

When dealing with problems in science and engineering, several special functions are employed for convenience. Among these, the **Dirac delta** function and the **Heaviside** step function hold particular importance.

The Dirac delta function is defined as follows:

1. **Spiky** property:

$$\delta(x - a) = \begin{cases} +\infty, & x = a \\ 0, & x \neq a \end{cases} \tag{4.34}$$

2. **Unit** property:

$$\int_{-\infty}^{\infty} \delta(x - a) dx = 1 \tag{4.35}$$

A schematic drawing of the Dirac delta function is shown in Fig. 4.15(a).

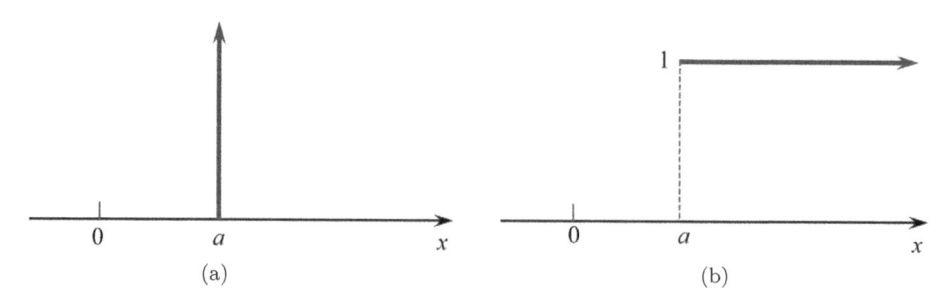

Figure 4.15.   A schematic drawing of (a) the Dirac delta function $\delta(x - a)$ and (b) the Heaviside step function $H(x - a)$.

The delta function is the derivative of the Heaviside function (see proof later):

$$\delta(x - a) = \frac{dH(x - a)}{dx} \tag{4.36}$$

in which the Heaviside function is defined as

$$H(x - a) = \begin{cases} 1, & x \geq a \\ 0, & x < a \end{cases} \tag{4.37}$$

A schematic drawing of the Heaviside step function is shown in Fig. 4.15(b). It is clear that the Heaviside step function is integrable over any finite interval. Note that there are a number of different definitions for the Heaviside step function. For example, in SymPy, $H(0)$ is set as $1/2$, which is the average of the values on the two sides of the jumping point.

As seen in Fig. 4.15(a), one cannot give a value from the delta function at $x = a$. It is a spike there. Since integration is a smooth operator, we can integrate the delta function (to smooth the spike), which gives

$$\int_{a^-}^{a^+} \delta(x - a)dx = \int_{a^-}^{a^+} \frac{dH(x - a)}{dx}dx = H(x - a)\big|_{a^-}^{a^+} = 1 \tag{4.38}$$

Here, the integration needs to be carried out in an infinitely small interval $[a^-, a^+]$ because the delta function is zero elsewhere. This result is an excellent "proof" that integration is a smoothing operator: It "smears" the spike to a constant! We note the following:

The Dirac delta function is integrable.

Using Eq. (4.38), we can derive the following important formula:

Consider a continuous function $f(x)$. We have

$$\int_{a^-}^{a^+} f(x)\delta(x-a)dx = f(a) \tag{4.39}$$

This means the following:

> The Dirac delta function "picks" up the function value at the spiky point of the delta function via an integration.

Equation (4.39) can be proven using the integration by parts. Letting $g(x) = H(x-a)$ and using Eq. (4.32) give

$$\int_{a^-}^{a^+} f(x)\underbrace{\delta(x-a)}_{\frac{dH(x-a)}{dx}}dx = f(x)H(x-a)\Big|_{a^-}^{a^+} - \int_{a^-}^{a^+} H(x-a)\frac{df(x)}{dx}dx \tag{4.40}$$

The right-hand side is

$$f(a)\cdot 1 - \int_a^{a^+} 1\cdot\frac{df(x)}{dx}dx = f(a) - \underbrace{f(x)\Big|_a^{a^+}}_{=0} = f(a) \tag{4.41}$$

Here, we used the fact that $H(x) = 0$ for $a^- \leq x < 0$, a^+ is infinitely close to a, and $f(x)$ is continuous at $x = a$. Equation (4.39) is thus proven.

This proof is an excellent application of the **integration by parts**. It shows the powerfulness of it in dealing with difficult integrations.

Equation (4.39) is useful in many applications, for example, in deriving nodal force vectors result from point forces in FEM formulations [6].

4.7.2 *Milder delta function*

4.7.2.1 *Definition*

We know that the Dirac delta is "spiky" and is infinite at $x = a$. Clearly, it is not differentiable at $x = a$ because it is not possible even to construct the definition of integration Eq. (2.1). We thus construct a "milder" delta function denoted as $\delta_\varepsilon(x-a)$:

$$\delta_\varepsilon(x-a) = \lim_{\varepsilon\to 0} \begin{cases} 0, & x < a - \varepsilon \\ \dfrac{1}{2\varepsilon}, & a - \varepsilon \leq x \leq a + \varepsilon \\ 0, & x > a + \varepsilon \end{cases}$$

$$= \lim_{\substack{\text{or } \varepsilon\to 0}} \begin{cases} \dfrac{1}{2\varepsilon}, & a - \varepsilon \leq x \leq a + \varepsilon \\ 0 & \text{elsewhere} \end{cases} \tag{4.42}$$

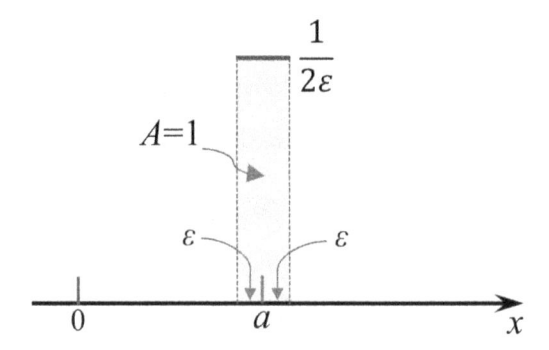

Figure 4.16. A milder delta function by construction. The area under the function $A(x)$ is always unit 1.

This milder delta was given in Example 3.5 in Ref. [7]. Before the limit is taken, it can be drawn in Fig. 4.16. The area under $\delta_\varepsilon(x - a)$, $A(x)$, is always unit 1.

4.7.2.2 *Properties*

Let us check whether the milder delta possesses the properties of the Dirac delta:

1. **Unit** property

$$\int_{-\infty}^{\infty} \delta_\varepsilon(x - a)dx = \int_{-\infty}^{\infty} \lim_{\varepsilon \to 0} \begin{cases} 0, & x < a - \varepsilon \\ \dfrac{1}{2\varepsilon}, & -\varepsilon \le x \le \varepsilon \\ 0, & x > a + \varepsilon \end{cases} dx$$

$$= \int_{-\varepsilon}^{\varepsilon} \lim_{\varepsilon \to 0} \frac{1}{2\varepsilon} dx = \lim_{\varepsilon \to 0} \int_{-\varepsilon}^{\varepsilon} \frac{1}{2\varepsilon} dx = \lim_{\varepsilon \to 0} \frac{1}{2\varepsilon} 2\varepsilon = \lim_{\varepsilon \to 0} 1 = 1$$

$$(4.43)$$

Thus, the unit property holds. Here, we swapped the order of the integral and limit operations. We hold ε constant while working out the integration and then take the limit (similar to the definition of the differentiation).

2. **Spiky** property

The spiky property is obvious:

$$\delta_\varepsilon(x - a) = \lim_{\varepsilon \to 0} \begin{cases} \dfrac{1}{2\varepsilon}, & a - \varepsilon \le x \le a + \varepsilon \\ 0 & \text{elsewhere} \end{cases} = \begin{cases} +\infty, & x = a \\ 0, & x \ne a \end{cases} \qquad (4.44)$$

3. Function value **picking-up**

Consider a continuous function $f(x)$. Using the milder delta, we have

$$\int_{a^-}^{a^+} f(x)\delta_\varepsilon(x-a)dx = \int_{a-\varepsilon}^{a+\varepsilon} f(x) \lim_{\varepsilon \to 0} \frac{1}{2\varepsilon}dx = \lim_{\varepsilon \to 0} \int_{a-\varepsilon}^{a+\varepsilon} f(x)\frac{1}{2\varepsilon}dx$$

$$= \lim_{\substack{(a-\varepsilon)\leq x^*\leq(a+\varepsilon) \\ \varepsilon \to 0}} f(x^*)(2\varepsilon)\frac{1}{2\varepsilon}$$

$$= \lim_{\substack{(a-\varepsilon)\leq x^*\leq(a+\varepsilon) \\ \varepsilon \to 0}} f(x^*) = f(a) \qquad (4.45)$$

where $(a-\varepsilon) \leq x^* \leq (a+\varepsilon)$. Here, we swapped the order of integration and limit, and used the Riemann integral definition for a continuous function. This proof is not only straightforward but also shows clearly how the milder delta picks up the function value. This means the following:

> The milder delta function picks up the function value at center point of the milder delta function via an integration.

4. Antiderivative of the milder delta

Let us now find the antiderivative of the milder delta. Using Eq. (4.26),

$$F(x) = \int_{-\infty}^x \delta_\varepsilon(t-a)dt = \underbrace{\int_{-\infty}^{a-\varepsilon} \delta_\varepsilon(t-a)dt}_{x<(a-\varepsilon),\ A(x)\ \text{always } 0} + \underbrace{\int_{a-\varepsilon}^x \delta_\varepsilon(t-a)dt}_{x\geq(a-\varepsilon),\ A(x)\ \text{always } 1}$$

$$\underset{\text{when } \varepsilon \to 0}{=} \begin{cases} 1, & x \geq a \\ 0, & x < a \end{cases} = H(x-a) \qquad (4.46)$$

This proves that the antiderivative of the milder delta is the Heaviside function! The is the fact we used in the proof of Eq. (4.38). We note the following:

> The derivative of the Heaviside function is the milder delta function, or the antiderivative of the milder delta function is the Heaviside function.

4.7.3 *A proof of square integrability of the delta function*

We proved that the Dirac delta function is integrable, and the integral is 1. However,

> Dirac delta function is not square integrable.

To prove this, we first prove that the milder delta function is not squarely integrable.

Proof. Let us examine the following limit of the integral of the squared milder delta:

$$\lim_{\varepsilon \to 0} \int_{a-\varepsilon}^{a+\varepsilon} [\delta_\varepsilon(x-a)]^2 dx = \lim_{\varepsilon \to 0} \int_{a-\varepsilon}^{a+\varepsilon} \left[\frac{1}{2\varepsilon}\right]^2 dx$$

$$= \lim_{\varepsilon \to 0} \frac{2\varepsilon}{4\varepsilon^2} = \lim_{\varepsilon \to 0} \frac{1}{2\varepsilon} \to \infty \qquad (4.47)$$

Therefore, the milder function is not squarely integrable.

Since the Dirac delta is more "spiky" than the milder function, we conclude that the Dirac delta function is not squarely integrable. This completes the proof.

This proof is crucial because it sheds light on the relationship between numerical model stability and the mathematical properties of functions used in solid and structure modeling. In these models, the strain energy is often computed using the squared strain function. However, when the assumed displacement field is discontinuous, the strain function's behavior mimics the Dirac delta function. This similarity to the delta function can cause the numerical solution for strain energy to not converge or converge to an inaccurate value [6–8]. Special operations, such as gradient smoothing, are needed to overcome such stability issues [7, 8].

4.7.4 *Property of the Heaviside function*

Based on the definition given in Eq. (4.37), the Heaviside function has the following properties:

1. The base antiderivative of $H(x)$ is

$$H_F = \int H(x)dx = \begin{cases} 0 & \text{for } x < 0 \\ x & \text{otherwise} \end{cases} \qquad (4.48)$$

A schematic drawing of the antiderivatives of $H(x)$ is shown in Fig. 4.17.

2. It is integrable over any finite interval, e.g.,

$$\int_{-b}^{0} H(x)dx = 0; \quad \int_{0}^{b} H(x)dx = b; \quad \int_{-b}^{b} H(x)dx = b \qquad (4.49)$$

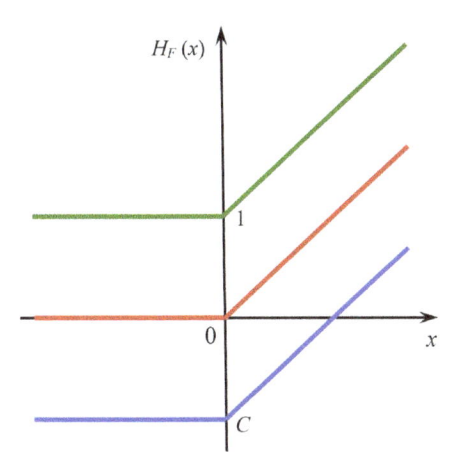

Figure 4.17. Schematic drawing of antiderivatives of the Heaviside function: red line: the base antiderivative; blue line: an arbitrary antiderivative with constant C; green line: an antiderivative with $C = 1$.

3. The base antiderivative of squared $H(x)$ is the same as the base antiderivative of $H(x)$:

$$H_F^2 = \int H^2(x)dx = \begin{cases} 0 & \text{for } x < 0 \\ x & \text{otherwise} \end{cases} \tag{4.50}$$

4. It is squarely integrable over any finite interval, e.g.,

$$\int_{-b}^{0} H^2(x)dx = 0; \quad \int_{0}^{b} H^2(x)dx = b; \quad \int_{-b}^{b} H^2(x)dx = b \tag{4.51}$$

Squaring has no effect on $H(x)$, which is akin to having no effect on 1. All these can be confirmed using the following SymPy code:

```
1  x = symbols('x', real=True)                        # variables
2  b = symbols('b', positive=True)
3  H = sp.Piecewise((0, x < 0), (1, x >= 0))           # define the Heaviside
4  display(Math(f" \\text{{The Heaviside function $H(x)$}}= {sp.latex(H)};\
5                              H(0)={H.subs(x,0)}"))
6  H_F = integrate(H,x)
7  display(Math(f" H_F = ∫ H(x) dx = {latex(integrate(H,x))}; \\;\\;\\;\
8                  ∫_{{-b}}^0 H(x)dx={latex(integrate(H,(x,-b,0)))}; \
9          \\;\\;\\; ∫_{{0}}^b H(x)dx={latex(integrate(H,(x,0,b)))}"))
10
11 display(Math(f" H^2_F=∫\\big(H(x)\\big)^2dx={latex(integrate(H**2,x))};\
12 \\;\\;\\; ∫_{{-b}}^0\\big(H(x)\\big)^2dx=\
13                  {latex(integrate(H**2,(x,-b,0)))};\
14 \\;\\;\\; ∫_{{0}}^b\\big(H(x)\\big)^2dx=\
15                  {latex(integrate(H**2,(x,0,b)))}"))
```

The Heaviside function $H(x) = \begin{cases} 0 & \text{for } x < 0 \\ 1 & \text{otherwise} \end{cases}$; $H(0) = 1$

$$H_F = \int H(x)dx = \begin{cases} 0 & \text{for } x < 0 \\ x & \text{otherwise} \end{cases} ; \quad \int_{-b}^{0} H(x)dx = 0; \quad \int_{0}^{b} H(x)dx = b$$

$$H_F^2 = \int (H(x))^2 dx = \begin{cases} 0 & \text{for } x < 0 \\ x & \text{otherwise} \end{cases} ; \quad \int_{-b}^{0} (H(x))^2 dx = 0;$$

$$\int_{0}^{b} (H(x))^2 dx = b$$

4.7.5 A case study: Integral of squared derivative of functions

4.7.5.1 Setting of the problem

In most real-life problems in science and engineering, functions that describe the behavior of a problem can be very complicated. The most commonly used approach is to approximate the function using a piecewise continuous function, allowing each piece to have a very simple form, such as linear or quadratic functions. In these cases, the continuity of the function becomes critical in obtaining stable and reliable solutions. This case study breaks down the problem to a very simple one: considering a function that consists of two linear functions, the green line and the blue line shown in Fig. 4.18.

We assume these two linear functions have a gap controlled by finite real constants, C_1 and C_2. The two lines are perfectly connected if $C_1 = C_2$. This piecewise function, denoted as $g_C(x)$, is expressed as follows:

$$g_C(x) = \begin{cases} k_1 x + C_1, & x < 0 \\ k_2 x + C_2, & x \geq 0 \end{cases} \tag{4.52}$$

It is well defined and bounded in $(-\infty, \infty)$. Hence, it is clearly integrable.

Consider a finite interval $[-b, b]$. Our task is to determine the conditions for

1. the derivative of this piecewise function being integrable and find the integral value over the given interval if it is integrable;
2. the derivative of this piecewise function being square-integrable, and find the integral value over the given interval if it is integrable;
3. consider $k_1 = 0$, $k_2 = 0$, $C_1 = 0$, $C_2 = 1$, repeat items 1 and 2, and discuss the results obtained.

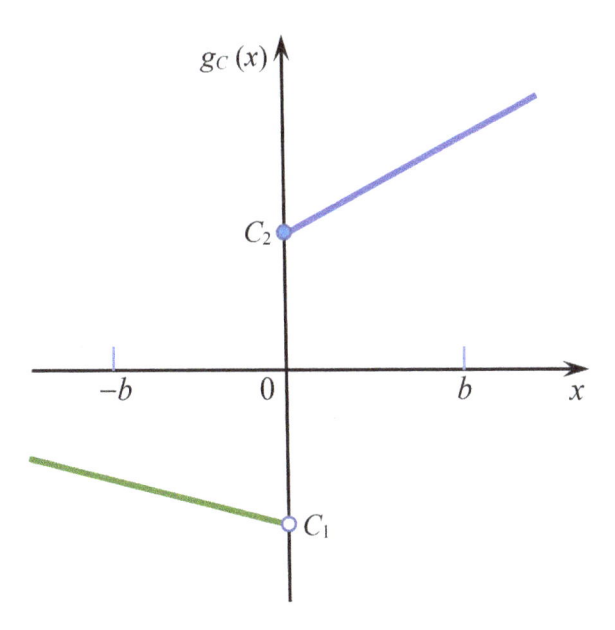

Figure 4.18. Schematic drawing of a piecewise function with two linear segments, one in green and the other in blue. There is a gap between these two lines at $x = 0$ if $C_1 \neq C_2$.

Solution to question 1: The derivative of this function becomes

$$g_C'(x) = \begin{cases} k_1, & x < 0 \\ \text{indeterminate}, & x = 0 \\ k_2, & x > 0 \end{cases} \tag{4.53}$$

It is very similar to the Heaviside function, except at $x = 0$, where it cannot be determined due to the jump of $g_C(x)$ at that point. The derivative at $x = 0$ experiences an infinite change: from a finite value to infinity (the vertical line) or vice versa. Therefore, we have an indeterminate scenario: $\Delta(g_C'(0))\Delta x = \infty \times 0$ at $x = 0$. The Riemann sum will not converge when integration is carried out over any finite interval that includes $x = 0$. We are stuck.

4.7.5.2 *A smoother piecewise continuous function*

To work around this problem, we construct a smoother piecewise continuous function by introducing a straight line (the shaded red line) to bridge the green and blue lines, passing through the origin. This forms a continuous three-piece function without any gaps, as shown in Fig. 4.19.

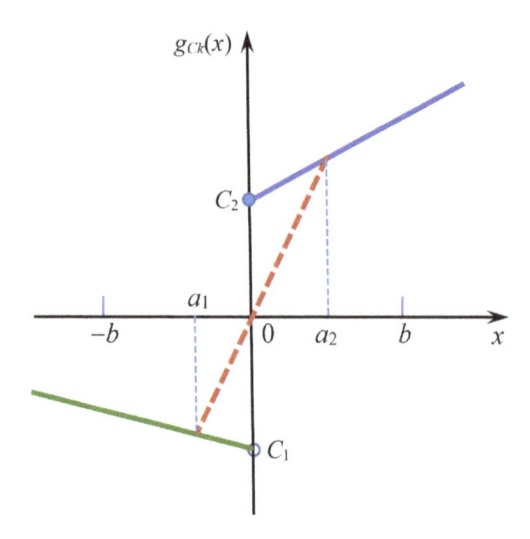

Figure 4.19. Construction of a piecewise continuous function with three lines. The shaded red line is used to bridge the green and blue lines, forming a continuous three-piece function without any gaps. As the slope of the dashed red line approaches infinity, the red line approaches vertical, and the three-piece function approaches the original two-piece function.

The expression for the smoother piecewise continuous function is given as follows:

$$g_{Ck}(x) = \begin{cases} k_1 x + C_1, & x < a_1 \\ kx, & a_1 \le x \le a_2 \\ k_2 x + C_2, & x > a_1 \end{cases} \qquad (4.54)$$

The smoother piecewise function has a new parameter k, which is the slope or gradient of the dashed red line. The intersection locations, a_1 and a_2, on this piecewise continuous function are determined through the following process:

First, let the red line intersect with the green line at $x = a_1$, which gives the following condition:

$$k a_1 = k_1 a_1 + C_1 \qquad (4.55)$$

Second, let the red line intersect with the blue line at $x = a_2$, which gives the following condition:

$$k a_2 = k_2 a_2 + C_2 \qquad (4.56)$$

The above two equations, (4.55) and (4.56), can be solved simultaneously for the two unknowns, a_1 and a_2. This is done using the following code:

```
1  # Python code to determine intersection points for the dashed red-line:
2
3  x, a1, a2 = sp.symbols('x, a1, a2', real=True)
4  C1, C2, k1, k2 = sp.symbols('C1, C2, k1, k2', real=True)
5  k = sp.symbols('k', positive=True)
6
7  f_green = k1*x + C1                      # function for the green line
8  f_blue  = k2*x + C2                      # function for the blue line
9  f_red   = k*x                # Assume a linear function for the red-line
10
11 expr1 = f_green.subs(x,a1) - k*a1              # Continuity at x=a1
12 expr2 = f_blue.subs(x, a2) - k*a2              # Continuity at x=a2
13
14 sln = sp.solve((expr1, expr2), (a1, a2))       # solve for a1, a2
15 display(Math(f" \\text{{solutions for $a_1$ and $a_2$:}}{latex(sln)}"))
```

Solutions for a_1 and a_2: $\left\{ a_1 : \dfrac{C_1}{k - k_1}, \quad a_2 : \dfrac{C_2}{k - k_2} \right\}$

4.7.5.3 *k-limit approach*

It is clear from Eq. (4.54) that when $k \to 0$, x on the dashed red line approaches vertical, and $g_{Ck}(x)$ approaches the original function $g_C(x)$. Therefore, our strategy is to examine $g_{Ck}(x)$, find the solutions, and then take the limit as $k \to \infty$. Since the parameter k controls the gradient of the shaded red line, we term this the gradient limit approach or the **k-limit approach**. This approach was proposed in Chapter 3 of Ref. [7].

The derivative of the smoother piecewise continuous function is found as follows:

$$g'_{Ck}(x) = \begin{cases} k_1, & x < a_1 \\ k, & a_1 < x < a_2 \\ k_2, & x > a_2 \end{cases} \tag{4.57}$$

Now, the derivative function still has three pieces and is discontinuous at $x = a_1$ and $x = a_2$. However, this case is very different from the case in Eq. (4.53). Our smoother function $g_{Ck}(x)$ is continuous at both $x = a_1$ and $x = a_2$. The derivative at either point has only a finite change: from one finite value to another (for a fixed k). Therefore, we have $\Delta\left(g'_{Ck}(0)\right)\Delta x = k \times 0$ at $x = a_1$ and $x = a_2$, where k is a finite number (before the limit is taken), however large it may be. Therefore, the Riemann sum will converge, and we can proceed with the integration piece by piece over $[-b, b]$. The integral for the first piece over $[-b, a_1]$ is simply $k_1(a_1 + b)$ and that for the third piece over $[a_2, b]$ is $k_2(b - a_2)$. The second piece over $[a_1, a_2]$ requires a little work and can be found using the following code:

```
1  df_red = f_red.diff(x)                    # Derivative of the dashed red-line
2  Isga1a2 = integrate(df_red , (x, sln[a1], sln[a2]))
3  display(Math(f" ∫_{{a_1}}^{{a_2}}{latex(df_red)}dx = {latex(Isga1a2)}"))
4
5  # Take the limit ε → 0:
6  display(Math(f"\\text{{limit}}_{{k→∞}} I_{{sg}} =\
7                {sp.latex(sp.limit(Isga1a2, k, oo))}"))
```

$$\int_{a_1}^{a_2} k\, dx = -\frac{C_1 k}{k - k_1} + \frac{C_2 k}{k - k_2}$$

$$\text{limit}_{k \to \infty} I_{sg} = -C_1 + C_2$$

Taking the limit of $k \to \infty$, which leads to $a_1 \to 0$ and $a_2 \to 0$, we obtain

$$\lim_{k \to \infty} \int_{-b}^{b} g'_{Ck}(x) = \lim_{k \to \infty} \left[\int_{-b}^{a_1} g'_{Ck}(x) + \int_{a_1}^{a_2} g'_{Ck}(x) + \int_{a_2}^{b} g'_{Ck}(x) \right]$$

$$= \lim_{k \to \infty} \left[k_1(a_1 + b) + \frac{C_2 k}{k - k_2} - \frac{C_1 k}{k - k_1} + k_2(b - a_2) \right]$$

$$= \underbrace{C_2 - C_1}_{\text{from jump}} + \underbrace{k_1 b}_{\text{green line}} + \underbrace{k_2 b}_{\text{blue line}} \tag{4.58}$$

As seen, the derivative of the piecewise function is integrable because both C_1 and C_2 are finite. The result of the integral over the interval $[-b, b]$ is $C_2 - C_1 + k_1 b + k_2 b$. We note the following:

1. The gap in the piecewise function affects the value of the integral of the derivative function. $C_2 - C_1$ accounts for the effect of the jump. This result is useful and frequently applied in handling functions with finite jumps in various computational methods.
2. The jump effect is independent of k_1 and k_2. In fact, the green and blue pieces can be any continuous functions; the parameters in these functions will not affect the jump effect. It is related only to the gap $C_2 - C_1$.

3. The contributions to the integral of the derivative of the function from the green and blue lines can be computed as per normal as if there is no jump.

This is the answer to question 1.

Compared to the integral result for the Heaviside function given in Eq. (4.48), the major difference is $C_2 - C_1$. This is caused by the indeterminacy of $g'_C(x)$ at $x = 0$. If $C_2 = C_1$, implying that the piecewise function $g_C(x)$ is continuous at $x = 0$, $g'_C(x)$ becomes the Heaviside function defined in Eq. (4.37) if setting $k_1 = 0$, $k_2 = 1$, $C_1 = C_2$. This implies that the base **antiderivative of the Heaviside function** is $g_C(x)$ with $C_2 = C_1$ (and $k_1 = 0$, $k_2 = 1$).

Solution to question 2: The squared derivative of this smoother piecewise function becomes

$$\left(g'_{Ck}(x)\right)^2 = \begin{cases} k_1^2, & x < a_1 \\ k^2, & a_1 < x < a_2 \\ k_2^2, & x > a_2 \end{cases} \tag{4.59}$$

It has three pieces with discontinuities at $x = a_1$ and $x = a_2$. In this case, we have $\Delta\left(g'_{Ck}(0)\right)^2 \Delta x = k^2 \times 0$ at $x = a_1$ or $x = a_2$, where k is a finite number, however large it may be. Therefore, the Riemann sum will still converge, and we can proceed with the integration piece by piece over $[-b, b]$. There are no issues with integrating the first and third pieces, and they can be computed as per normal. For the second integral over $[a_1, a_2]$, we first find the solution using the following code for a fixed k:

```
1  #Img2a1a2 = k**2*(sln[a2]-sln[a1])                    # Alternative
2  #display(Math(f" I_{{\\text{{2nd piece}}}}={sp.latex(Img2a1a2)}"))
3
4  Img2a1a2 = integrate(df_red**2, (x, sln[a1], sln[a2]))
5  display(Math(f" ∫_{{a_1}}^{{a_2}}{latex(df_red**2)}dx = \
6                  {latex(Img2a1a2)}"))
```

$$\int_{a_1}^{a_2} k^2 dx = -\frac{C_1 k^2}{k - k_1} + \frac{C_2 k^2}{k - k_2}$$

Next, taking the limit of $k \to \infty$, we obtain the following:

```
1  display(Math(f"\\text{{limit}}_{{k→∞}} I_{{mg2}}=\
2                  {sp.latex(sp.limit(Img2a1a2, k, oo))}"))
```

$$\text{limit}_{k\to\infty} I_{mg2} = \infty \, \text{sign}\left(-C_1 + C_2\right)$$

The entire integral becomes

$$\lim_{k\to\infty} \int_{-b}^{b} (g'_{Ck}(x))^2 dx = \lim_{k\to\infty} [\int_{-b}^{a_1} k_1^2 dx + \int_{a_1}^{a_2} k^2 dx + \int_{a_2}^{b} k_2^2 dx]$$

$$= \lim_{k\to\infty} [k_1^2 dx(a_1 + b) + k^2 dx(a_2 - a_1) + k_2^2 dx(b - a_2)]$$

$$= \underbrace{k_1^2 b}_{\text{green line}} + \underbrace{\infty\,\text{sign}\,(C_2 - C_1)}_{\text{from jump}} + \underbrace{k_2^2 b}_{\text{blue line}} \qquad (4.60)$$

Note that when taking the limit as $k \to \infty$, it leads to $a_1 \to 0$ and $a_2 \to 0$. As seen from Eq. (4.60), the square of the derivative of the piecewise function is integrable if and only if $C_1 = C_2$. In this case, the result over the interval $[-b, b]$ is $(k_1^2 + k_2^2)b$. This is the answer to question 2.

Solution to question 3: The conditions given are $k_1 = 0$, $k_2 = 0$, $C_1 = 0$, $C_2 = 1$.

First, on the integrability of the derivative of the function, we found the following. When $k_1 = 0$, $k_2 = 0$, $C_1 = 0$, $C_2 = 1$, the original function $g_C(x)$ is the Heaviside function defined in Eq. (4.37) (with $a = 0$). Using Eq. (4.57), the derivative $g'_C(x)$ becomes

$$g'_{Ck}(x) = \begin{cases} 0, & x < a_1 \\ k, & a_1 < x < a_2 \\ 0, & x > a_2 \end{cases} \qquad (4.61)$$

Its integral over $[a_1, a_2]$ is found using the following code, which gives 1:

```
1  df_red = f_red.diff(x)                # Derivative of the dashed red-line
2  Isga1a2 = integrate(df_red , (x, sln[a1], sln[a2]))
3  display(Math(f" ∫_{{a_1}}^{{a_2}}{{latex(df_red)}}dx =\
4          {latex(Isga1a2.subs({k1:0,k2:0,C1:0,C2:1}))}"))
```

$$\int_{a_1}^{a_2} k dx = 1$$

When $k \to \infty$, $a_1 \to 0$, $a_2 \to 0$, and $g'_{Ck}(x) \to \infty$ (spiky). Therefore, Eq. (4.61) is the same as the "milder" delta function $\delta_\varepsilon(x - a)$ defined in Eq. (4.42)!

In conclusion, $g'_C(x)$ is integrable, and the result can be obtained using Eq. (4.58), which simply gives 1. This finding supports our earlier proof that the integral of the delta function is 1.

Second, on the square integrability of the derivative function, Eq. (4.60) shows that $(g'_C(x))^2$ cannot be bounded if $C_1 \neq C_2$, and hence it is not square-integrable. This finding also supports our earlier proof that the delta function is not square-integrable. This completes the entire case study.

This case study is lengthy and may be difficult to comprehend. If it can be comprehended, we should have quite a good understanding of both derivatives and integration of functions at least of piecewise functions.

4.8 Major techniques to find antiderivatives

Finding antiderivatives, which are functions whose derivative is the original function, is a cornerstone of calculus. As mentioned earlier, it's essentially an inverse operation of differentiation and can be challenging. Over time, mathematicians have developed techniques to find antiderivatives for various types of functions. This section introduces the major techniques, providing a deeper understanding of the concepts and also equipping you with some tools in Python code. We'll concentrate on finding the base antiderivative (without the constant of integration) in this discussion.

4.8.1 *Use of the known antiderivatives*

Using differentiation techniques, we can obtain the derivatives of a given function. Using these formulas in an inverse manner, we can found antiderivatives of many elementary functions. Here are some common examples:

```
1  # antiderivatives of power functions:
2  x, n = symbols('x, n', real=True)                        # variables
3  f = x**n
4  display(Math(f" f(x) ={sp.latex(f)}"))
5  dfdx = f.diff(x).simplify()
6  display(Math(f" f'(x) ={sp.latex(dfdx)}"))                # n ≠-1
7  # for n=-1, x>0:
8  dfdx = (sp.log(x)).diff(x)
9  display(Math(f" \\frac{{d}}{{dx}}\\ln(x)={sp.latex(dfdx)}; n=-1,x>0"))
```

$$f(x) = x^n$$
$$f'(x) = nx^{n-1}$$
$$\frac{d}{dx}\ln(x) = \frac{1}{x}; \quad n = -1, \quad x > 0$$

Inverse the operations, we shall have

$$\int x^n \, dx = \frac{x^{n+1}}{n+1} \qquad \text{for } n \neq -1$$

$$\int \frac{1}{x} \, dx = \ln(x) \qquad \text{for } n = -1, \quad x > 0 \tag{4.62}$$

This is the first two formulas given in Eq. (4.29). A large number of antiderivatives are found in this manner.

4.8.2 *Variable substitution*

If the integrand has the form of $f(g(x))g'(x)$, let $u = g(x)$, then $du = g'(x)dx$. We thus have

$$\int f(g(x))g'(x) \, dx = \int f(u) \, du \tag{4.63}$$

This technique is known as the ***u*-substitution**. Antiderivatives of many composite functions can be found in this way. The following is an example:

```
1  x, θ, u = symbols('x, θ, u', real=True)
2
3  f = sin(θ)**3 * cos (θ)                        # The integrand
4
5  display(Math(f" f(x) ={sp.latex(f)}"))
6  u_ = sin(θ)                                    # u-substitution
7  dudθ = u_.diff(θ)
8
9  fu = u**3
10 Fu = integrate(fu,u)                           # antiderivative in u
11 display(Math(f" F(u) ={sp.latex(Fu)}"))
12 Fθ = Fu.subs(u,u_)                             # substitute back
13 display(Math(f" F(θ) ={sp.latex(Fθ)}"))       # final antiderivative in θ
14 #sp.integrate(f, θ).simplify()                 # direct integratoin using SymPy
```

$$f(x) = \sin^3 (\theta) \cos (\theta)$$

$$F(u) = \frac{u^4}{4}$$

$$F(\theta) = \frac{\sin^4 (\theta)}{4}$$

4.8.3 *Partial fraction decomposition*

If a rational integrand has a form like $\frac{1}{(x+c_1)(x+c_2)}$, we can decompose it into simpler fractions that can be integrated separately with ease. For example,

$$\int \frac{1}{(x+c_1)(x+c_2)}\,dx = \int \frac{1}{(c_1-c_2)(c_2+x)} - \frac{1}{(c_1-c_2)(c_1+x)}\,dx$$

(4.64)

The splitting can be done by hand, but SymPy is an excellent help. The code get this done is as follows:

```
1  c1, c2 = symbols('c1, c2', real=True)
2
3  f = 1/((x+c1)*(x+c2))                      # The original integrand
4
5  f_apart = f.apart(x)
6  display(f_apart)                           # split it to 2 parts
7  f1, f2 = f_apart.args                      # get the 1st and 2nd part
8  F1 = integrate(f1, x)                      # Integrate the 1st part
9  F2 = integrate(f2, x)                      # Integrate the 2nd part
10 display(Math(f" ∫{latex(f1)}dx = {latex(F1)}"))
11 display(Math(f" ∫{latex(f2)}dx = {latex(F2)}"))
12 #display(Math(f" ∫{latex(f)}dx = {latex(integrate(f, x))}")) # no split
```

$$\frac{1}{(c_1-c_2)(c_2+x)} - \frac{1}{(c_1-c_2)(c_1+x)}$$

$$\int \frac{1}{(c_1-c_2)(c_2+x)}dx = \frac{\log\left(c_1 c_2 - c_2^2 + x\left(c_1 - c_2\right)\right)}{c_1-c_2}$$

$$\int -\frac{1}{(c_1-c_2)(c_1+x)}dx = -\frac{\log\left(c_1^2 - c_1 c_2 + x\left(c_1 - c_2\right)\right)}{c_1-c_2}$$

For this example, SymPy can produce the same results without splitting. However, if the function is too complicated, SymPy might take a long time to integrate, produce lengthy output, or may not integrate it at all. Therefore, it is a good idea to split the function first and then use SymPy. Additionally, since we often do not know how many terms will be produced, a for-loop comes in handy. Following is an example:

```
1  f = (x**2+c1)/((x+c2)**2*(x**2+2*x+1))    # The original integrand
2
3  f_apart = f.apart(x)                       # split it to n parts
4  for fi in f_apart.args:                    # use a for-loop to integrate one by one
5      Fi = integrate(fi, x)                  # Integrate the 1st part
6      display(Math(f" ∫{latex(fi)}dx = {latex(Fi)}"))
```

$$\int \frac{c_1 + 1}{(c_2 - 1)^2 (x + 1)^2} dx = -\frac{c_1 + 1}{c_2^2 - 2c_2 + x\left(c_2^2 - 2c_2 + 1\right) + 1}$$

$$\int \frac{c_1 + c_2^2}{(c_2 - 1)^2 (c_2 + x)^2} dx = -\frac{c_1 + c_2^2}{c_2^3 - 2c_2^2 + c_2 + x\left(c_2^2 - 2c_2 + 1\right)}$$

$$\int -\frac{2\left(c_1 + c_2\right)}{(c_2 - 1)^3 (x + 1)} dx$$

$$= \frac{(-2c_1 - 2c_2)\log\left(c_2^3 - 3c_2^2 + 3c_2 + x\left(c_2^3 - 3c_2^2 + 3c_2 - 1\right) - 1\right)}{c_2^3 - 3c_2^2 + 3c_2 - 1}$$

$$\int \frac{2\left(c_1 + c_2\right)}{(c_2 - 1)^3 (c_2 + x)} dx$$

$$= \frac{(2c_1 + 2c_2)\log\left(c_2^4 - 3c_2^3 + 3c_2^2 - c_2 + x\left(c_2^3 - 3c_2^2 + 3c_2 - 1\right)\right)}{c_2^3 - 3c_2^2 + 3c_2 - 1}$$

```
1  # One may try without splitting, it can produce a very long formula:
2  #display(Math(f" ∫{latex(f)}dx = {latex(integrate(f, x))}"))
```

4.8.4 *Use of trigonometric identities*

If the integrand is products of trigonometric functions, one may use trigonometric identities to simplify the integrand. For example,

$$\int \cos^2(\theta)\, d\theta = \int \frac{1 + \cos(2\theta)}{2}\, d\theta = \frac{\theta}{2} + \frac{\sin(2\theta)}{4} \qquad (4.65)$$

As seen, the double angle function is much easier to integrate. The code to get this done is as follows:

```
1  from sympy.simplify.fu import TRpower
2
3  f0 = cos(θ)**2
4
5  display(Math(f" f(θ) ={sp.latex(f0)}"))
6  f = TRpower(cos(θ)**2)                        # convert to double angle
7  display(Math(f" {sp.latex(f0)} ={sp.latex(f)}"))
8  F = integrate(f, θ)                           # get the antiderivative
9  display(Math(f" F(θ) ={sp.latex(F)}"))
```

$$f(\theta) = \cos^2(\theta)$$

$$\cos^2(\theta) = \frac{\cos(2\theta)}{2} + \frac{1}{2}$$

$$F(\theta) = \frac{\theta}{2} + \frac{\sin(2\theta)}{4}$$

4.8.5 Use of trigonometric substitution

If the integrand has terms like $\sqrt{c^2 - x^2}$, one may use trigonometric functions to simplify it via substitution. Let $x = c\sin\theta$, and then $dx = c\cos\theta d\theta$. For example, in the following integral, we have

$$\int \sqrt{c^2 - x^2}dx = \int \sqrt{c^2 - (c\sin\theta)^2}dx = |c|\int \cos\theta(c\cos\theta)d\theta$$

$$= |c|c \int \left(\frac{1}{2} + \frac{\cos(2\theta)}{4}\right) d\theta = |c|c \left(\frac{\theta}{2} + \frac{\sin(\theta)\cos(\theta)}{2}\right)$$

$$(4.66)$$

After the integration is done, we can then substitute back:

$$\int \sqrt{c^2 - x^2}dx = |c|c \left(\frac{x\sqrt{1 - x^2}}{2} + \frac{\operatorname{asin}(x)}{2}\right) \qquad (4.67)$$

The code to get this done is as follows:

```
1  x, c, θ = symbols('x, c, θ', real=True)
2  f = sqrt(c**2 - x**2)
3  display(Math(f" f(x) = {sp.latex(f)}"))
4
5  x_ = c*sin(θ)                              # trigonometric substitution
6  dx_ = x_.diff(θ)
7  display(Math(f"dx = {sp.latex(dx_)}dθ"))
8
9  fθ = sp.Abs(c)*cos(θ)*dx_                  # function after substitution
10 display(Math(f" f(θ) = {sp.latex(fθ)}"))
11
12 Fθ = sp.integrate(fθ, θ)        # get the antiderivative in terms of θ
13 display(Math(f" F(θ) = {sp.latex(Fθ)}"))
14
15 F = Fθ.subs(θ,sp.asin(x))           # substitute back in terms of x
16 display(Math(f" F(x) = {sp.latex(F)}"))       # final antiderivative
17 #sp.integrate(f, x).simplify()          # direct integratoin using SymPy
```

$$f(x) = \sqrt{c^2 - x^2}$$

$$dx = c\cos(\theta)d\theta$$

$$f(\theta) = c\cos^2(\theta)|c|$$

$$F(\theta) = \begin{cases} c^2\left(\dfrac{\theta}{2} + \dfrac{\sin(\theta)\cos(\theta)}{2}\right) & \text{for } c \geq 0 \\[2ex] -c^2\left(\dfrac{\theta}{2} + \dfrac{\sin(\theta)\cos(\theta)}{2}\right) & \text{otherwise} \end{cases}$$

$$F(x) = \begin{cases} c^2\left(\dfrac{x\sqrt{1-x^2}}{2} + \dfrac{\operatorname{asin}(x)}{2}\right) & \text{for } c \geq 0 \\[2ex] -c^2\left(\dfrac{x\sqrt{1-x^2}}{2} + \dfrac{\operatorname{asin}(x)}{2}\right) & \text{otherwise} \end{cases}$$

As another example, if the integrand has terms like $c^2 + x^2$, one may use substitution $x = c\tan\theta$, and then $dx = c(1 + \tan^2\theta)d\theta$. For example,

$$\int \frac{1}{c^2 + x^2}dx = \int \frac{c(\tan^2\theta + 1)}{c^2(\tan^2\theta + 1)}d\theta = \int \frac{1}{c}d\theta = \frac{\theta}{c} \tag{4.68}$$

As seen, the integral become extremely simple after the substitution. Once the integration is done, we can then substitute back to obtain the base antiderivative:

$$F(x) = \int \frac{1}{c^2 + x^2}dx = \frac{\operatorname{atan}\left(\frac{x}{c}\right)}{c} \tag{4.69}$$

The code to get this done is as follows:

```
1  x, c, θ = symbols('x, c, θ', real=True)
2  f = 1/(c**2+x**2)
3  display(Math(f" f(x) = {sp.latex(f)}"))
4
5  x_ = c*sp.tan(θ)                        # trigonometric substitution
6  dx_ = x_.diff(θ)
7  display(Math(f"dx = {sp.latex(dx_)}dθ"))
8
9  fθ = f.subs(x,x_)*dx_                   # function after substitution
10 display(Math(f" f(θ) = {sp.latex(fθ)}"))
11
12 Fθ = sp.integrate(fθ, θ)         # get the antiderivative in terms of θ
13 display(Math(f" F(θ) = {sp.latex(Fθ)}"))
14
15 F = Fθ.subs(θ,sp.atan(x/c)).simplify()  # substitute back in terms of x
16 display(Math(f" F(x) = {sp.latex(F)}"))          # final antiderivative
17 #sp.integrate(f, x).simplify()          # direct integratoin using SymPy
```

$$f(x) = \frac{1}{c^2 + x^2}$$

$$dx = c\left(\tan^2(\theta) + 1\right)d\theta$$

$$f(\theta) = \frac{c\left(\tan^2(\theta) + 1\right)}{c^2\tan^2(\theta) + c^2}$$

$$F(\theta) = \frac{\theta}{c}$$

$$F(x) = \frac{\operatorname{atan}\left(\frac{x}{c}\right)}{c}$$

4.8.6 *Use of integration by parts*

As mentioned earlier, integration by parts is extremely useful in handling complicated integrands, and hence it is widely used. Typical examples are given in Eqs. (4.33) and (4.40).

In addition, we can also use coordinate transformation to simple integrations, which will be discussed in the following chapter.

These techniques given above can be combined when handling complicated integrands.

Note that most of these techniques (and many others) are implemented in SymPy. If one uses SymPy to find an antiderivative, almost no manual derivation is needed. Additionally, SymPy utilizes a large database of known solutions, allowing many definite and indefinite solutions to be directly retrieved from there. The author worked intensively during his university years on deriving antiderivatives manually, but now uses SymPy most of the time. However, it's still valuable for readers to be aware of the techniques listed above.

4.9 Symbolic integration using SymPy

Let us take a look at some examples set to demonstrate the formulas presented earlier.

4.9.1 *Integration then differentiation*

```
1  x, a, b, C = sp.symbols('x, a, b, C')        # define symbolic variables
2  f = sp.Function('f')(x)                       # define a sympy function
3
4  F = sp.integrate(f, x) + C                    # define indefinite integration
5  display(Math(f" F = {sp.latex(F)}"))
6  display(Math(f" F'(x) = {sp.latex(F.diff(x))}"))  # differentiation of F
```

$$F = C + \int f(x)\,dx$$

$$F'(x) = f(x)$$

```
1  I = sp.integrate(f, (x, a, b))              # definite integral, a to b
2  display(Math(f" I = {sp.latex(I)}"))
3  display(Math(f" I'(x) = {sp.latex(I.diff(x))}")) # differentiation of I
```

$$I = \int_a^b f(x)\,dx$$

$$I'(x) = 0$$

We got zero because a definite integral gives a constant. Its derivative must be zero:

```
1  I = sp.integrate(f, (x, a, a))              # the trivial case: [a, a]
2  display(Math(f" ∫_a^af(x)dx = {sp.latex(I)}"))
```

$$\int_a^a f(x)dx = 0$$

These confirm some of the formulas given in the theoretical study section.

4.9.2 *Elementary functions with closed-form integrals*

We present closed-form integrals of some often used elementary functions using SymPy. Readers can modify the codes and find out this rich list of closed-form integrals or antiderivatives.

We define some symbolic variables for later use:

```
1  x,a,b,c,d = sp.symbols('x,a,b,c,d')              # symbolic variables
2  a0,a1,a2,a3,a4,C0,C1,C2 = sp.symbols('a:5,C:3')
3  f = sp.Function('f')(x)                           # define a sympy function
```

4.9.2.1 *Polynomial functions*

```
1  # Define a polynomial funciton:
2  f = a0 + a1*x + a2*x**2 + a3*x**3
3  F = sp.integrate(f, x) + C0              # add in an integral constant
4  display(Math(f" ∫({sp.latex(f)})dx = {sp.latex(F)}"))
```

$$\int (a_0 + a_1 x + a_2 x^2 + a_3 x^3)dx = C_0 + a_0 x + \frac{a_1 x^2}{2} + \frac{a_2 x^3}{3} + \frac{a_3 x^4}{4}$$

Since the antiderivative is also a function, there is nothing preventing it to be integrated again. Each integration brings in another integral constant.

```
1  F2 = sp.integrate(F, x) + C1          # add in another integral constant
2  display(Math(f" ∫\\bigg(∫({sp.latex(f)})dx\\bigg)dx = {sp.latex(F2)}"))
```

$$\int \left(\int (a_0 + a_1 x + a_2 x^2 + a_3 x^3) dx \right) dx$$

$$= C_0 x + C_1 + \frac{a_0 x^2}{2} + \frac{a_1 x^3}{6} + \frac{a_2 x^4}{12} + \frac{a_3 x^5}{20}$$

We obtained the antiderivative of an integral of an integral.

4.9.2.2 *Trigonometric functions*

```
1  print(sp.integrate(sp.sin(x), x) + C0)     # add in an integral constant
2  print(sp.integrate(sp.cos(x), x) + C0)
3  print(sp.integrate(sp.tan(x), x) + C0)
4  print(sp.integrate(sp.cot(x), x) + C0)
```

```
C0 - cos(x)
C0 + sin(x)
C0 - log(cos(x))
C0 + log(sin(x))
```

4.9.2.3 *Inverse trigonometric functions*

```
1  print(sp.integrate(sp.asin(x), x) + C0)    # add in an integral constant
2  print(sp.integrate(sp.acos(x), x) + C0)
3  print(sp.integrate(sp.atan(x), x) + C0)
4  print(sp.integrate(sp.acot(x), x) + C0)
```

```
C0 + x*asin(x) + sqrt(1 - x**2)
C0 + x*acos(x) - sqrt(1 - x**2)
C0 + x*atan(x) - log(x**2 + 1)/2
C0 + x*acot(x) + log(x**2 + 1)/2
```

4.9.2.4 *Trigonometric composite functions*

Integrals of composite functions involving trigonometric functions are important in many applications, including wave propagation and vibration problems [9]. A sinusoidal term in such an integrand can oscillate violently when its argument is large. Integrating it numerically by discretizing the interval requires dense sampling and can be extremely costly. The use of antiderivatives is often particularly useful because integrals can be done analytically using Eq. (4.8) without dense sampling, as reported in Ref. [10].

The following lists some examples of the antiderivatives of composite integrands involving trigonometric functions:

```
1  f = (a0 + a1*x + a2*x**2)*sp.sin(a*x)
2  display(Math(f" f(x) = {sp.latex(f)}"))
3  F = sp.integrate(f, x) + C0
4  F                        # F(x)
```

$$f(x) = \left(a_0 + a_1 x + a_2 x^2 \right) \sin (ax)$$

$$C_0 + \begin{cases} \begin{aligned} & -\frac{a_0 \cos(ax)}{a} - \frac{a_1 x \cos(ax)}{a} - \frac{a_2 x^2 \cos(ax)}{a} + \frac{a_1 \sin(ax)}{a^2} \\ & + \frac{2a_2 x \sin(ax)}{a^2} + \frac{2a_2 \cos(ax)}{a^3} \end{aligned} & \text{for } a \neq 0 \\ \\ 0 & \text{otherwise} \end{cases}$$

```
1  f = sp.sin(sp.ln(x))
2  F = sp.integrate(f, x).doit() + C0        # add in an integral constant
3  display(Math(f" ∫{sp.latex(f)}dx = {sp.latex(F)}"))
4
5  F2 = sp.integrate(F,x) + C1              # add in another integral constant
6  display(Math(f" ∫\\bigg(∫{sp.latex(f)}dx\\bigg) dx = {sp.latex(F2)}"))
```

$$\int \sin(\log(x)) dx = C_0 + \frac{x \sin(\log(x))}{2} - \frac{x \cos(\log(x))}{2}$$

$$\int \left(\int \sin(\log(x)) dx \right) dx = C_0 x + C_1 + \frac{x^2 \sin(\log(x))}{10} - \frac{3x^2 \cos(\log(x))}{10}$$

```
1  f = a*x**2*sp.sin(x)+sp.exp(x)*sp.cos(x)        # a composite function
2  F = sp.integrate(f, x) + C0
3  display(Math(f" ∫{sp.latex(f)}dx =\\\\ \\hspace{{4.5cm}}{sp.latex(F)}"))
4  display(Math(f" dF/dx = {sp.latex(F.diff(x))}")) # This gives back to f
```

$$\int ax^2 \sin(x) + e^x \cos(x) dx$$

$$= C_0 + a\left(-x^2 \cos(x) + 2x \sin(x) + 2\cos(x)\right) + \frac{e^x \sin(x)}{2} + \frac{e^x \cos(x)}{2}$$

$$dF/dx = ax^2 \sin(x) + e^x \cos(x)$$

```
1  x, l = sp.symbols('x, l ', real=True)
2  m = sp.symbols('m ', integer = True)
3  fsin = sp.sin(m*sp.pi*x/l)
4  f = x*fsin**2
5  F = sp.integrate(x*fsin**2, x)          # ignore the integral constant
6  display(Math(f" ∫{sp.latex(f)}dx =\\\\ \\hspace{{2.2cm}}{sp.latex(F)}"))
```

$$\int x \sin^2\left(\frac{\pi m x}{l}\right) dx$$

$$= \begin{cases} \dfrac{l^2 \sin^2\left(\frac{\pi m x}{l}\right)}{4\pi^2 m^2} - \dfrac{lx \sin\left(\frac{\pi m x}{l}\right) \cos\left(\frac{\pi m x}{l}\right)}{2\pi m} + \dfrac{x^2 \sin^2\left(\frac{\pi m x}{l}\right)}{4} \\ \quad + \dfrac{x^2 \cos^2\left(\frac{\pi m x}{l}\right)}{4} & \text{for } m \neq 0 \\ \\ 0 & \text{otherwise} \end{cases}$$

4.10 Non-integrable (closed-form) functions

There are many (infinite) functions that do not have closed-form integrals. The following list includes just some examples that have a simple form:

Function name	Expression
Sine integral function	$\frac{\sin(x)}{x}$
Sine integral function	$\cos\left(\frac{1}{x}\right)$
Cosine integral function	$\frac{\cos(x)}{x}$
Cosine integral function	$\sin\left(\frac{1}{x}\right)$
Logarithmic integral function	$\frac{1}{\ln(x)}$
Logarithmic function	$\frac{x}{\ln(x)}$
Generalized Gaussian function	e^{ax^2}

These functions look simple, but their integrals do not have closed-form expressions in terms of elementary functions. The first six cases have singularities within the domain of the function. Therefore, if the denominator of any of these six functions involves a power, its integral may not have a closed-form solution. The generalized Gaussian function exhibits extreme unboundedness. If the independent variable x has an exponent greater than 1, its integral will not have a closed-form expression.

Let's take a look at the outputs from SymPy for these functions:

```
1  f = sp.exp(a*x**2)                    # or sp.exp(a*x**3), sp.exp(a*x**4)...
2  F = sp.integrate(f, x).doit() + C0     # add in an integral constant
3  display(Math(f" ∫{sp.latex(f)}dx = {sp.latex(F)}"))      # help(sp.erf)
```

$$\int e^{ax^2}\,dx = C_0 + \begin{cases} \dfrac{\sqrt{\pi}\,\mathrm{erf}\left(x\sqrt{-a}\right)}{2\sqrt{-a}} & \text{for } a \neq 0 \\ \\ x & \text{otherwise} \end{cases}$$

The integral function involves a Gauss error function erf() that needs numerical integration. Readers may find details about it using 'help(sp.erf)'.

```
1  f = sp.sin(x)/x                    # sp.sin(x)/x**2, (sp.sin(x)/x)**2, ...
2  F = sp.integrate(f, (x, 1,2)).doit() + C0
3  display(Math(f" ∫{sp.latex(f)}dx = {sp.latex(F)}"))        # help(sp.Si)
```

$$\int \frac{\sin(x)}{x}\,dx = C_0 - \mathrm{Si}\,(1) + \mathrm{Si}\,(2)$$

The integral function involves a Sine integral function Si(). For a given arbitrary argument c, Si(c) needs to be computed numerically:

```
1  c = np.random.random(1)[0]                       # a random number
2  print(f"Si({c}) = {sp.Si(c)}")          # Compute Si(x) numerically
```

```
Si(0.18442922727895616) = 0.184081071506254
```

The similar situation is for the Ci() function:

```
1  f = sp.cos(x)/x                    # sp.cos(x)/x**2, (sp.cos(x)/x)**2, ...
2  F = sp.integrate(f, (x, 1,2)).doit() + C0
3  display(Math(f" ∫{sp.latex(f)}dx = {sp.latex(F)}"))
4  print(f"Ci({c}) = {sp.Ci(c)}")                # Compute Si(x) numerically
```

$$\int \frac{\cos(x)}{x}\,dx = C_0 - \mathrm{Ci}\,(1) + \mathrm{Ci}\,(2)$$

```
Ci(0.18442922727895616) = -1.12176530856376
```

The following gives more examples of integrals expressed in the Sine integral of Cosine integral functions:

```
1  f = sp.sin(1/x)
2  F = sp.integrate(f, x).doit() + C0      # add in an integral constant
3  display(Math(f" ∫{sp.latex(f)}dx = {sp.latex(F)}"))
4
5  f = sp.cos(1/x)
6  F = sp.integrate(f, x).doit() + C0
7  display(Math(f" ∫{sp.latex(f)}dx = {sp.latex(F)}"))
```

$$\int \sin\left(\frac{1}{x}\right) dx = C_0 + x\sin\left(\frac{1}{x}\right) - \frac{\log\left(\frac{1}{x^2}\right)}{2} + \log\left(\frac{1}{x}\right) - \mathrm{Ci}\left(\frac{1}{x}\right)$$

$$\int \cos\left(\frac{1}{x}\right) dx = C_0 + x\cos\left(\frac{1}{x}\right) + \mathrm{Si}\left(\frac{1}{x}\right)$$

It is clear that this function is not defined at $x = 0$.

Another special function is the classical logarithmic integral, $\mathrm{li}(x)$, resulting from integrand $\frac{1}{\ln(x)}$:

```
1  f = 1/sp.log(x)                    # The classical logarithmic integral: li(x)
2  F = sp.integrate(f, x).doit() + C0          # add in an integral constant
3  display(Math(f" ∫{sp.latex(f)}dx = {sp.latex(F)}"))        # help(sp.li)
4  print(f"li({c}) = {sp.li(c)}")               # Compute li(x) numerically
```

$$\int \frac{1}{\log(x)} dx = C_0 + \mathrm{li}(x)$$

```
li(0.18442922727895616) = -0.0756844164011224
```

The following is the classical exponential integral, $\mathrm{Ei}(x)$, resulting from integrand $\frac{x}{\ln(x)}$:

```
1  f = x/sp.log(x)                    # The classical exponential integral: Ei(x)
2  F = sp.integrate(f, x).doit() + C0          # add in an integral constant
3  display(Math(f" ∫{sp.latex(f)}dx = {sp.latex(F)}"))        # help(sp.Ei)
4  print(f"Ei({c}) = {sp.Ei(c)}")               # Compute Ei(x) numerically
```

$$\int \frac{x}{\log(x)} dx = C_0 + \mathrm{Ei}(2\log(x))$$

```
Ei(0.18442922727895616) = -0.919980126097073
```

Sympy has a number of built-in methods to deal with these special functions efficiently. Interested readers may check its documentation.

4.11 Improper integrals

In the section on integrability, we examined examples of power functions with singularities within a finite interval, as shown in Fig. 4.5. However, power functions are not the only cases; there are many types of functions with singular points. Additionally, intervals can be infinite, such as $[a, \infty)$. A power function that is integrable over $[0, 1)$ may not be integrable over $[1, \infty)$. Integrals with a singular or unbounded integrand, or where the interval extends to infinity, are called improper integrals, which will be discussed in this section.

Our discussion extends the integration of the power function from Section 4.12.1, as shown in Fig. 4.5, but now considers the infinite interval $[1, \infty)$.

4.11.1 *Example: Power integrand over an infinite interval*

The antiderivatives of the power function $\frac{1}{x^p}$ are elementary functions:

```
1  x, p = sp.symbols("x, p")
2  f = 1/x**p
3  F = sp.integrate(f, x)
4  display(Math(f" ∫{sp.latex(f)}dx = {sp.latex(F)}"))
```

$$\int x^{-p}dx = \begin{cases} \dfrac{x^{1-p}}{1-p} & \text{for } p \neq 1 \\[2mm] \log(x) & \text{otherwise} \end{cases}$$

which is given in Eq. (4.29). Here, we consider the definite integral of $\frac{1}{x^p}$ over $[1, \infty)$. Depending on p, this integrand may diverge as $x \to \infty$. Also, because the interval extends to infinity, the integrand must converge to zero and do so rapidly enough to ensure the integral remains bounded, even though these functions are continuous. To gain a clearer understanding of these issues, let's plot the power function for values of x beyond 1:

```
1  colors = ['b', 'r', 'g', 'c', 'm']
2  xL = 0.001; xR = 3.0; N = 200
3  ps = [-0.5, 0, 0.5, 1, 1.5]
4  x  = np.linspace(xL, xR, N)                              # on x-axis
5
6  plt.figure(); plt.ioff()
7  fig, ax = plt.subplots(1,1,figsize=(4,2.8))
8  ax.grid(c='r', linestyle=':', linewidth=0.3)
9
10 for k, p in enumerate(ps):
11     ax.plot(x, [1/xi**p for xi in x],c=colors[k],label='p='+str(p),lw=1)
12     plt.fill_between(x,[1/xi**p for xi in x],[0]*len(x),\
13                      where=(x>=xL)&(x<=xR), color=colors[k], alpha=0.1)
14
15 ax.scatter(1, 1, c='r' )
16 ax.set_xlabel('x')
17 ax.set_ylabel(r"$1/x^p$"); ax.set_ylim(0, 6)
18 ax.axvline(x=0, c="k", lw=0.6); ax.axhline(y=0, c="k", lw=0.6)
19 ax.legend()
20
21 plt.savefig('imagesDI/powerf0oo.png', dpi=500)
22 #plt.show();
```

It is clear from Fig. 4.20 that when $p < 0$, the function is bounded and continuous on $[0, 1]$, but it becomes unbounded as $x \to \infty$. Conversely, when $p > 0$, the function is unbounded on $[0, 1]$, but converges to zero as $x \to \infty$. The behavior is opposite.

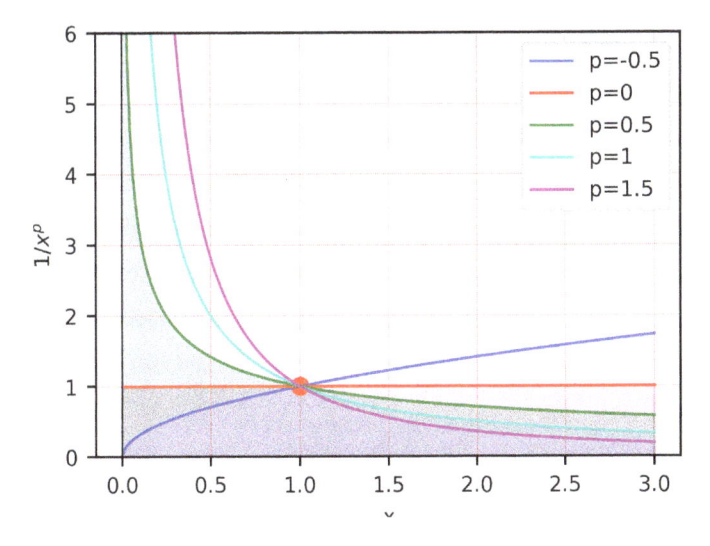

Figure 4.20. Distribution of the power functions with different p values over $(0, 3]$ and the area under these functions.

However, an integrand that converges to zero does not automatically imply that it is integrable. The integrability depends on the rate of convergence. Let's examine this case by case.

Case 0: First, let's consider the case where $p = 0$. This results in a horizontal straight line. When integrated over $[x, \infty)$, the area is obviously infinite. For all cases where $p < 0$, the power function is larger than the case where $p = 0$ throughout $[x, \infty)$, and therefore all these functions are not integrable over $[x, \infty)$.

Case I: When $(0 < p < 1)$, the decay rate of the function $\frac{1}{x^p}$ is less than 1. The definite integral over the domain $[1, \infty)$ is not Riemann integrable:

```
1  x = sp.symbols("x")
2  R = sp.Rational
3  print(f"p=3/10, over [0, 1], I={sp.integrate(1/x**R(3,10),(x,0, 1))}")
4  print(f"p=3/10, over [1,oo], I={sp.integrate(1/x**R(3,10),(x,1,oo))}")
```

```
p=3/10, over [0, 1], I=10/7
p=3/10, over [1,oo], I=oo
```

```
1  print(f"p=1/2, over [0, 1], I={sp.integrate(1/x**R(1,2),(x,0,1))}")
2  print(f"p=1/2, over [1,oo], I={sp.integrate(1/x**R(1,2),(x,1,oo))}")
```

```
p=1/2, over [0, 1], I=2
p=1/2, over [1,oo], I=oo
```

Case II: When $(p > 1)$, the definite integral over domain $[0, 1]$ is not Riemann integrable, but these are all integrable over $[x, \infty)$:

```
1  print(f"p=3/2, over [0, 1], I={sp.integrate(1/x**R(3,2),(x,0, 1))}")
2  print(f"p=3/2, over [1,oo], I={sp.integrate(1/x**R(3,2),(x,1,oo))}")
```

```
p=3/2, over [0, 1], I=oo
p=3/2, over [1,oo], I=2
```

Case III: When $(p = 1)$, the power function is not integrable over both $[0, 1]$ and $[x, \infty)$:

```
1  print(f"p=1, over [0, 1], I={sp.integrate(1/x,(x,0, 1))}")
2  print(f"p=1, over [1,oo], I={sp.integrate(1/x,(x,1,oo))}")
```

```
p=1, over [0, 1], I=oo
p=1, over [1,oo], I=oo
```

The power function $1/x$ corresponds to the cyan line in Fig. 4.20. It grows too fast in $[0, 1]$ and decays too slowly in $[x, \infty)$. Therefore, it is not integrable over either interval.

Note that in all these examples, we use 1 as the endpoint of the intervals, which is represented by the red dot in Fig. 4.20. As far as integrability is concerned, this value can be replaced by any positive non-zero real number. Using $x = 1$ as a reference point is convenient when comparing the behavior of these functions.

Case IV: If our integration interval is $(0, b]$ where b is a finite positive real number, the power functions will not have an opportunity to grow to infinity. In this case, the function is Riemann integrable regardless of the value of p:

```
1  x = sp.symbols("x")
2  ps = [-1, -0.5, 0, 0.5, 0.8, 1.5]                          # any p
3  a = 0.00001; b = 1000.                                     # a > 0, b > a
4  Is = np.array([sp.integrate(1/x**p, (x, a, b)) for p in ps],dtype=float)
5  F = np.array([(b**(1-p)-a**(1-p))/(1-p) for p in ps], dtype=float)
6  print(f"Definite integrals over (0, b] are all finite:\n{Is}")
7  print(f"Same results by antiderivatives over (0, b] :\n{F}")
```

```
Definite integrals over (0, b] are all finite:
[5.0000e+05 2.1082e+04 1.0000e+03 6.3239e+01 1.9405e+01 6.3239e+02]
Same resutls by antiderivatives over (0, b] :
[5.0000e+05 2.1082e+04 1.0000e+03 6.3239e+01 1.9405e+01 6.3239e+02]
```

The same analysis and results can also be obtained using antiderivatives of the power function.

4.11.2 *Splitting rules*

With the concepts established through the above examples, we now present the general rules for handling improper integrals:

1. **Exclude the singular point and then take limits:** To ensure integrability, any improper integral over a finite interval $[a, b]$ that contains a singular point $c \in [a, b]$ is split into two integrals, each covering an interval that excludes the singular point. Perform these integrals separately and then take the limits:

$$\int_a^b f(x)\, dx = \lim_{\varepsilon \to 0} \left(\int_a^{c-\varepsilon} f(x)\, dx + \int_{c+\varepsilon}^b f(x)\, dx \right) \qquad (4.70)$$

If the limit is finite, it is the result of the original integral. Otherwise, the original integral is not integrable. For convenience, let's term this as ε-limit approach. It is used for integrands with singularities.

2. **Change the interval limit to finite and then take the limit:** For any improper integral over an infinite interval $[a, \infty]$, change the interval to a finite one $[a, b]$. Perform the integral over this finite interval and then take the limit as b approaches infinity:

$$\int_a^\infty f(x)\, dx = \lim_{b \to \infty} \int_a^b f(x)\, dx \qquad (4.71)$$

If the limit is finite, it is the result of the original integral. Otherwise, the original integral is not integrable. For convenience, let's term this as b-limit approach. It is used for integrals with an infinite interval.

4.11.2.1 *Example: Convert to bounded integral, singular case*

Let's see some examples applying the rules given above: first, integral with a singularity:

```
 1  x, a, b = sp.symbols("x, a, b", positive=True)
 2  b = 1
 3  f = 1/x**R(1,2)                                        # integrand
 4  I = sp.integrate(f, (x, a, b))
 5  print(f"Expression of the bounded integral result = {I}")
 6  print(f"At the limit, I of 1/x^0.5 = {I.limit(a,0)}") # integrable case
 7
 8  f = 1/x                                                # integrand
 9  I = sp.integrate(f, (x, a, b))
10  print(f"Expression of the bounded integral result = {I}")
11  print(f"At the limit, I of 1/x = {I.limit(a,0)}") # not integrable case
```

```
Expression of the bounded integral result = 2 - 2*sqrt(a)
At the limit, I of 1/x^0.5 = 2
Expression of the bounded integral result = -log(a)
At the limit, I of 1/x = oo
```

We have obtained the same results as those obtained earlier.

4.11.2.2 *Example: Convert to bounded integral, infinity case*

Let us now look at a case with infinite interval:

```
 1  x, a, b = sp.symbols("x, a, b", positive=True)
 2  a = 1
 3  f = 1/x**2                                          # integrand
 4  I = sp.integrate(f, (x, a, b))
 5  print(f"Expression of the bounded integral result = {I}")
 6  print(f"At the limit, I of 1/x^0.5 = {I.limit(b,oo)}") #integrable case
 7
 8  f = 1/x                                             # integrand
 9  I = sp.integrate(f, (x, a, b))
10  print(f"Expression of the bounded integral result = {I}")
11  print(f"At the limit, I of 1/x = {I.limit(b,oo)}") #not integrable case
```

```
Expression of the bounded integral result = 1 - 1/b
At the limit, I of 1/x^0.5 = 1
Expression of the bounded integral result = log(b)
At the limit, I of 1/x = oo
```

We have obtained the same results as those obtained earlier. The major takeaway from these examples is that any improper integral can be handled by splitting it into multiple integrals, each of which is bounded, continuous, and over finite intervals.

Using SymPy, improper integrals can be solved directly without manual domain splitting. SymPy performs the splitting and takes the limit automatically. The following are some examples.

4.11.3 *Indefinite integration intervals via SymPy*

For integrations with infinite intervals like $[a, \infty)$

$$\int_a^\infty f(x)dx \tag{4.72}$$

Sympy can be used directly to obtain the solution if it exists. This following code gets this done in bulk:

```
1  # power function family x^p:
2
3  x = sp.symbols('x', real=True)
4
5  fxs = [x,  1/x, 1/x**(11/10), 1/x**2]          # put f(x) in a list
6  Fxs = [sp.integrate(fi,(x, 1, oo)) for fi in fxs]      # compute F(x)
7                              # The integral constant C is omitted
8  results = {}               # Empty dictionary for holding the results
9  for key, value in zip(fxs, Fxs):
10     results[key] = value
11
12 results
```

$$\left\{ \frac{1}{x^2} : 1, \ x^{-1.1} : 9.99999999999999, \ \frac{1}{x} : \infty, \ x : \infty \right\}$$

It is seen that when the risen power $p < -1$, the integral has a converged value. Otherwise, it diverges:

```
1  # Exponential function family e^(ax):
2
3  x, a = sp.symbols('x, a', real=True)
4  a = [-1, -1/10, 0, 1/10, 1]
5  fxs = [sp.exp(ai*x) for ai in a]
6  Fxs = [sp.integrate(fi,(x, 1, oo)) for fi in fxs]
7
8  results = {}               # Empty dictionary for holding the results
9  for key, value in zip(fxs, Fxs):
10     results[key] = value
11
12 results
```

$$\left\{ 1 : \infty, \ e^{-x} : e^{-1}, \ e^{-0.1x} : 9.04837418035959, \ e^{0.1x} : \infty, \ e^x : \infty \right\}$$

For this family of exponential functions, a must be smaller than 0 when integrated over $[a, \infty)$.

The following examples involve sinusoidal functions:

```
1  # Trigonometric function family, sine:
2  x = sp.symbols('x', real=True)
3  fxs = [sp.sin(x), sp.sin(x)/x, sp.sin(x)/x**sp.Rational(1,28)]
4  Fxs = [sp.integrate(fi,(x, 1, oo)).evalf() for fi in fxs]
5
6  results = {}               # Empty dictionary for holding the results
7  for key, value in zip(fxs, Fxs):
8      results[key] = value.evalf(8)
9
10 results
```

$$\left\{ \frac{\sin(x)}{x} : 0.62471326, \ \frac{\sin(x)}{\sqrt[28]{x}} : 0.55189029, \right.$$

$$\left. \sin(x) : \langle -1 + \cos(1), \cos(1) + 1 \rangle \right\}$$

For sine functions, the indefinite integral gives oscillatory values due to its sinusoidal nature. The integral values have multiple bounds and are called "accumulated bounds" denoted in $\langle \rangle$. When the sine function is modulated with $\frac{1}{x^p}$ and $p > 0$, the integral converges because the oscillation dies off as $x \rightarrow \infty$, compounded with cancellations (see Fig. 4.10). This example further reinforces the signed area definition of a definite integral.

4.11.4 *On residue integrals*

In the real domain, we know the integral

$$\int_{-1}^{1} \frac{1}{x}\, dx \qquad (4.73)$$

is not integrable due to the singularity of order one at $x = 0$. However, when integrals extend into the complex domain, they can become integrable and significant. In such cases, this type of integral can be multivalued, as the variables enter the complex domain, as discussed in Chapter 3 of Ref. [4]. These integrations involve complex contours that need to be properly chosen. This is a rich topic known as residue integration. We need to consider the physics of the actual problem to determine a physically meaningful solution. Residue integration is important in many practical problems, including wave propagation [9], where complex analysis is involved. This topic is extensive and beyond the scope of this volume, so we will stop here.

4.12 Major properties of integrals

Integrals of any integrable functions over closed intervals have the following major properties.

4.12.1 *Linearity*

First, because the integration of any integrable real function on a closed interval $[a, b]$ produces a scalar in $\mathbb{R}$, the collection of integrable functions forms a real space. It is closed under the standard arithmetic operations similar to the closure property of real numbers discussed in Chapter 2 of Ref. [4].

Second, since the integration of any integrable function on a closed interval $[a, b]$ is essentially a summation operation, any two integrable functions,

$f(x)$ and $g(x)$, satisfy

$$\int_a^b \left(\alpha f(x) + \beta g(x)\right) dx = \alpha \int_a^b f(x)\, dx + \beta \int_a^b g(x)\, dx \qquad (4.74)$$

where α and β are arbitrary scalars in $\mathbb{R}$. This means that the integral of a linear combination of functions is the linear combination of the integrals of these functions.

Additionally, the definite integration of a zero function gives zero, thus there is a zero element.

Therefore, the collection of integrable functions forms a real vector space [4].

The following code demonstrates this property:

```
1  x, a, b, α, β = sp.symbols('x, a, b, α, β ')        # symbolic variables
2  f = sp.Function('f')(x)
3  g = sp.Function('g')(x)                    # define another sympy function
4  fNg_int = sp.integrate(α*f+β*g, x)                  # integral of αf+βg
5  display(Math(f" {sp.latex(fNg_int)} = {sp.latex(fNg_int.expand())}"))
```

$$\int (\alpha f(x) + \beta g(x))\, dx = \int \alpha f(x)\, dx + \int \beta g(x)\, dx$$

4.12.2 *Inequalities*

There are quite a number of inequalities for integrals of functions in a closed interval when they are integrable. This is possible because such an integration results in a real number that is ordered and hence can be compared.

Upper and lower bounds: If a real function $f(x)$ is bounded and integrable in the interval $[a, b]$, its integral is bounded on that interval. The inequality is

$$m(b - a) \leq \int_a^b f(x)\, dx \leq M(b - a) \qquad (4.75)$$

where m and M are the bounds for $f(x)$: $m \leq f(x) \leq M$ for all x in $[a, b]$. The equality holds when $f(x)$ is a constant:

```
1  f = sp.sin(x)
2  a = 0; b = sp.pi
3  I = sp.integrate(f, (x, a, b))
4  I
```

It is found that the integration of the sine function over the interval $[0, \pi]$ is 2. Let us now find the minimum and maximum of the integrand in the same interval and then check whether or not the inequality is satisfied. This requires some coding as follows:

```
1  from scipy.optimize import minimize
2  domain = (0., sp.pi)                         # define the domain
3  len_domain = domain[1] - domain[0]           # Length of the domain
4  x_guess = [1.5]                              # set the initial guess
5
6  np_f = lambdify(x, f)                        # convert to numpy func.
7  results = minimize(np_f, x_guess, bounds = [domain], tol=1e-6)
8  min_f = results.fun                          # function valuse at min.
9
10 np_f = lambdify(x, -f)                        # flip sign of f for max.
11 results = minimize(np_f, x_guess, bounds = [domain], tol=1e-6)
12 max_f = -results.fun                          # flip-back to get max.
13
14 l_b = min_f*len_domain
15 u_b = max_f*len_domain
16
17 print(f'm={min_f:.4f}, M={max_f:.4f}')
18 print(f'Integral lower bound={l_b:.4f}; Integral upper bound={u_b:.4f}')
19 print(f'Is the integral inequality satisfied? {l_b <= I <= u_b}')
```

```
m=0.0000, M=1.0000
Integral lower bound=0.0000; Integral upper bound=3.1416
Is the integral inequality satisfied? True
```

Subinterval: If integrable $f(x)$ is non-negative for all x in $[a, b]$, and $[c, d]$ is a subinterval of $[a, b]$, then

$$\int_c^d f(x)dx \leq \int_a^b f(x)dx \tag{4.76}$$

```
1  x, a, b = sp.symbols('x, a, b ')            # symbolic variables
2  random_value = np.random.random()*10  # use random number for the tests
3  a = 0; b = random_value*sp.pi                # interval a to b
4  c = 0; d = 0.9*b                             #  subinterval c to d
5  f_a_b = sp.integrate(sp.sin(x)**2, (x, a, b))
6  f_c_d = sp.integrate(sp.sin(x)**2, (x, c, d))
7
8  print(f'Integral from a to b = {f_a_b:.4f}, from c to d={f_c_d:.4f}')
9  print(f'Is the inequality satisfied? {f_c_d <= f_a_b}')
```

```
Integral from a to b = 10.9950, from c to d=9.5949
Is the inequality satisfied? True
```

The above code uses a random number generator, so it produces different results each time it is executed. Readers may run it multiple times to see if a counterexample can be found.

Absolute inequality: Any integrable $f(x)$ in $[a, b]$ satisfies

$$\left| \int_a^b f(x)\, dx \right| \le \int_a^b |f(x)|\, dx \tag{4.77}$$

The equality holds only if $f(x)$ is non-negative over $[a, b]$.

```
1  random_value = np.random.random()*10   # use random number for the tests
2  a = 0; b = random_value*sp.pi                      # interval a to b
3  abs_I = abs( sp.integrate(sp.sin(x), (x, a, b)))
4  I_abs_f = sp.integrate(abs(sp.sin(x)), (x, a, b))
5
6  print(f'abs_I = {abs_I:.4f}, I_abs_f={I_abs_f:.4f}')
7  print(f'Is the inequality satisfied? {abs_I <= I_abs_f}')
```

```
abs_I = 0.2363, I_abs_f=3.7637
Is the inequality satisfied? True
```

Inequality for function difference: Consider two integrable functions $f(x)$ and $g(x)$. If $f(x) \le g(x)$ for x in $[a, b]$, then

$$\int_a^b \underbrace{[f(x) - g(x)]}_{<0 \text{ everywhere}} dx < 0 \tag{4.78}$$

Thus,

$$\int_a^b f(x)\, dx \le \int_a^b g(x)\, dx \tag{4.79}$$

```
1   random_value = np.random.random()*10   # use random number for the tests
2   a = 0; b = random_value*sp.pi                      # interval a to b
3   f = sp.sin(x)
4   g = 1.1*sp.sin(x)
5   f_I = sp.integrate(f, (x, a, b))
6   g_I = sp.integrate(g, (x, a, b))
7   solution = sp.solve(sp.simplify(f) <= sp.simplify(g), x)
8   print(f'Where f <= g? {solution}')
9
10  print(f'f_I = {f_I:.4f}, g_I={g_I:.4f}')
11  print(f'Is the inequality satisfied? {f_I <= g_I}')
```

```
Where f <= g? (0 <= x) & (x <= pi)
f_I = 0.0470, g_I=0.0517
Is the inequality satisfied? True
```

Minkowski inequality for function addition: For a real number $p \ge 1$, if $|f(x)|^p$, $|g(x)|^p$, and $|f(x) + g(x)|$ are integrable, we have

$$\left(\int_a^b |f(x) + g(x)|^p\, dx \right)^{1/p} \le \left(\int_a^b |f(x)|^p\, dx \right)^{1/p} + \left(\int_a^b |g(x)|^p\, dx \right)^{1/p} \tag{4.80}$$

Note that when $p < 1$, the above inequality reverses, known as the reverse inequality.

When $p = 2$, we have

$$\left(\int_a^b |f(x) + g(x)|^2 \, dx\right)^{1/2} \leq \left(\int_a^b |f(x)|^2 \, dx\right)^{1/2} + \left(\int_a^b |g(x)|^2 \, dx\right)^{1/2}$$

(4.81)

This is the integral version of the well-known **triangle inequality** for integrals, which is a useful inequality for proving properties of computational methods [7, 8]. This is expected because integrals form linear vector space, and hence they should have all these inequalities for a linear vector space:

```
1   p = 1.5                                  # 2 or any real >=1
2   a = 0; b = sp.pi
3   f = sp.sin(x); g = sp.cos(x)
4   fNg2_I = (sp.integrate(abs(f+g)**p, (x, a, b))**(1/p)).evalf(6)
5   f2_I   = (sp.integrate(abs(f)**p,   (x, a, b))**(1/p)).evalf(6)
6   g2_I   = (sp.integrate(abs(g)**p,   (x, a, b))**(1/p)).evalf(6)
7
8   print( fNg2_I, f2_I+g2_I )
9   print(f'fNg2_I={fNg2_I:.4f}, f2_I + g2_I={(f2_I+g2_I):.4f}')
10  print(f'Is the inequality satisfied? {fNg2_I <= f2_I+g2_I}')
```

```
2.05218 2.90222
fNg2_I=2.0522, f2_I + g2_I=2.9022
Is the inequality satisfied? True
```

Cauchy–Schwarz inequality for function product: Any two *squarely* integrable functions $f(x)$ and $g(x)$ in $[a, b]$ satisfy

$$\left(\int_a^b f(x)g(x) \, dx\right)^2 \leq \left(\int_a^b f(x)^2 \, dx\right)\left(\int_a^b g(x)^2 \, dx\right)$$

(4.82)

To prove this, we need to define an inner product for our vector space of square-integrable functions on closed intervals. The Cauchy–Schwarz inequality is a useful tool for proving properties of computational methods [7, 8]. It is also one of the inequalities for a linear vector space. We omit the proof here but demonstrate it using the following code for some functions:

```
1   a = 0; b = 0.5*sp.pi; p = 2  # p may changed to
2   f = sp.sin(x); g = sp.cos(x)
3   fg_I2 = sp.integrate(f*g,   (x, a, b))**2
4   f_I2  = sp.integrate(f**2, (x, a, b))
5   g_I2  = sp.integrate(g**2, (x, a, b))
6
7   print(f'fg_I2={fg_I2:.4f}, f_I2*g_I2 ={f_I2*g_I2:.4f}')
8   print(f'Is the inequality satisfied? {fg_I2 <= f_I2*g_I2}')
```

```
fg_I2=0.2500, f_I2*g_I2 =0.6169
Is the inequality satisfied? True
```

Hölder's inequality for function product: The Cauchy–Schwarz inequality can be extended to Hölder's inequality:

$$\left| \int_a^b f(x)g(x)\,dx \right| \leq \left(\int_a^b |f(x)|^p\,dx \right)^{1/p} \left(\int_a^b |g(x)|^q\,dx \right)^{1/q} \tag{4.83}$$

where p and q are real numbers, $1 \leq p, q \leq \infty$, and $\frac{1}{p} + \frac{1}{q} = 1$. This inequality requires that $f(x)g(x)$, $|f(x)|^p$, and $|g(x)|^q$ are integrable. When $p = q = 2$, Hölder's inequality reduces to the Cauchy–Schwarz inequality:

```
 1  a = 0; b = 0.5*sp.pi
 2  p = 3                                        # one may try other p
 3  q = p/(p-1)
 4  print(f'1/p + 1/q = {1/p + 1/q}')
 5  f = sp.sin(x); g = sp.cos(x)
 6  abs_fg_I = abs(sp.integrate(f*g, (x, a, b)))
 7  fp_Ip  = sp.integrate(abs(f)**p, (x, a, b))**(1/p)
 8  gq_Iq  = sp.integrate(abs(g)**q, (x, a, b))**(1/q)
 9
10  print(f'abs_fg_I={abs_fg_I}, fp_Ip*gq_Iq={fp_Ip*gq_Iq:.4f}')
11  print(f'Is the inequality satisfied? {abs_fg_I <= fp_Ip*gq_Iq}')
```

```
1/p + 1/q = 1.0
abs_fg_I=1/2, fp_Ip*gq_Iq=0.7986
Is the inequality satisfied? True
```

4.13 Integration using legend polynomials

In Chapter 6 of Ref. [4], we presented the theory, techniques, and codes for approximating functions using basis functions including monomials and the legend orthogonal polynomials (Chebyshev, Legendre, Laguerre and Hermite). Utilizing the linearity property of integrals, functions approximated with any of these polynomials can be integrated easily, as these basis functions have simple analytical antiderivatives. Therefore, integrating the original function becomes a linear combination of the approximation coefficients and the integrals of the basis functions.

This section outlines the detailed procedure for integrating functions approximated using polynomial basis functions. Three examples are provided: 1. using the monomial basis, 2. using the Chebyshev basis, and 3. using the Legendre basis.

4.13.1 *Example: Integration using monomials*

4.13.1.1 *Integrals of monomials*

The monomials basis can be given as [4]

$$1, \ x^1, \ x^2, \ldots, x^n \tag{4.84}$$

The domain of these monomial basis functions is $(-\infty, \infty)$. The formulas for the antiderivatives are given in Eq. (4.29).

4.13.1.2 *Formula for integration of polynomials*

Assuming a given function $f(x)$ has been approximated using these monomials up to order n, the expression for the approximated function becomes a polynomial. Its general form is given as

$$\langle f(x) \rangle = c_0 x^0 + c_1 x^1 + c_2 x^2 + \cdots + c_n x^n = \sum_{i=0}^{n} c_i x^i \tag{4.85}$$

where $\langle\rangle$ stands for approximation and c_i are the coefficients found in the process of approximation [4].

The integration of the approximated function can be done exactly using the formulas for the power function (monomials) for each of the terms. We have the antiderivative

$$F_{\langle f(x) \rangle} = C + c_0 x + \frac{c_1}{2} x^2 + \frac{c_2}{3} x^3 + \cdots + \frac{c_n}{n+1} x^{n+1}$$

$$= \sum_{i=0}^{n} \frac{c_i}{i+1} x^{i+1} + C \tag{4.86}$$

where C is the integration constant.

4.13.1.3 *Approximation of functions in polynomials*

The approximation of a given function follows exactly the procedure given in Chapter 6 of Ref. [4]. The following code is copied from Ref. [4]:

```
 1  import numpy.polynomial.polynomial as poly
 2
 3  # Define the true function to be approximated:
 4  x = symbols('x', real=True)
 5  sp_f = 0.2*x**3+x*sp.sin(3*x)              # sympy func, Non-polynomial
 6
 7  np_f = lambdify(x, sp_f, 'numpy')             # convert to numpy func.
 8
 9  Ns = [5]                                    # order of approximation
10  np_a, np_b = 0, 1                       # integration interval [a, b]
11
12  x_nodes = np.linspace(np_a, np_b, Ns[-1]+1) # Sample in interval [a, b]
13  # Evenly spaced nodes on x, for lower orders, use least square fitting.
14
15  Pfit = poly.polyfit                         # instance of fitting tool
16  Px = poly.Polynomial                   # Use monomials. Create an object
17
18  P_ap_x = gr.PolyAprox(np_f, Ns, x_nodes, Pfit, Px)  # fitted Polynomial
19  # The coefficients botained are given in P_ap_x.coef
20
21  P_ap_x_coef = [float(f"{c:.7f}") for c in P_ap_x.coef]  # Limits digits
22
23  Px(P_ap_x_coef)                        # approximated f(x) in monomial bases
```

$$x \mapsto 0.0 + (5.1019\text{e-}03)\, x + 2.9071\, x^2 + 0.828\, x^3 - 6.4938\, x^4 + 3.0948\, x^5$$

4.13.1.4 *Integration of the approximation polynomial*

P_ap_x is the polynomial approximating the given original function, which can be complicated and may lack a closed-form antiderivative. We thus proceed by integrating the polynomial using Eq. (4.86). First, integrate the monomial basis functions to obtain their antiderivatives:

```
 1  n1 = Ns[-1]+1                               # n+1 (order + 1, terms)
 2  xk_int = [integrate(x**k,(x,np_a,np_b)) for k in range(n1)] # monomials
 3  print(xk_int)                  # integrated value of these basis functions
 4
 5  xk_int = [integrate(x**k, (x)) for k in range(n1)]          # monomial
 6  xk_int                     # integrated basis functions, antiderivatives
```

[1, 1/2, 1/3, 1/4, 1/5, 1/6]

$$\left[x,\ \frac{x^2}{2},\ \frac{x^3}{3},\ \frac{x^4}{4},\ \frac{x^5}{5},\ \frac{x^6}{6} \right]$$

Next, we apply the linear property of integrals to integrate the approximated function. This involves straightforwardly combining the integrals of each monomial basis function, weighted by the coefficients derived during the original function's approximation process:

```
1  F_f = 0
2  for i in range(len(P_ap_x.coef)):            # Use basis antiderivatives
3      F_f += P_ap_x.coef[i]*xk_int[i]    # P_ap_x.coef is obtained earlier
4                                         # F(b) - F(a):
5  print(f'Definite integral, I={F_f.subs({x:np_b})-F_f.subs({x:np_a})}\n')
6
7  # Get the coefficients of the integrated polynomial:
8  coeffs_ = [sp.Poly(F_f).coeff_monomial((i,)) for i in range(n1)]
9
10 F_coef = [float(f"{c:.7f}") for c in coeffs_]  # Limit the print digits
11 print(f"The integrated polynomial, approximated antiderivative F:")
12 Px(F_coef)                        # form numpy polynomial using coefficients
```

```
Definite integral, I=0.395607356711189
```

```
The integrated polynomial, approximated antiderivative F:
```

$$x \mapsto 0.0 + 0.0\,x + (2.551\mathrm{e}\text{-}03)\,x^2 + 0.969\,x^3 + 0.207\,x^4 - 1.2988\,x^5$$

Let us use SymPy to compute the definite integral value of the original function, which gives the exact integral result:

```
1  sp_I = integrate(sp_f, (x, np_a, np_b))               # definite int.
2  print(f'Exact dfinite integral result={sp_I:.8f}')
```

```
Exact definite integral result=0.39567750
```

It is observed that using polynomial approximation up to order five provides four correct digits for the given interval. We note the following points:

1. **Accuracy dependence:** The accuracy of the approximation depends on the order of the polynomial and the length of the integration interval. Readers are encouraged to experiment with these parameters using the code provided above to examine the accuracy.

2. **Exact solution for polynomials:** The technique yields the exact solution only when the original function is a polynomial and the order of approximation matches the order of the original polynomial. In such cases, direct analytical integration is possible without the need for approximation.

3. **Utility for complicated functions:** This technique is particularly useful when an analytical antiderivative for a complicated function cannot be determined, which is often the scenario in real-life applications.

4. **Computational cost and error source:** The primary computational expense lies in the function approximation. Once the approximation is completed, the integration process is quick and straightforward. The main source of error stems from the function approximation, while the integration step itself is exact.

Given above is the detailed step-by-step procedure for computing the integral of function that are approximated using monomial bases.

Alternatively, one may directly use the numpy.polynomial.polynomial. polyint() to obtain the definite integral value after the approximation polynomial is found. The code is as follows:

```
1  # Integrate and then get its coefficients:
2  int_coef = poly.polyint(P_ap_x.coef)
3
4  # Compute F(b)-F(a):
5  F_b_a = poly.polyval(np_b, int_coef)-poly.polyval(np_a, int_coef)
6  print(f'Definite integral result using numpy = {F_b_a:.8f}')
```

```
Definite integral result using numpy = 0.39560736
```

4.13.2 *Case study: Integration using Chebychev bases*

This study uses Chebyshev basis functions to approximate a non-polynomial function and then carries out the integration. The procedure is the same as in the previous example, with the only change being the basis functions.

4.13.2.1 *Integrals of Chebychev bases*

First, let us generate Chebychev basis functions using the code given in Chapter 6 of Ref. [4]:

```
1  x = symbols('x', real=True)          # define a symbolic variable
2  n = 5                                # generate the first n+1 terms
3  T = [gr.cheby_T(i, x).expand() for i in range(n + 1)]   # gr module
4  T
```

$$\left[1,\ x,\ 2x^2 - 1,\ 4x^3 - 3x,\ 8x^4 - 8x^2 + 1,\ 16x^5 - 20x^3 + 5x\right]$$

The domain of these basis functions is $(-\infty, \infty)$. The integral of the Chebychev bases can be obtained easily:

```
1 np_a, np_b = 0, 1                              # integration interval [a, b]
2 I_T = [integrate(T[k], (x,np_a,np_b)) for k in range(n+1)]   # definite
3 print(I_T)                            # integrated value of these basis functions
4
5 F_T = [integrate(T[k], (x)) for k in range(n + 1)]     # polynomial form
6 F_T                              # indefinite integral or base antiderivatives
```

[1, 1/2, -1/3, -1/2, -1/15, 1/6]

$$\left[x, \ \frac{x^2}{2}, \ \frac{2x^3}{3} - x, \ x^4 - \frac{3x^2}{2}, \ \frac{8x^5}{5} - \frac{8x^3}{3} + x, \ \frac{8x^6}{3} - 5x^4 + \frac{5x^2}{2}\right]$$

4.13.2.2 *Formula for integration of Chebyshev bases*

Assume that a given function $f(x)$ has been approximated using these Chebyshev bases up to order n. The expression for the approximated function becomes a polynomial:

$$\langle f(x)\rangle = c_0 T_0(x) + c_1 T_1(x) + c_2 T_2(x) + \cdots + c_n T_n(x) = \sum_{i=0}^{n} c_i T_i(x)$$

$$(4.87)$$

where c_i are the coefficients found in the process of approximation using the Chebyshev bases [4].

The integration of the approximated function can be done exactly and has the following form:

$$F_{\langle f \rangle}(x) = c_0 T_{F0}(x) + c_1 T_{F1}(x) + c_2 T_{F2}(x) + \cdots + c_n T_{Fn}(x)$$

$$= \sum_{i=0}^{n} c_i T_{Fi}(x) + C \qquad\qquad (4.88)$$

where T_{Fi} is the antiderivative of the Chebyshev basis T_i obtained in the previous code cell and C is the integral constant.

4.13.2.3 *Approximation in Chebyshev polynomials*

The approximation of a given function follows exactly the procedure given in Chapter 6 of Ref. [4]. The following code is a copy from Ref. [4]:

```
 1  import numpy.polynomial as P
 2  import numpy.polynomial.chebyshev as Che
 3  import numpy.polynomial.polynomial as poly
 4
 5  # Define the true function to be approximated:
 6  x = symbols('x', real=True)
 7  sp_f = 0.2*x**3+x*sp.sin(3*x)               # sympy func, Non-polynomial
 8
 9  np_f = lambdify(x, sp_f, 'numpy')           # convert to numpy func.
10
11  Ns = [5]                                    # order of approximation
12  np_a, np_b = 0, 1                           # integration interval [a, b]
13
14  x_nodes = np.linspace(np_a, np_b, Ns[-1]+1) # Sample in interval [a, b]
15  # Evenly spaced nodes on x, for lower orders, use least square fitting.
16
17  Pfit = Che.chebfit                          # instance of fitting tool
18  Px = P.Chebyshev                       # Use Chebyshev. Create an object
19
20  P_ap_x = gr.PolyAprox(np_f, Ns, x_nodes, Pfit,Px) # fitted in Chebyshev
21  # The coefficients botained are given in P_ap_x.coef
22
23  P_ap_x_coef = [float(f"{c:.3f}") for c in P_ap_x.coef]  # Limits digits
24
25  Px(P_ap_x_coef)                      # approximated f(x) in Chebyshev bases
```

$$x \mapsto -0.982\, T_0(x) + 2.56\, T_1(x) - 1.793\, T_2(x) + 1.174\, T_3(x) - 0.812\, T_4(x) + 0.193\, T_5(x)$$

P_ap_x is the Chebyshev polynomial that approximates the given original function. We can convert it to the polynomial if so required:

```
 1  Px(P_ap_x_coef).convert(kind=P.Polynomial)
```

$$x \mapsto -1.0e - 03 + (3.0e\text{-}03)\, x + 2.91\, x^2 + 0.836\, x^3 - 6.496\, x^4 + 3.088\, x^5$$

4.13.2.4 *Integration in Chebyshev polynomial*

The integral of the approximated function is a linear combination of the antiderivatives of all these Chebyshev bases with the fitted coefficients obtained during the approximation of the original function:

```
1  F_f = 0
2  for i in range(len(P_ap_x.coef)):              # antiderivative in bases
3      F_f += P_ap_x.coef[i]*F_T[i]        # P_ap_x.coef is obtained earlier
4                                    # F(b) - F(a):
5  print(f'Definite integral, I={F_f.subs({x:np_b})-F_f.subs({x:np_a})}\n')
6
7  # Get the coefficients of the integrated Chebyshev polynomial:
8  coeffs_ = [sp.Poly(F_f).coeff_monomial((i,)) for i in range(Ns[0])]
9
10 F_coef = [float(f"{c:.5f}") for c in coeffs_]         # limit the digits
11 print(f"The integrated polynomial, approximated antiderivative F:")
12 Px(F_coef)                    # form Chebyshev polynomial using coefficients
```

Definite integral, I=0.395607356711188

The integrated polynomial, approximated antiderivative F:

$$x \mapsto 0.0\,T_0(x) + 0.0\,T_1(x) + (2.55\text{e-}03)\,T_2(x) + 0.969\,T_3(x) + 0.207\,T_4(x)$$

Let us use SymPy to compute the definite integral value of the original function, which gives the exact result:

```
1  sp_I = integrate(sp_f, (x, np_a, np_b))              # definite int.
2  print(f'Exact dfinite integral result={sp_I:.8f}')
```

Exact definite integral result=0.39567750

We can use directly the numpy.polynomial.chebint() to obtain the definite integral value. The code is as follows:

```
1  # Integrate and then get its coefficients:
2  int_coef = Che.chebint(P_ap_x.coef)
3
4  # Compute F(b)-F(a):
5  F_b_a = Che.chebval(np_b, int_coef)-Che.chebval(np_a, int_coef)
6  print(f'Definite integral result using numpy = {F_b_a:.8f}')
```

Definite integral result using numpy = 0.39560736

It is seen that the result through Chebyshev basis approximation with order five gives four correct digits for the given interval. The accuracy depends on the number of terms of Chebyshev bases used and the size of the integration interval.

4.13.3 *Example: Integration of the Gaussian function*

The Gaussian function $\exp(-x^2)$ looks simple, but its antiderivative cannot be expressed in an explicit formula:

```
1  x = symbols('x', real=True)
2  sp_f = sp.exp(-x**2)
3  sp_F = integrate(sp.exp(-x**2), x)    # attempts to obtain an expression
4  display(Math(f" ∫{sp.latex(sp_f)}}dx = {sp.latex(sp_F)}"))
5  print(f"erf(1.1) = {sp.erf(1.1)}")          # Compute erf(x) numerically
```

$$\int e^{-x^2}\,dx = \frac{\sqrt{\pi}\,\mathrm{erf}\,(x)}{2}$$

```
erf(1.1) = 0.880205069574082
```

As seen, the SymPy integral result can only be given in the Gauss error function form, which can only be evaluated numerically.

In this section, we first approximate the original Gaussian function using Legendre polynomial bases and then integrate these bases exactly to obtain an approximate definite integral for the function. The procedure mirrors that of the previous example. Additionally, monomials or Chebyshev bases can be employed for this task. Our choice of Legendre this time is merely to demonstrate an alternative approach.

4.13.3.1 *Integrals of Legendre bases*

First, generate Legendre basis functions using the code given in Chapter 6 of Ref. [4]:

```
1  n = 5                              # generate the first n+1 terms
2  legP = [gr.LegendreR(i, x) for i in range(n+1)]          # gr module
3  legP
```

$$\left[1,\ x,\ \frac{3x^2}{2} - \frac{1}{2},\ \frac{5x^3}{2} - \frac{3x}{2},\ \frac{35x^4}{8} - \frac{15x^2}{4} + \frac{3}{8},\ \frac{63x^5}{8} - \frac{35x^3}{4} + \frac{15x}{8}\right]$$

The domain of these basis functions is $(-\infty, \infty)$. The integrals of the Chebychev bases are obtained easily:

```
1  I_P = [integrate(legP[k], (x,np_a,np_b)) for k in range(n+1)] # definite
2  print(I_P)                      # integrated value of these basis functions
3
4  F_P = [integrate(legP[k], (x)) for k in range(n + 1)]   # polynomial form
5  F_P                            # indefinite integral, base antiderivatives
```

```
[1, 1/2, 0, -1/8, 0, 1/16]
```

$$\left[x,\ \frac{x^2}{2},\ \frac{x^3}{2} - \frac{x}{2},\ \frac{5x^4}{8} - \frac{3x^2}{4},\ \frac{7x^5}{8} - \frac{5x^3}{4} + \frac{3x}{8},\ \frac{21x^6}{16} - \frac{35x^4}{16} + \frac{15x^2}{16}\right]$$

4.13.3.2 *Approximation in Legendre polynomials*

The approximation of a given function follows exactly the procedure given in Chapter 6 of Ref. [4]. The following code is copied from Ref. [4]:

```
1  import numpy.polynomial.legendre as Leg
2
3  np_f = lambdify(x, sp_f, 'numpy')                    # convert to numpy func.
4
5  Ns = [5]                                             # order of approximation
6  np_a, np_b = 0, 1                                    # integration interval [a, b]
7
8  x_nodes = np.linspace(np_a, np_b, Ns[-1]+1) # Sample in interval [a, b]
9  # Evenly spaced nodes on x, for lower orders, use least square fitting.
10
11 Pfit = Leg.legfit                                    # instance of fitting tool
12 Px = P.Legendre                                      # Use Legendre. Create an object
13
14 P_ap_x = gr.PolyAprox(np_f, Ns, x_nodes, Pfit, Px)  # fitted Polynomial
15 # The coefficients obtained are given in P_ap_x.coef
16
17 P_ap_x_coef = [float(f"{c:.3f}") for c in P_ap_x.coef]  # limits digits
18
19 Px(P_ap_x_coef)                                      # approximated f(x) in monomial bases
```

$$x \mapsto 0.798\,P_0(x) - 0.131\,P_1(x) - 0.292\,P_2(x) - 0.125\,P_3(x) + 0.148\,P_4(x) - 0.031\,P_5(x)$$

P_ap_x is the Legendre polynomial that approximates the given original function.

4.13.3.3 *Integration in Legendre polynomial*

In this example, we skip the detailed procedure and use directly the numpy.polynomial.legint() to obtain the definite integral value. The code is as follows:

```
1  # Integrate and then get its coefficients:
2  int_coef = Leg.legint(P_ap_x.coef)
3
4  # Compute F(b)-F(a):
5  F_b_a = Leg.legval(np_b, int_coef)-Leg.legval(np_a, int_coef)
6  print(f'Definite integral result using numpy = {F_b_a:.8f}')
```

```
Definite integral result using numpy = 0.74682936
```

This integration technique can also be applied to functions approximated using other legend polynomials, such as Laguerre and Hermite polynomials. Readers are encouraged to try it using the codes provided in Chapter 6 of Ref. [4] and the code provided above.

4.13.3.4 *Numerical integration of functions*

Let's integrate the original function directly. Since an analytical antideriva-tive for the Gaussian function is not available, we'll use a numerical tool provided in scipy.integrate to compute the result for comparison. Numeri-cal integration is a rich, useful, and important topic, deserving a separate volume for detailed presentation. Here, we simply utilize the existing tool provided by Python modules. The following code snippet employs scipy.integrate.quad(), which also provides an accuracy estimate for the numerical integral value:

```
1  from scipy.integrate import quad
2  I_direct = quad(np_f, np_a, np_b)            # integrate numerically
3  print(f" The area under y(x) in [{np_a},{np_b:.2f}]={I_direct}")
```

```
The area under y(x) in [0,1.00]=(0.7468241328124271, 8.291413475940725e-15)
```

The comparison reveals that the Legendre basis approximation with order five provides accuracy to five significant digits over the given interval.

The advantage of polynomial approximation lies in its ability to provide an approximate formula for the antiderivative in polynomial form. In con-trast, direct numerical integration yields only the numerical value of the integral.

4.14 Line integration

Line integration involves integrating an integrand along a curve $\mathcal{C}$. Its defini-tion is akin to one-dimensional integration, as given in Eq. (4.1), where the integral variable is singular. However, the integrand and curve $\mathcal{C}$ are defined in a multi-dimensional space. The integral can be expressed as

$$I = \int_{\mathcal{C}} f(x, y, z) \, ds = \lim_{\substack{m \to \infty \\ s_{i-1} \leq s_i^* \leq s_i \\ \text{all } (s_i - s_{i-1}) \to 0}} \left(\sum_{i=1}^{m} f(s_i^*)(s_i - s_{i-1}) \right) \tag{4.89}$$

where s is a curve coordinate defined along $\mathcal{C}$ and s_i denotes the location at x_i, y_i, z_i that is on $\mathcal{C}$. The length of s_i is the arc length on $\mathcal{C}$ between the starting point a at x_a, y_a, z_a and the point at x_i, y_i, z_i. The end point b of $\mathcal{C}$ corresponds to x_b, y_b, z_b.

All the theory, rules, and techniques for 1D integration apply to line integrals with s as the coordinate. In a specific scenario, when the curve $\mathcal{C}$ lies directly on any of the x-, y-, or z-axes, the line integration is equivalent to 1D integration.

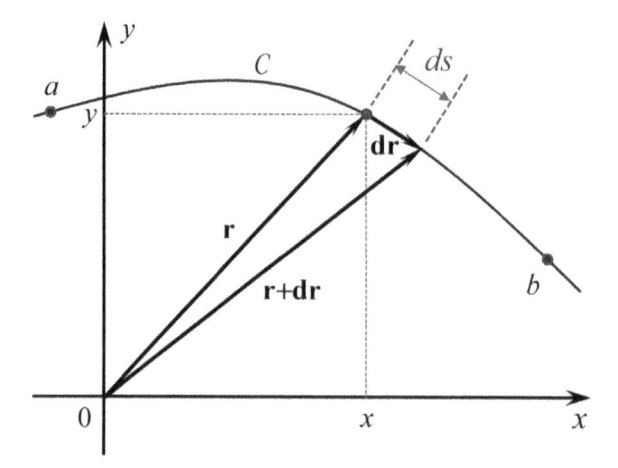

Figure 4.21. Position vector **r** introduced to create the relationship between (x, y) on a curve and a parameter t for line integral along $\mathcal{C}$.

For proper formulation, curve $\mathcal{C}$ is expressed as a bijective mapping $\mathbf{r}(t) \colon [a, b] \to \mathcal{C}$, where $\mathbf{r}(t)$ is typically a continuous position vector parametrized by a new variable t, denoted as $\mathbf{r}(x(t), y(t), z(t))$. In 2D, this simplifies to $\mathbf{r}(x(t), y(t))$, as illustrated in Fig. 4.21.

As shown in Fig. 4.21, $\mathbf{r}(t)$ at point (x, y) is simply

$$\mathbf{r}(t) = \begin{bmatrix} x(t) \\ y(t) \end{bmatrix} \tag{4.90}$$

When a point on the curve undergoes an infinitesimal movement along the curve, the corresponding change in differential length becomes ds. Similarly, the position vector $\mathbf{r}(t)$ experiences a corresponding change of $d\mathbf{r}(t)$, as depicted in Fig. 4.21. It is evident that ds represents the L2 norm of $d\mathbf{r}(t)$: $ds = |d\mathbf{r}(t)| = |\mathbf{r}'(t)dt|$.

Using Eq. (4.90), we have

$$ds = \|d\mathbf{r}\| = \left\| \frac{d\mathbf{r}}{dt} dt \right\| = \sqrt{\left(\frac{dx}{dt} \right)^2 + \left(\frac{dy}{dt} \right)^2} \, dt = \|\mathbf{r}'(t)\| dt \tag{4.91}$$

Equation (4.89) becomes

$$I = \int_{\mathcal{C}} f\left(\mathbf{r}(t)\right) ds = \int_{t_a}^{t_b} f\left(\mathbf{r}(t)\right) \underbrace{\|\mathbf{r}'(t)\|}_{\text{length scalar}} dt \tag{4.92}$$

where t_a and t_b are values of t corresponding, respectively, to x_a, y_a, z_a and x_b, y_b, z_b on $\mathcal{C}$. This integral is nothing but a 1D integral over t similar to

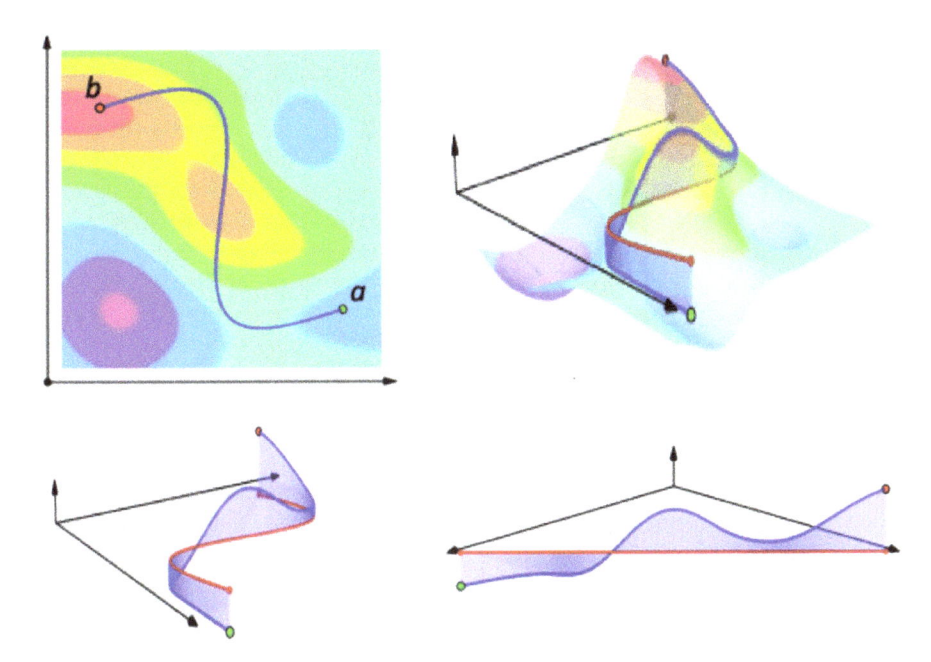

Figure 4.22. The line integral of a function (represented by the blue curve) along a curve $\mathcal{C}$ (shown in red) occurs within a 2D domain. It represents the signed area under the function along the curve. To convert the integral along a curve into a standard 1D integral, a bijective mapping, ensuring one-to-one correspondence, is essential.

Source: Images created based on movie from en.wikipedia Wikimedia Commons by Lucas Vieira to the public domain.

that given in Eq. (4.1). We have successfully converted the line integral to a form of standard 1D integral.

The length scalar becomes

$$\|\mathbf{r}'(t)\| = \sqrt{\left(\frac{d\xi}{dt}\right)^2 + \left(\frac{d\eta}{dt}\right)^2} \qquad (4.93)$$

The mapping process is schematically shown in Fig. 4.22.

Equation (4.92) can be used to compute the length of a curve by setting $f(\mathbf{r}) = 1$.

4.14.1 *Length of a planar curve via integration*

In a 2D plane, the curve can be defined as a function $y(x)$ shown in Fig. 4.23.

At any location on the curve in the 2D plane, consider a small line segment Δs on the curve. The corresponding increments in x- and y-axis become,

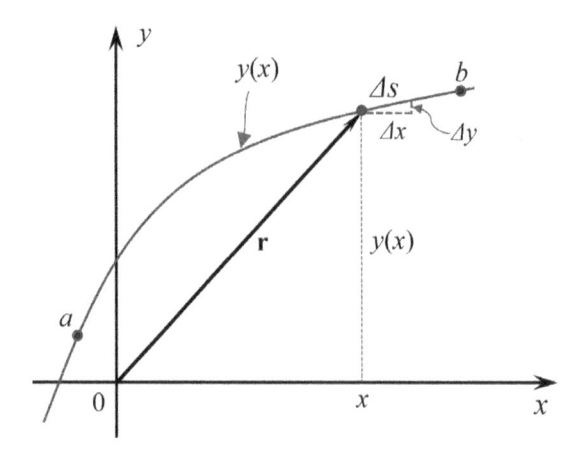

Figure 4.23. A curve C in a 2D plane defined by function $y(x)$ and its relation to the position vector $\mathbf{r}$ for mapping.

respectively, Δx and Δy. The length of Δs is computed using

$$\|\Delta s\| = \sqrt{(\Delta x)^2 + (\Delta y)^2} \tag{4.94}$$

At $\Delta x \to 0$, the length of the infinitely small line segment ds becomes

$$ds = \sqrt{(dx)^2 + (dy)^2} = \sqrt{1 + \left(\frac{dy}{dx}\right)^2}\, dx \tag{4.95}$$

We now find a simpler mapping: t is simply x, and vector $\mathbf{r}$ is

$$\mathbf{r} = \begin{bmatrix} x \\ y(x) \end{bmatrix} \tag{4.96}$$

The L2 norm of $\mathbf{r}$ is

$$\|\mathbf{r}'(x)\| = \sqrt{\left(\frac{dx}{dx}\right)^2 + \left(\frac{dy}{dx}\right)^2} = \sqrt{1 + \left(\frac{dy}{dx}\right)^2} \tag{4.97}$$

Equation (4.92) becomes

$$I = \int_C f(x, y)ds = \int_{x_a}^{x_b} f(x, y(x)) \underbrace{\sqrt{1 + \left(\frac{dy}{dx}\right)^2}\, dx}_{\text{length scalar}} \tag{4.98}$$

where x_a and x_b are the x-coordinates at points a and b on C. The formula for computing the length of the curve l_s between a and b becomes

$$l_s = \int_C ds = \int_{x_a}^{x_b} \sqrt{1 + \left(\frac{dy}{dx}\right)^2}\, dx \tag{4.99}$$

It also simplifies to a standard 1D integral. This integrand may appear simple but can pose challenges in finding an antiderivative function, especially due to squared derivatives in the square root.

Let's consider some examples: first, a simple $y(x)$ that is a linear function.

4.14.2 *Example: Length of an inclined straight line*

We can now use Eq. (4.99) to compute the length of any given curve defined by $y(x)$. Let's employ SymPy for both indefinite and definite integrals to accomplish this. First, consider a simple straight line defined by a linear function:

```
1  # Use indefinite integral:
2  x, k, c, C = symbols('x, k, c, C', real=True)
3  y = k*x  + c                          # a straight line
4  ls = integrate(sp.sqrt(1+(y.diff(x))**2), x) + C   # integral constant C
5  display(Math(f" \\text{{Length of a straight line formula: }} \
6                          {sp.latex(ls)}"))
```

Length of a straight line formula: $C + x\sqrt{k^2 + 1}$

```
1  # Set parameters for a specific line:
2  k_ = 1                                # Slope of the line
3  c_ = 0                          # The intersection of the line
4  a_ = 0; b_ = 1
5  dict_ = {k:k_, c:c_}
6
7  # at x=a_, length should be 0. Use it to solve for integral constant C:
8  sln_C = sp.solve(ls.subs(dict_).subs(x,a_), C)
9  ls_ax = ls.subs(dict_).subs(C,sln_C[0])        # Length in [a, x]
10 display(Math(f" \\text{{formula for computing the length at $x$: }} \
11                          {sp.latex(ls_ax)}"))
```

Formula for computing the length at x: $\sqrt{2}x$

This formula is correct: The length of a straight line inclined at 45 degrees is $\sqrt{2}$ times x. By setting $x = b$, we can determine the length within the interval $[a, b]$:

```
1  ls_ab = ls_ax.subs(x,b_)              # compute the length in [a, b]
2  print(f"The length of a straight line in [{a_},{b_}] = {ls_ab}")
```

The length of a straight line in [0,1] = sqrt(2)

On the other hand, we can use the definite integration to directly compute the final length in $[a, b]$:

```
1  ls_ab = integrate(sp.sqrt(1+(y.diff(x))**2), (x, a_, b_)).subs(dict_)
2  print(f"The length of a straight line in [{a_},{b_}]={ls_ab}")
```

```
The length of a straight line in [0,1]=sqrt(2)
```

We obtained the same result using both the indefinite and definite integrals. The indefinite integral provides a formula to compute the length at any location x. The same formula can also be derived using the definite integral, where $b = x_b$, with x_b defined as a symbolic variable. Readers are encouraged to try this.

4.14.3 *Example: Arc length of a power function*

Let consider a curve defined by $y = x^{(3/2)}$. The code is as follows:

```
1  y = x**(3/2)
2  a_ = 0; b_ = 4
3  ls = integrate(sp.sqrt(1+(y.diff(x))**2), x) + C
4  gr.printM(ls,"General formula for computing length of the curve:")
```

```
General formula for computing length of the curve:
```

$$C + 1.0 \left(x^{1.0} + 0.4444 \right)^{\frac{3}{2}}$$

```
1  # at x=a_, the length should be 0:
2  sln_C = sp.solve(ls.subs(dict_).subs(x,a_),C) #solve for int. constant C
3  ls_ax = ls.subs(dict_).subs(C,sln_C[0])              # length in [a, x]
4  gr.printM(ls_ax, 'The formula for computing the length at x:')
```

```
The formula for computing the length at x:
```

$$1.0 \left(x^{1.0} + 0.4444 \right)^{\frac{3}{2}} - 0.2963$$

Letting $x = b$, we shall obtain the length in $[a, b]$ interval:

```
1  ls_ab = ls_ax.subs(x,b_)              # compute the arc length in [a, b]
2  print(f"The length of a straight line in [{a_},{b_}] = {ls_ab}")
```

```
The length of a straight line in [0,4] = 9.07341528938779
```

Now, we use the definite integration to directly compute the arc length in $[a, b]$:

```
1  ls_ab = integrate(sp.sqrt(1+(y.diff(x))**2), (x, a_, b_)).subs(dict_)
2  print(f"The length of a straight line in [{a_},{b_}] = {ls_ab}")
```

```
The length of a straight line in [0,4] = 9.07341528938779
```

The results obtained via infinite and definite integral are the same.

4.14.4 *Example: Arc length of a sine function*

Next, let us compute the arc length of the $\sin(x)$ curve over the interval $[0, 2\pi]$. Since it is sinusoidal, it exhibits negative values within this interval. The code for this computation is as follows. First, we plot the curve:

```
1  def y(x): return np.sin(x)              # define the function in numpy
2
3  A = 0.0                    # Start of x for the function to be drawn [A, B]
4  B = 2.0*np.pi              # End of x for the function to be drawn [A, B]
5  a_  = A; b_  = B                            # interval [a, b]
6  f0= lambda x: 0.0*x                              # the x-axis
7
8  plot_f_shaded(y, f0, A, B, a_, b_, p_name='sin2pi',shade=False)
9  #plt.show()
```

The curve of $\sin(x)$ in $[0, 2\pi]$ is plotted in Fig. 4.24.

Using Eq. (4.99), the arc length becomes

$$l_s = \int_C ds = \int_a^b \sqrt{1 + \left(\frac{dy}{dx}\right)^2}\, dx = \int_a^b \sqrt{1 + \cos^2 x}\, dx \qquad (4.100)$$

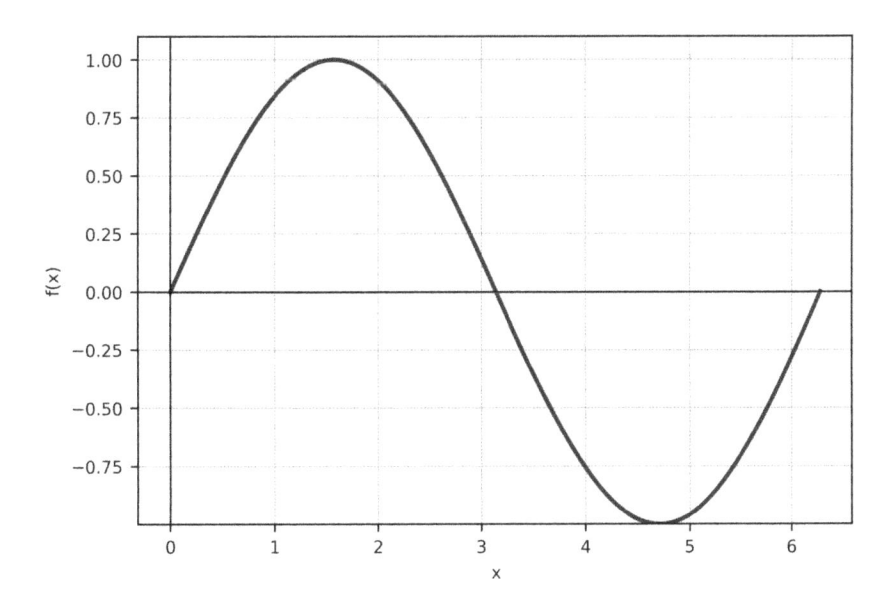

Figure 4.24. The curve of $\sin(x)$ in $[0, 2\pi]$. Its arc length is computed via line integration.

The code is given as follows:

```
1  x = symbols('x')
2  y = sp.sin(x)                                          # sympy function
3  ls_ab = integrate(sp.sqrt(1+(y.diff(x))**2), (x, a_, b_))
4  print(f" In [{a_},{b_:.4f}]={ls_ab}")
```

```
In [0.0,6.2832]=Integral(sqrt(cos(x)**2 + 1), (x, 0.0, 6.28318530717959))
```

This time, analytical integration will not work because no closed-form antiderivative can be found for our integrand, $\sqrt{1 + \cos^2 x}$, which falls under the **elliptic type**. Although it appears simple, no closed-form antiderivative exists. Therefore, we will resort to numerical integration using Scipy. The code is as follows:

```
1  from scipy.integrate import quad
2  def np_y(x):                                           # use numpy
3      return np.sqrt(1+(np.cos(x))**2)                   # y'=cos(x)
4
5  ls_ab = quad(np_y, a_, b_)
6  print(f"Length of sin(x) in [{a_},{b_:.6f}]={np.array(ls_ab)}")
```

```
Length of sin(x) in [0.0,6.283185]=[7.6404e+00 2.6031e-13]
```

The result is correct with a numerical error of 2.6031×10^{-13}. The numerical computation takes only a few milliseconds. This example shows that the arc length measure is not affected, regardless of whether or not the function has negative values. Note that the signed area under this function, however, is zero:

```
1  A = integrate(y, (x, a_, b_))
2  print(f" The area under sin(x) interval [{a_},{b_:.4f}]={A}")
```

```
The area under sin(x) in terval [0.0,6.2832]=0
```

4.14.5 *Example: Arc length of a composite function*

Let us consider a more challenging problem: Computing the arc length of the following composite function,

$$y(x) = \frac{\sin(x^2)}{x+1} \tag{4.101}$$

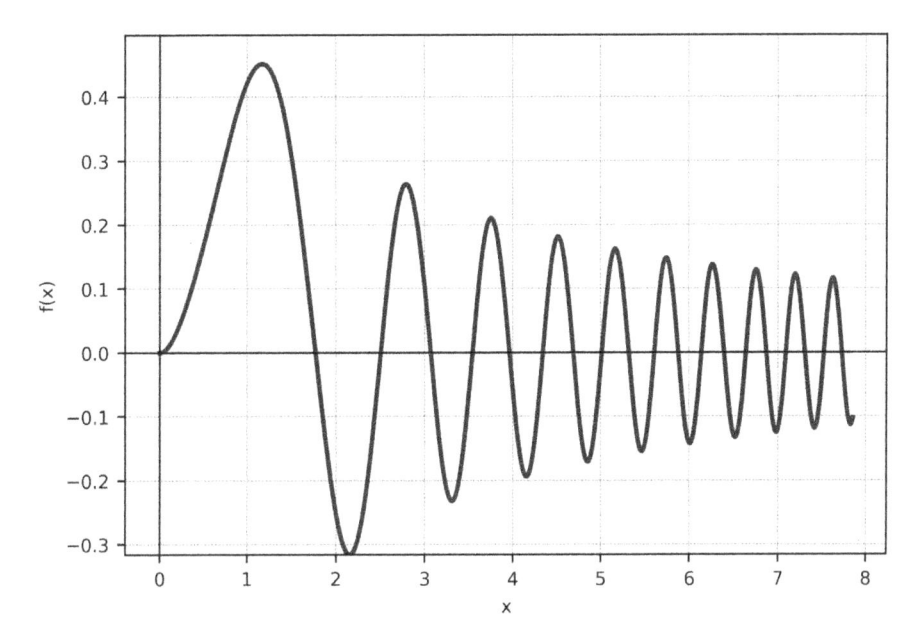

Figure 4.25. The curve of the composite function in $[0, 2.5\pi]$. Its arc length is computed via numerical line integration.

over the interval $[0, 2.5\pi]$. The function is sinusoidal with increasing frequency and decays as x increases. It has negative values within this interval. The code for this computation is as follows:

```
1  def np_y(x): return np.sin(x**2)/(x+1)
2
3  A = 0.0                  # Start of x for the function to be drawn [A, B]
4  B = 2.5*np.pi            # End of x for the function to be drawn [A, B]
5  a_ = A; b_ = B                      # interval [a, b]
6  f0= lambda x: 0.0*x              # Start of y for shading the area
7
8  plot_f_shaded(np_y, f0, A, B, a_, b_, p_name='sin2d5pi',shade=False)
9  #plt.show()
```

The curve of the composite function in $[0, 2.5\pi]$ is plotted in Fig. 4.25.

This time, there is little hope that analytical integration might work. We proceed directly to numerical integration using Scipy. To apply Eq. (4.99), we first need to find the formula for $\frac{dy}{dx}$. The code for this computation is as follows:

```
1  x = symbols('x')
2  f = sp.sin(x**2)/(x+1)                          # The composite function
3  dfdx = f.diff(x)
4  display(Math(f" \\frac{{d}}{{dx}}({sp.latex(f)}) = {sp.latex(dfdx)}"))
```

$$\frac{d}{dx}\left(\frac{\sin\left(x^2\right)}{x+1}\right) = \frac{2x\cos\left(x^2\right)}{x+1} - \frac{\sin\left(x^2\right)}{(x+1)^2}$$

```
1  from scipy.integrate import quad
2
3  def np_ds(x):                                    # define the function in numpy
4      np_dydx = 2*x*np.cos(x**2)/(x + 1) - np.sin(x**2)/(x + 1)**2
5      return np.sqrt(1+np_dydx**2)
6
7  ls_ab = quad(np_ds, a_, b_)
8  print(f"Length of sin(x) in [{a_},{b_:.4f}]={np.array(ls_ab)}")
```

```
Length of sin(x) in [0.0,7.8540]=[1.1077e+01 1.4760e-07]
```

The result is correct with numerical error at 1.476×10^{-07}. The area under $y(x)$ is very small because of the oscillation resulting in area cancellation:

```
1  A = quad(np_y, a_, b_)
2  print(f" The area under y(x) in [{a_},{b_:.4f}]={np.array(A)}")
```

```
The area under y(x) in [0.0,7.8540]=[3.3922e-01 3.6093e-09]
```

4.14.6 *Example: Line integration along a circular*

Consider a circular curve with radius R. This example computes the arc length of the curve and the area between a given function and the circular curve. Assume that the function is given as

$$f(x, y) = x^2 + y^2 \tag{4.102}$$

Solution: Using an angular parameter θ, and Cartesian–polar coordinate transformation, a circle can be expressed as

$$x = R\cos\theta, \quad y = R\sin\theta \tag{4.103}$$

In this case, the mapping is found as follows: t is θ, and vector $\mathbf{r}$ is

$$\mathbf{r} = \begin{bmatrix} R\cos\theta \\ R\sin\theta \end{bmatrix} \tag{4.104}$$

The L2 norm of $\mathbf{r}'$ is

$$|\mathbf{r}'(t)| = \sqrt{(-R\sin\theta)^2 + (R\cos\theta)^2} = R \tag{4.105}$$

Equation (4.92) becomes

$$I = \int_{\mathcal{C}} f(x, y)ds = \int_{\theta_a}^{\theta_b} f(R\cos\theta, R\sin\theta)R d\theta \qquad (4.106)$$

where θ_a and θ_b are the θ-coordinates at points a and b on the circle. The formula for computing the length of the curve l_s becomes

$$l_s = \int_{\mathcal{C}} ds = \int_{\theta_a}^{\theta_b} R d\theta \qquad (4.107)$$

This time, the length scalar is R.

The circumferential length of the circle is known as $2\pi R$. Our task is to compute the length of the circle via line integration. The code is as follows:

```
1  R = symbols('R ', positive=True)
2  θ = symbols('θ ', real=True)
3  x_ = R*sp.cos(θ)                          # Cartesian-polar transformation
4  y_ = R*sp.sin(θ)                    # center of the circle is at (x,y)=(0,0)
5
6  θa = 0; θb = 2*sp.pi
7  ds = sp.sqrt(x_.diff(θ)**2+y_.diff(θ)**2)      # the differential length
8  ls = sp.integrate(ds, (θ, θa, θb))            # The length of the circle
9  #ls = sp.integrate(R, (θ, 0, 2*sp.pi))        # Or simply use this
10 display(Math(f" \\text{{Arc length}} = {sp.latex(ls)}"))
```

Arc length $= 2\pi R$

We obtained the expected results. One can easily compute the path length when θ completes multiple circles:

```
1  n = 5                                                   # n circles
2  ls = sp.integrate(ds, (θ, 0, n*2*sp.pi))
3  display(Math(f" {n} \\text{{-circle arc length}} = {sp.latex(ls)}"))
```

5-circle arc length $= 10\pi R$

The area between the function $f(x, y)$ and the circular curve can be computed using the following:

```
1  x, y = symbols('x, y', real=True)
2  f = x**2 + y**2                        # if f=x+y (anti-symmetric); A = 0
3  A = sp.integrate(f.subs({x:x_, y:y_})*R, (θ, θa, θb)).simplify()
4  display(Math(f" ∫({sp.latex(f)})R dx = {sp.latex(A)}"))
```

$$\int (x^2 + y^2)R d\theta = 2\pi R^3$$

One can also easily compute the area under the circle placed at $z = 1$, which gives $f(x, y) = 1$:

```
1  f = 1
2  A = sp.integrate(f*R, (θ, θa, θb)).simplify()
3  display(Math(f" ∫({sp.latex(f)})R dθ = {sp.latex(A)}"))
```

$$\int (1) R d\theta = 2\pi R$$

4.14.7 *Example: Circumference and area of an ellipse*

Consider an ellipse defined as

$$\frac{x^2}{a} + \frac{y^2}{b} = 1 \tag{4.108}$$

where a is the major radius and b is the minor radius. Compute the circumference and area of the ellipse, assuming $a = 2$, and $b = 1$.

Solution: The curve of an ellipse can be parameterized using

$$x = a\cos(\theta); \quad y = b\sin(\theta) \tag{4.109}$$

The computation is the same as that for the circle. The code is as follows:

```
1  a, b = symbols('a, b', positive=True)
2  θ = symbols('θ', real=True)
3  x_ = a*sp.cos(θ)                          # parameterization for ellipse
4  y_ = b*sp.sin(θ)
5
6  ds = sp.sqrt(x_.diff(θ)**2+y_.diff(θ)**2)  # Differential circumference
7  ls = sp.integrate(ds, (θ, 0, 2*sp.pi))              # The circumference
8  display(Math(f" \\text{{Ellipse circumference}} = {sp.latex(ls)}"))
```

$$\text{Ellipse circumference} = \int_0^{2\pi} \sqrt{a^2 \sin^2(\theta) + b^2 \cos^2(\theta)} \, d\theta$$

This integral simplifies to the elliptic integral of the second kind. Despite its apparent simplicity, it does not have a closed-form antiderivative. Therefore, we need to resort to numerical integration to obtain the solution.

```
1  a_ = 2; b_ = 1
2  dict_ = {a:a_, b:b_}
3  np_ds = lambdify(θ, ds.subs(dict_), 'numpy')          # convert to numpy
4  ls = si.quad(np_ds, 0, 2*np.pi)
5  print(f"Circumference of ellipse [a={a_},b={b_}]={np.array(ls)}")
6
7  # Use the Ramanujan's second approximation:
8  h = ((a-b)/(a+b))**2
9  ls_R = (sp.pi*(a+b)*(1+3*h/(10+sqrt(4-3*h)))).subs(dict_).evalf(12)
10 print(f"Ramanujan's second approximation = {ls_R}")
11 print(f"The simplest estimation = {(sp.pi*(a_+b_)).evalf(6)}")
```

```
Circumference of ellipse [a=2,b=1]=[9.6884e+00 3.4673e-10]
Ramanujan's second approximation = 9.68844821613
The simplest estimation = 9.42478
```

Our numerical solution has an error of 3.467×10^{-10}. For comparison, we used Ramanujan's second approximation, which achieves accuracy up to eight significant digits with an h^5 order accuracy.

Interestingly, the simplest estimation of $\pi(a + b)$ (mimicking the formula for circles) has less than 3% error in this case.

The area for the ellipse is much easier to compute. We obtained it symbolically, providing the formula for computing the area:

```
1  x= symbols('x', real=True)
2  y = b*sp.sqrt(1-(x/a)**2)
3  A = 2*sp.integrate(y, (x, -a, a))
4  display(Math(f" \\text{{Formula of ellipse area}} = {sp.latex(A)}"))
```

Formula of ellipse area $= \pi ab$

4.14.8 *Parameterized curves in 3D*

In many applications, 3D curves can be parameterized using a common parameter. For instance, when describing the trajectory of a spacecraft, we introduce a parameter, typically time t, and express the spacecraft's coordinates as $(x(t), y(t), z(t))$. Here, t serves as a physical parameter that determines the spacecraft's position, reducing the problem to 1D along the parameter.

By parameterizing, we gain additional insights into the spacecraft, such as its speed and acceleration, obtained by differentiating its positional variables with respect to t.

Consider an arc length increment Δs at a point on the 3D curve. Through parameterization and the application of the chain rule of differentiation, we derive

$$\Delta x = \frac{dx}{dt}\Delta t, \quad \Delta y = \frac{dy}{dt}\Delta t, \quad \Delta z = \frac{dz}{dt}\Delta t \tag{4.110}$$

This was also given in Eq. (2.70). The arc length increment is computed using

$$\Delta s = \sqrt{\left(\frac{dx}{dt}\right)^2 + \left(\frac{dy}{dt}\right)^2 + \left(\frac{dz}{dt}\right)^2}\,\Delta t \tag{4.111}$$

At $\Delta t \to 0$, $\Delta x \to 0, \Delta y \to 0$ and $\Delta z \to 0$, the arc length increment becomes ds. The position vector $\mathbf{r}$ for the mapping can be defined as

$$\mathbf{r} = \begin{bmatrix} x(t) \\ y(t) \\ z(t) \end{bmatrix} \tag{4.112}$$

The L2 norm of $\mathbf{r}'$ is

$$|\mathbf{r}'(t)| = \sqrt{\left(\frac{dx}{dt}\right)^2 + \left(\frac{dy}{dt}\right)^2 + \left(\frac{dz}{dt}\right)^2} \tag{4.113}$$

Equation (4.92) becomes

$$I = \int_C f(x, y, z)ds = \int_{t_a}^{t_b} f(x, y, z)\frac{ds}{dt}dt$$

$$= \int_{t_a}^{t_b} f\left(x(t), y(t), z(t)\right) \underbrace{\sqrt{\left(\frac{dx}{dt}\right)^2 + \left(\frac{dy}{dt}\right)^2 + \left(\frac{dz}{dt}\right)^2}}_{\text{length scalar}} dt \tag{4.114}$$

This is a 1D integral along t. The formula for computing the length of the curve l_s becomes

$$l_s = \int_C ds = \int_{t_a}^{t_b} \frac{ds}{dt}dt = \int_{t_a}^{t_b} \sqrt{\left(\frac{dx}{dt}\right)^2 + \left(\frac{dy}{dt}\right)^2 + \left(\frac{dz}{dt}\right)^2}\,dt \tag{4.115}$$

Let us look at an example.

4.14.9 *Case study: A traveling spacecraft in 3D space*

Let us use a spherical coordinate system: (r, θ, φ), where r is the radial, θ is the polar angle, and φ is the azimuthal angle, as shown in Fig. 4.26.

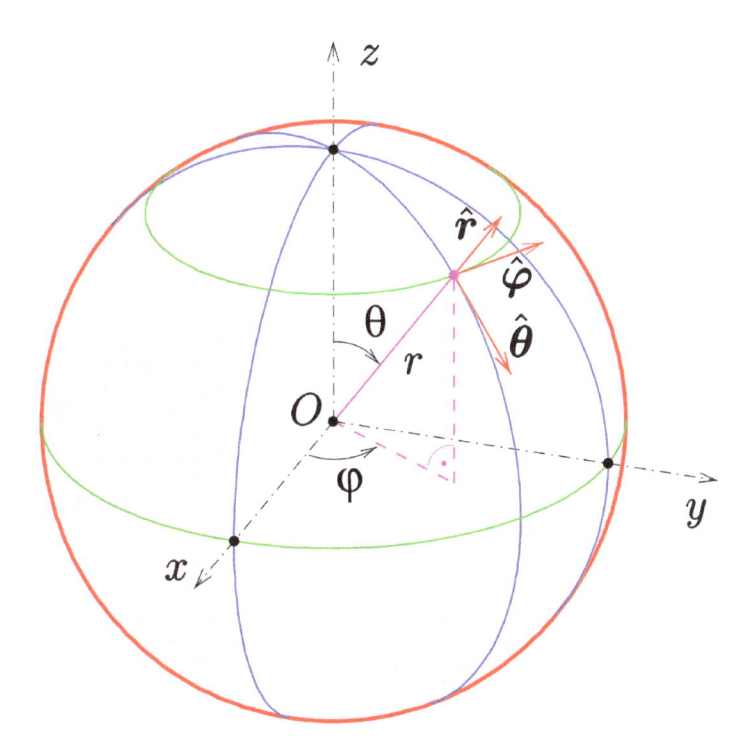

Figure 4.26. Spherical coordinates with radial distance r, polar angle θ, and azimuthal angle φ.

Source: Images from en.wikipedia by Ag2gaeh under the CC BY-SA 4.0 license.

All these three coordinates can be parameterized using parameter time t. The coordinates of a spacecraft can be expressed as

$$x = r(t) \sin\big(\theta(t)\big) \cos\big(\varphi(t)\big)$$
$$y = r(t) \sin\big(\theta(t)\big) \sin\big(\varphi(t)\big)$$
$$z = r(t) \cos\big(\theta(t)\big) \tag{4.116}$$

Assume that r, θ and φ all change with time in the following manner:

$$r(t) = c_{r1}t + c_{r2}t^2$$
$$\theta(t) = c_\theta t$$
$$\varphi(t) = c_\varphi t \tag{4.117}$$

where c_{r1}, c_{r2}, c_θ and c_φ are all given coefficients.

Our task is to derive formulas for the velocity and acceleration of the spacecraft as functions of time t as well as calculate the distance traveled by the spacecraft in the time interval $[0.0, 5.0]$. The code to accomplish these tasks is as follows:

```
1  t = symbols('t', positive=True)
2  cr2, cr1, cθ, cφ = symbols('c_r2, c_r1, c_θ, c_φ',real=True)
3  r = cr2*t**2 + cr1*t
4  θ = cθ*t
5  φ = cφ*t
6
7  dict_ = {cr2:1, cr1:1, cθ:0, cφ:0}        # coefficients, vertical travel
8  #dict_= {cr2:0, cr1:1, cθ:0, cφ:0}        # contant velocity for testing
9
10 x = r*sp.sin(θ)*sp.cos(φ)
11 y = r*sp.sin(θ)*sp.sin(φ)
12 z = r*sp.cos(θ)
13
14 dsdx = x.diff(t).subs(dict_)              # velocity in the x-direction
15 dsdy = y.diff(t).subs(dict_)              # velocity in the y-direction
16 dsdz = z.diff(t).subs(dict_)              # velocity in the z-direction
17 dsdr = sp.sqrt(dsdx**2+dsdy**2+dsdz**2)
18 print(f'Velocity of the spacecraft in the x-direction = {dsdx}')
19 print(f'Velocity of the spacecraft in the y-direction = {dsdy}')
20 print(f'Velocity of the spacecraft in the z-direction = {dsdz}')
21 print(f'Absolute velocity of the spacecraft = {dsdr}')
```

```
Velocity of the spacecraft in the x-direction = 0
Velocity of the spacecraft in the y-direction = 0
Velocity of the spacecraft in the z-direction = 2*t + 1
Absolute velocity of the spacecraft = 2*t + 1
```

```
1  d2sdz2 = dsdz.diff(t)
2  gr.printM(d2sdz2, 'Acceleration of the spacecraft in the x-direction:')
```

```
Acceleration of the spacecraft in the x-direction:

2
```

Let us compute the distance traveled by the spacecraft in the time interval $[0, 1]$. The integral is of elliptic type, so we use numerical integration:

```
1  a_ = 0.0; b_ = 5.0
2
3  ds = sp.sqrt(x.diff(t)**2+y.diff(t)**2+z.diff(t)**2)
4  #ls = sp.integrate(ds, (t, a_, b_))        # not analytically integrable
5
6  np_ds = lambdify(t, ds.subs(dict_), 'numpy')
7  ls = np.array(si.quad(np_ds, a_, b_))
8  print(f"Distance of the spacecraft traveled in [{a_},{b_:.1f}]={ls}")
```

```
Distance of the spacecraft traveled in [0.0,5.0]=[3.0000e+01 3.3307e-13]
```

4.15 Remarks

This chapter thoroughly examines 1D function integrals, covering theory, properties, techniques, and numerous examples with Python codes. We conclude this chapter with the following key points:

1. A definite integral of a function sums small areas under the function, yielding the signed area between the function and the x-axis over a given interval. Integrability requires the convergence of the Riemann sum to the signed area. An indefinite integral is the antiderivative of the integrand, which is unique up to an additive constant.
2. Not all functions are integrable. Generally, a bounded, continuous function with a finite number of discontinuities within a closed interval is Riemann integrable. Functions may still be integrable if they are unbounded at singular points inside the interval, provided their growth rate is slower than linear; otherwise, they are unlikely to be integrable.
3. The k-limit approach is useful for handling integrands with jumps where the gradient is infinite, the ε-limit approach is useful for integrands with singularities, and the b-limit approach is useful for integrals with infinite intervals.
4. Finding an antiderivative of a function is essentially an inverse problem and can be challenging. While antiderivatives exist for many functions, they remain undiscovered for others. Once found, the definite integral is simply the difference of the antiderivative values at the interval endpoints. The derivative of the antiderivative yields the original integrand.
5. Integrating polynomials is straightforward. Functions can be approximated using polynomials, facilitating easy integration and anti-differentiation. Similar approximations using other basis functions are also feasible.
6. Line integrals are valuable and are approached through parameterization, yielding a scalar length. Only one parameter is necessary because line integrals are inherently one-dimensional.

The following chapter extends these discussions to integrals of nD functions.

References

[1] Y.C. Fan, *Advanced Mathematics*, 1979.

[2] L.X. Yang and B.Y. Bi, *Analytic Mathematics Practices*, Boris Demidovich, 2005.

[3] Gilbert Strang, *Calculus*, 1991. Available online (http://ocw.mit.edu/OcwWeb/resources/RES-18-001Spring-2005/ResourceHome/index.htm).

[4] G.R. Liu, *Numbers and Functions: Theory, Formulation, and Python Codes*, World Scientific, 2024.

[5] G.R. Liu and X. Han, *Computational Inverse Techniques in Nondestructive Evaluation*, Taylor and Francis Group, New York, 2003.

[6] G.R. Liu and Siu Sin Quek, *The Finite Element Method: A Practical Course*, Butterworth-Heinemann, 2013.

[7] G.R. Liu and T.T. Nguyen, *Smoothed Finite Element Methods*, Taylor and Francis Group, New York, 2010.

[8] G.R. Liu, *Mesh Free Methods: Moving Beyond the Finite Element Method*, Taylor and Francis Group, New York, 2010.

[9] G.R. Liu and Z.C. Xi, *Elastic Waves in Anisotropic Laminates*, 2001.

[10] G.R. Liu, K.Y. Lam, and T. Ohyoshi, *A Technique for Analyzing Elastodynamic Responses of Anisotropic Laminated Plates to Line Loads*, Composites Part B: Engineering, Vol. 28, number 5, pp. 667–677, 1997. Available online (https://www.sciencedirect.com/science/article/pii/S1359836896000807).

Chapter 5

Integration of Multi-Dimensional Functions

```
 1  # Place cursor in this cell, and press Ctrl+Enter to import dependences.
 2  import sys                        # For accessing the computer system
 3  sys.path.append('../grbin/')   # Add in the path to your system
 4
 5  from commonImports import *        # Import dependences from '../grbin/'
 6  import grcodes as gr                # Import the module of the author
 7  importlib.reload(gr)                # When grcodes is modified, reload it
 8
 9  init_printing(use_unicode=True)       # For latex-like quality printing
10  np.set_printoptions(precision=4,suppress=True,
11                      formatter={'float_kind': '{:.4e}'.format})
```

This chapter extends the discussion of integrals from one-dimensional (1D) functions to multi-dimensional (nD) functions. We focus more on the issues resulting from this dimensional extension. As mentioned in Chapter 3, any problem in nD becomes more complex in terms of formulation and calculation. The key idea is often to convert it to an eventual 1D problem, which we are familiar with solving. Readers should keep this in mind while studying this chapter to avoid confusion caused by the complex formulations.

This chapter is written with reference to textbooks in Refs. [1–4]. Many of the proofs can be found in the first three textbooks. Wikipedia pages, particularly those on integral (https://en.wikipedia.org/wiki/Integral), serve as valuable additional references.

Both NumPy and SymPy are used in the development of code for the demonstration examples. "Discussions" with ChatGPT, Gemini, and Bing have also greatly helped in coding and in the preparation of this chapter.

5.1 General definition

Integration of nD functions is a dimensional extension from the Riemann integration for 1D functions discussed in Chapter 4. Consider an nD integrand function $f(\mathbf{x})$ defined in n-dimension Cartesian coordinate space $\mathbf{x} = (x_1, x_2, \ldots, x_n)$. The **definite integral** of $f(\mathbf{x})$ is expressed as

$$I = \int_{\mathcal{D}^n} f(\mathbf{x})d\mathbf{x} = \int \cdots \int f(x_1, x_2, \ldots, x_n)\, dx_1 \cdots dx_n$$

$$= \lim_{\substack{\mathbf{x}_i^* \in \Delta \mathbf{x}_i \\ m \to \infty \\ \text{all } \Delta \mathbf{x}_i \to 0}} \left(\sum_{i=1}^{m} f(\mathbf{x}_i^*)(\Delta \mathbf{x}_i) \right) \tag{5.1}$$

where $\mathcal{D}^n$ is the domain of integration and each $\Delta \mathbf{x}_i$ is a small sub-domain of $\mathcal{D}^n$. The union of these sub-domains forms $\mathcal{D}^n$ without any overlapping or gaps. In general, $\Delta \mathbf{x}_i$ can be different in shape, but all have a finite aspect ratio.

When $n = 2$, the definite integral is often called **double integral**. It gives the signed volume under the integrand function $f(\mathbf{x})$. When $n \geq 3$, it is the hyper-volume under the integrand. If the limit given in Eq. (5.1) converges to the signed hyper-volume under the function $f(\mathbf{x})$, it is said to be integrable.

A schematic drawing for 2D case is shown in Fig. 5.1.

When $n = 3$, the definite integral is often call **triple integral**, which is more difficult to visualize.

Since nD integration relies on repeated summation, the fundamental theories, rules, and properties developed for 1D integrals (in Chapter 4) generally apply to nD functions as well. The key idea is that we must hold constant for all other independent variables while examining a single variable. Therefore, we do not repeat those discussions here. Instead, we leverage these existing concepts to facilitate our study of nD functions. Naturally, this can be done significantly easier if the nD domain has a regular (rectangular) shape.

5.2 Integration in rectangular domain

Consider an integrand function $f(x, y)$ defined in 2D Cartesian coordinate space (x, y) in a rectangular domain, $\mathcal{D} = [x_L, x_U] \times [y_L, y_U]$, and it is continuous with only a finite number of jumping points in $\mathcal{D}$. The definite double

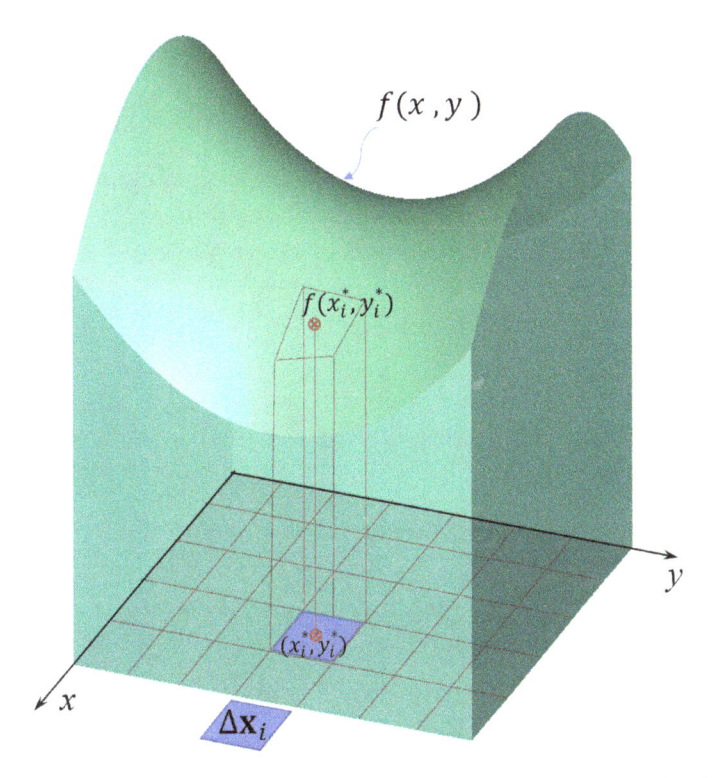

Figure 5.1. Schematic drawing of definite integration of a function $f(x, y)$ over a 2D integration domain. It gives the volume under the function.
Source: Images modified based on the one from en.wikipedia Wikimedia Commons by Oleg Alexandrov to the public domain.

integral can now be expressed as

$$I = \int_{\mathcal{D}^2} f(\mathbf{x}) d\mathbf{x} = \int_{[x_L, x_U] \times [y_L, y_U]} f(\mathbf{x}) d\mathbf{x}$$

$$= \int_{x_L}^{x_U} \left[\int_{y_L}^{y_U} f(x, y) dy \right] dx$$

$$\underset{\text{or}}{=} \int_{y_L}^{y_U} \left[\int_{x_L}^{x_U} f(x, y) dx \right] dy \qquad (5.2)$$

Here, we use the following notation:

$$x_L: \text{Lower end of the interval along } x\text{-axis}$$

$$x_U: \text{Upper end of the interval along } x\text{-axis}$$

$$y_L: \text{Lower end of the interval along } y\text{-axis}$$

$$y_U: \text{Upper end of the interval along } y\text{-axis} \qquad (5.3)$$

Subscripts L and U denote the lower and upper ends of the integration limits, respectively.

Equation (5.2) allows us to perform the integration in two steps:

1. For a fixed value of x within the interval $[x_L, x_U]$, integrate $f(x, y)$ over the interval $[y_L, y_U]$. This yields an intermediate antiderivative, $F(x)$, which is a function of x only. Here, we assume $f(x, y)$ is integrable with respect to y for any fixed value of x. Since integration acts as a smoothing operator (as discussed earlier), the intermediate antiderivative $F(x)$ is expected to be smoother.

2. Assuming that $F(x)$ is integrable with respect to x, integrate $F(x)$ over the interval $[x_L, x_U]$. This results in the final integral value, I, which represents the signed volume under the surface defined by $f(x, y)$.

Equation (5.2) essentially extends the techniques and code used for 1D integrals to 2D integrations by performing two nested 1D integrations. The order of these integrations can sometimes be swapped depending on the function's behavior and computational ease.

Let us look at some examples using Python.

5.2.1 *Example: 2D integral of a linear function*

In this example, we consider a simple linear function in 2D, which allows us to reveal the detailed integration procedure:

```
1  x, y, a, b, c = symbols('x, y, a, b, c', real=True)  # define variables
2
3  f = a*x+b*y+c                         # linear function to be integrated
4  xL, xU = 0, 1.;  yL, yU = 0, 1.            # for 2D integral domain
5  # xL: x lower end, xU: x upper end
6  # yL: y lower end, yU: y upper end
7  A_base = (xU-xL)*(yU-yL)                  # area of the integral domain
8
9  Fx = integrate(f, (y, yL, yU)) # integral along y over [yL,yU], x fixed
10 display(Math(f" F(x) =∫_0^1({sp.latex(f)})dy = {sp.latex(Fx)}"))
11
12 I = integrate(Fx, (x, xL, xU))       # integral along x over [xL, xU]
13 display(Math(f" I = ∫_0^1({sp.latex(Fx)})dx = {sp.latex(I)}"))
14 print(f' For the constant f, I = {I.subs({a:.0, b:.0, c:1.})}')
```

$$F(x) = \int_0^1 (ax + by + c)dy = 1.0ax + 0.5b + 1.0c$$

$$I = \int_0^1 (1.0ax + 0.5b + 1.0c)dx = 0.5a + 0.5b + 1.0c$$

```
For the constant f, I = 1.000000000000000
```

When the function is a constant with $c = 1$, and $a = b = 0$, we obtain $I = 1$, which is the volume under the surface defined by the constant function. If it is a linear function, we have the following:

```
1  dict_ = {a:3., b:2., c:1.}                          # a linear function
2  print(f' For the linear f(x,y), I = {I.subs(dict_)}')
3  print(f' f@center*A_base        = {f.subs(dict_|{x:.5, y:.5})}')
```

```
For the linear f(x,y), I = 3.50000000000000
f@center*A_base        = 3.50000000000000
```

When the function is linear in the integral domain, the volume obtained via integration is the same as that obtained using the function value at the centroid of the domain multiplied with the area of the domain. This is true for all linear functions. Intuitively, this can be understood because linear functions have a constant rate of change throughout the domain. The centroid represents the "average" location within the domain. By integrating a linear function, we essentially calculate the average value of the function across the entire domain. Multiplying this average value (function value at the centroid) by the domain area gives the total volume under the linear surface. This discussion reveals the essence of integration.

5.2.2 *Example: Volume under a composite function*

Let us compute the volume under the surface of a composite function over a square domain. The function is defined as

$$f(x, y) = ax^3 \cos(y) + bx^2 \exp(x \mid y) + cy \sin(x) \tag{5.4}$$

where a, b, and c are given constants:

```
1  x, y = symbols('x, y', real=True)                    # variables
2  a, b, c = symbols('a, b, c', positive=True)          # given constants
3
4  f = a*x**3*cos(y) + b*x**2*exp(x+y) + c*y*sin(x)    # f to be integrated
5  xL, xU = -1, 1;  yL, yU = -1, 1                      # for 2D integral domain
6
7  Fx = integrate(f, (y, yL,yU))                        #  along y over [yL,yU]
8  display(Math(f" F(x) =∫_{{-1}}^1({sp.latex(f)})dy = {sp.latex(Fx)}"))
9  I = integrate(Fx, (x, xL, xU))        # integral along x over [xL, xU]
10 display(Math(f" I = ∫_{{-1}}^1({sp.latex(Fx)})dx = {sp.latex(I)}"))
```

$$F(x) = \int_{-1}^{1} (ax^3 \cos(y) + bx^2 e^{x+y} + cy \sin(x)) dy = 2ax^3 \sin(1) - bx^2 e^{x-1}$$
$$+ bx^2 e^{x+1}$$

$$I = \int_{-1}^{1} (2ax^3 \sin(1) - bx^2 e^{x-1} + bx^2 e^{x+1}) dx = -b + be^2 - \frac{-5b + 5be^2}{e^2}$$

The integration result, I, is the formula for computing the volume under the surface of the composite function. Note that the formula does not depend on a and c. This is because of the symmetric property of integrals discussed in Section 4.3:

- The corresponding parts of the integrand are odd functions of x.
- The interval along the x-axis is symmetric.

For a given set of constants, the value of the volume is computed using the following:

```
1  dict_ = {b:2}                              # a composite function
2  print(f' Volume under the function, I = {I.subs(dict_).evalf(8)}')
```

```
Volume under the function, I = 4.1314650
```

5.3 Coordinate transformation

5.3.1 *Issues for integration over irregular domains*

As the previous examples demonstrate, definite integrals are straightforward to solve when the domain has a regular shape like a rectangle. However, in science and engineering, we often encounter problems with irregular (non-rectangular) domains, $\mathcal{D}$, in Cartesian coordinates (x, y, z). In these cases, unlike Eq. (5.2), we lack convenient formulas for the integral limits within the inner layers. This is because the limits become functions of the coordinates themselves.

To overcome this challenge, the most effective approach is to perform a **coordinate transformation**. This process, also called **mapping** (discussed in Chapter 3 for differentiation), transforms the irregular domain into a regular one, often a square in 2D or a cube in 3D. For example, when dealing with circular or spherical domains, using polar or spherical coordinates can significantly simplify the integration. The key is to choose an appropriate coordinate transformation that allows integration over the preferred coordinate system so that the nD integral becomes n-layer 1D integrals.

The formulations and techniques for coordinate transformation in integration are essentially the same as those discussed in Section 3.10. In integration, however, we also need to compute the determinant of the Jacobian matrix. This determinant acts as a **scaling factor**, accounting for the area or volume of the irregular domain under the transformation. We begin our discussion with 3D domains, as the formulations can then be directly applied to 2D domains as a special case.

5.3.2 *Differential vectors in 3D*

Suppose that the relationships between the Cartesian coordinates (x, y, z) and another coordinate system (ξ, η, ζ) are given in the form of

$$x = x(\xi, \eta, \zeta); \quad y = y(\xi, \eta, \zeta); \quad z = z(\xi, \eta, \zeta) \tag{5.5}$$

Assume that both coordinate systems are orthogonal, and these coordinate functions are differentiable. We thus have

$$dx = \frac{\partial x}{\partial \xi} d\xi + \frac{\partial x}{\partial \eta} d\eta + \frac{\partial x}{\partial \zeta} d\zeta$$

$$dy = \frac{\partial y}{\partial \xi} d\xi + \frac{\partial y}{\partial \eta} d\eta + \frac{\partial y}{\partial \zeta} d\zeta$$

$$dz = \frac{\partial z}{\partial \xi} d\xi + \frac{\partial z}{\partial \eta} d\eta + \frac{\partial z}{\partial \zeta} d\zeta \tag{5.6}$$

These equations can be rewritten in a matrix form, which gives the relationship of the differential vectors in (x, y, z) and (ξ, η, ζ):

$$\begin{bmatrix} dx \\ dy \\ dz \end{bmatrix} = \underbrace{\begin{bmatrix} \frac{\partial x}{\partial \xi} & \frac{\partial x}{\partial \eta} & \frac{\partial x}{\partial \zeta} \\ \frac{\partial y}{\partial \xi} & \frac{\partial y}{\partial \eta} & \frac{\partial y}{\partial \zeta} \\ \frac{\partial z}{\partial \xi} & \frac{\partial z}{\partial \eta} & \frac{\partial z}{\partial \zeta} \end{bmatrix}}_{\mathbf{J}} \begin{bmatrix} d\xi \\ d\eta \\ d\zeta \end{bmatrix} = \mathbf{J} \begin{bmatrix} d\xi \\ d\eta \\ d\zeta \end{bmatrix} \tag{5.7}$$

where the Jacobian matrix $\mathbf{J}$ bridges these two vectors.

5.3.3 *Volume scalar*

In Section 4.14, we encountered the concept of a length scalar when working with 1D line integrals. Similarly, for 3D integrals, a new concept emerges: the **volume scalar**. The determinant of the Jacobian matrix plays a crucial role here. It relates the differential volume (a tiny unit of volume) in the original coordinate system to that in the transformed coordinate system. In essence, the determinant acts as a scaling factor that accounts for this volume change under the transformation.

To derive this relationship, let's define three vectors:

$$d\mathbf{x} = \begin{bmatrix} \frac{\partial x}{\partial \xi} d\xi \\ \frac{\partial x}{\partial \eta} d\eta \\ \frac{\partial x}{\partial \zeta} d\zeta \end{bmatrix} ; \quad d\mathbf{y} = \begin{bmatrix} \frac{\partial y}{\partial \xi} d\xi \\ \frac{\partial y}{\partial \eta} d\eta \\ \frac{\partial y}{\partial \zeta} d\zeta \end{bmatrix} ; \quad d\mathbf{z} = \begin{bmatrix} \frac{\partial z}{\partial \xi} d\xi \\ \frac{\partial z}{\partial \eta} d\eta \\ \frac{\partial z}{\partial \zeta} d\zeta \end{bmatrix} \tag{5.8}$$

These three vectors are schematically shown in Fig. 5.2.

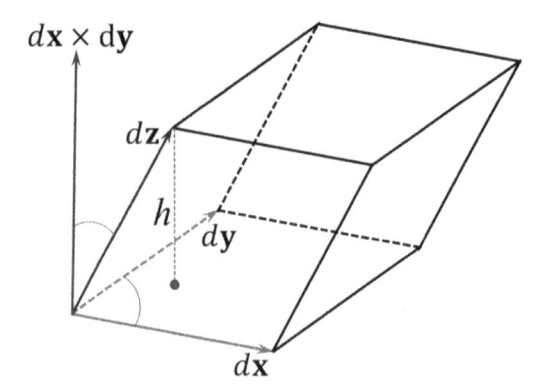

Figure 5.2. Three differential vectors form a differential volume: a parallelepiped.

We can calculate the differential volume, dV, in two steps:

1. **Area of the base:** First, take the cross product of vectors $d\mathbf{x}$ and $d\mathbf{y}$. This results in a new vector perpendicular to the plane formed by $d\mathbf{x}$ and $d\mathbf{y}$. The magnitude (or length) of this new vector represents the area of the base of the parallelepiped, which is analogous to the bottom surface of a parallelogram.
2. **Height of the parallelepiped:** Next, take the dot product of the new vector obtained in step 1 with $d\mathbf{z}$. This incorporates the height of the parallelepiped into the calculation.

The following formula expresses this concept, as detailed in Ref. [5]:

$$dV = (d\mathbf{x} \times d\mathbf{y}) \cdot d\mathbf{z} \tag{5.9}$$

While the derivation is somewhat lengthy, it is conceptually straightforward (we perform it using SymPy in the following section). By substituting Eq. (5.8) into Eq. (5.9) and finally factoring out $d\xi d\eta d\zeta$, we obtain

$$dV = \det(\mathbf{J})\underbrace{d\xi d\eta d\zeta}_{d\mathcal{B}} = \underbrace{\det(\mathbf{J})}_{\text{volume scaling}} d\mathcal{B} \tag{5.10}$$

where $d\mathcal{B} = d\xi d\eta d\zeta$ is the differential volume in the ξ, η, ζ coordinates.

The determinant of the Jacobian matrix is a **volume scalar** that relates the differential volumes between the original and transformed coordinate systems.

If we can find a suitable mapping function (like Eq. (5.5)) that transforms the regular shape (e.g., a cube) domain in the ξ, η, ζ coordinates into the original irregular shape in x, y, z, then the 3D integral defined in Eq. (5.1) becomes

$$I = \int_{\mathcal{D}^3} f(x, y, z)\,dxdydz$$

$$= \int_{\xi_L}^{\xi_U} \int_{\eta_L}^{\eta_U} \int_{\zeta_L}^{\zeta_U} f(\xi, \eta, \zeta) \underbrace{\det(\mathbf{J})}_{\text{volume scalar}} d\zeta d\eta d\zeta \tag{5.11}$$

As we discussed in Chapter 3 for differentiations, Eq. (5.5) can be given using Lagrange nodal shape functions. We can do exactly the same here for integration.

Extending this derivation to even higher dimensions would involve new concepts like the exterior product or tensor algebra, which fall outside the scope of this discussion. Therefore, we focus on 3D integrals for now.

5.3.4 *Python code proof of the volume scalar*

While manually deriving these formulas can be tedious and error-prone, SymPy can be a valuable tool to automate the process and ensure accuracy. Let's leverage SymPy to complete the verification of these relationships:

```
1  # Define the Jacobian matrix with symbolic variables:
2
3  j11, j12, j13 = symbols("j11, j12, j13", real=True)    # elements for J
4  j21, j22, j23 = symbols("j21, j22, j23", real=True)
5  j31, j32, j33 = symbols("j31, j32, j33", real=True)
6
7  J = sp.Matrix([[j11, j12, j13],                        # the Jacobian matrix
8                 [j21, j22, j23],
9                 [j31, j32, j33]])
10 J
```

$$\begin{bmatrix} j_{11} & j_{12} & j_{13} \\ j_{21} & j_{22} & j_{23} \\ j_{31} & j_{32} & j_{33} \end{bmatrix}$$

```
1  # Form these three vectors dx, dx, dx:
2
3  dξ, dη, dζ = symbols("dξ, dη, dζ", real=True)
4  dξηζ = sp.Matrix([dξ, dη, dζ])
5  dx = sp.Matrix([J[0,j]*dξηζ[j] for j in range(len(dξηζ))])
6  dy = sp.Matrix([J[1,j]*dξηζ[j] for j in range(len(dξηζ))])
7  dz = sp.Matrix([J[2,j]*dξηζ[j] for j in range(len(dξηζ))])
8  [dx, dy, dz]
```

$$\begin{bmatrix} \begin{bmatrix} d\xi j_{11} \\ d\eta j_{12} \\ d\zeta j_{13} \end{bmatrix}, \begin{bmatrix} d\xi j_{21} \\ d\eta j_{22} \\ d\zeta j_{23} \end{bmatrix}, \begin{bmatrix} d\xi j_{31} \\ d\eta j_{32} \\ d\zeta j_{33} \end{bmatrix} \end{bmatrix}$$

To perform cross product, we write the following code (see Chapter 2 of Ref. [5]) for later use:

```
1  def cross2vectors(a, b, sympy=True):
2      '''Cross-product of two 3D vectors: axb, returns h pependicular to
3          plane formed by a and b.
4          if sympy=True: a, b, and h are all sympy Matrixes.
5          otherwise: all in numpy arrays
6      '''
7      h1 = (a[1]*b[2] - a[2]*b[1])
8      h2 = (a[2]*b[0] - a[0]*b[2])
9      h3 = (a[0]*b[1] - a[1]*b[0])
10     if sympy:
11         h = sp.Matrix([h1, h2, h3])
12     else:
13         h = np.array([h1, h2, h3])
14     return h
```

Let us now apply it to our problem:

```
1  # Cross-product of vector dx with dy:
2  dx_cross_dy = cross2vectors(dx, dy)
3  dx_cross_dy
```

$$
\begin{bmatrix}
d\zeta\, d\eta\, j_{12} j_{23} - d\zeta\, d\eta\, j_{13} j_{22} \\
-d\zeta\, d\xi\, j_{11} j_{23} + d\zeta\, d\xi\, j_{13} j_{21} \\
d\eta\, d\xi\, j_{11} j_{22} - d\eta\, d\xi\, j_{12} j_{21}
\end{bmatrix}
$$

```
1  # Dot-product of (dxxdy)· dz, which gives dV:
2  dx_cross_dy.dot(dz).simplify()                                    # dV
```

$$
d\zeta\, d\eta\, d\xi\, (j_{31} (j_{12} j_{23} - j_{13} j_{22}) - j_{32} (j_{11} j_{23} - j_{13} j_{21}) + j_{33} (j_{11} j_{22} - j_{12} j_{21}))
$$

```
1  # Get the coefficient of dζdηdξ:
2  J_dxCdydz = dx_cross_dy.dot(dz).simplify().args[3]
3  J_dxCdydz
```

$$
j_{31} (j_{12} j_{23} - j_{13} j_{22}) - j_{32} (j_{11} j_{23} - j_{13} j_{21}) + j_{33} (j_{11} j_{22} - j_{12} j_{21})
$$

```
1  # Check if it is det(J):
2  print(f"Is (dxxdy)·dz = |J|dζdηdξ? {J_dxCdydz.expand() == sp.det(J)}")
```

Is $(dx \times dy) \cdot dz$ = $|J| d\zeta d\eta d\xi$? True

This proves that $|\mathbf{J}|$ is the volume scalar in Eq. (5.10).

5.4 Integration in 2D coordinates

Using these formulas for 3D, the corresponding formulas for 2D can be easily derived as a special case.

Assume that a coordinate mapping in 2D is done between the Cartesian coordinates (x, y) and another coordinate system (ξ, η):

$$x = x(\xi, \eta); \quad y = y(\xi, \eta) \tag{5.12}$$

Then, the relationship between the differential areas $dxdy$ and $d\xi d\eta$ is given as follows:

$$dxdy = \underbrace{\begin{vmatrix} \dfrac{\partial x}{\partial \xi} & \dfrac{\partial x}{\partial \eta} \\ \dfrac{\partial y}{\partial \xi} & \dfrac{\partial y}{\partial \eta} \end{vmatrix}}_{|\mathbf{J}|} d\xi d\eta \tag{5.13}$$

where $\mathbf{J}$ is the Jacobian matrix for 2D cases and $|\mathbf{J}|$ is the **area scalar**.

Note that the Jacobian matrix is computed using the differential relations of the two coordinate systems involved. When performing line/curve integrals in Section 4.14, we introduced a length scalar, which is also computed using the differential relations of the two coordinate systems involved, as shown in Eq. (4.93).

5.4.1 *Polar to Cartesian coordinate mapping*

Consider now a mapping from the polar coordinates to the Cartesian coordinates shown in Fig. 5.3.

In this case, (ξ, η) becomes (r, θ), and Eq. (5.5) can be obtained by orthogonal projections [5], as shown in Fig. 5.3. We thus obtain

$$x = r \cos \theta; \quad y = r \sin \theta \tag{5.14}$$

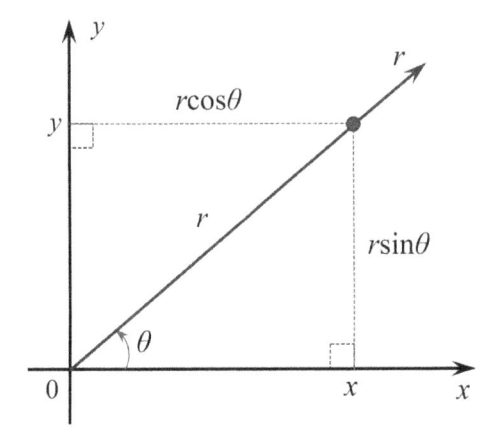

Figure 5.3. Polar and Cartesian coordinates.

The Jacobian matrix becomes

$$\mathbf{J} = \begin{bmatrix} \frac{\partial x}{\partial r} & \frac{\partial x}{\partial \theta} \\ \frac{\partial y}{\partial r} & \frac{\partial y}{\partial \theta} \end{bmatrix} = \begin{bmatrix} \cos\theta & -r\sin\theta \\ \sin\theta & r\cos\theta \end{bmatrix} \tag{5.15}$$

Its determinant is $|\mathbf{J}| = r(\cos^2\theta + \sin^2\theta) = r$. Thus, the area scalar is simply r. Equation (5.13) becomes

$$dxdy = rdrd\theta \tag{5.16}$$

Using Eq. (5.11), the integral can now be carried out in the polar coordinates:

$$I = \int_{\mathcal{D}^2} f(x,y)dxdy = \int_{\xi_L}^{\xi_U} \int_{\eta_L}^{\eta_U} f(\xi,\eta)\det(\mathbf{J})d\zeta d\eta$$

$$= \int_{r_L}^{r_U} r\left[\int_{\theta_L}^{\theta_U} f(r,\theta)d\theta\right] dr = \int_{\theta_L}^{\theta_U}\left[\int_{r_L}^{r_U} f(r,\theta)rdr\right] d\theta \tag{5.17}$$

The integration along r and θ can be done in the same way as that over a 2D regular domain. All the integration limits are fixed values. It is a nested two-layer 1D integral.

Let us now look at an example.

5.4.2 *Example: Computing the area of circular regions*

The following is a SymPy code to do this:

```
1  r, rL, rU = symbols('r, r_L,r_U',real=True) # interval along r: [rL,rU]
2  θ, θL, θU = symbols('θ, θ_L,θ_U',real=True) # interval along r: [θL,θU]
3  f = 1      # use unit function to compute the area of the integral domain
4  A = integrate( integrate(f*r, (r, rL, rU)), (θ, θL, θU))
5  display(Math(f" \\text{{Formula for area of a circular region: }} \
6                                  {sp.latex(A)}"))
```

Formula for area of a circular region: $-\theta_L\left(-\frac{r_L^2}{2}+\frac{r_U^2}{2}\right)+\theta_U\left(-\frac{r_L^2}{2}+\frac{r_U^2}{2}\right)$

```
1  # Use the formula to compute some cases with specified parameters:
2
3  print(f' Area of a full circle = {A.subs({rL:0,rU:1,θL:0,θU:2*sp.pi})}')
4  print(f' Area of half circle   = {A.subs({rL:0,rU:1,θL:0,θU:sp.pi})}')
5  print(f' Area of a ring        = {A.subs({rL:1,rU:2,θL:0,θU:2*sp.pi})}')
```

```
Area of a full circle = pi
Area of half circle   = pi/2
Area of a ring        = 3*pi
```

5.4.3 *Example: Polar moment of inertia of a ring*

In the previous example, we set $f(r,\theta)$ to be a constant value of 1 (unit function) because we were calculating the area. However, in many applications, $f(r,\theta)$ will vary over the polar coordinates. For instance, when computing the polar moment of inertia of a ring region, we encounter the following function [5]:

$$f(r,\theta) = r^2 \tag{5.18}$$

Equation (5.17) can now be used to compute the polar moment of inertia of a ring. The Python snippet is given as follows:

```
1  # Compute the polar moment of inertia of a ring:
2  f = r**2                        # function for polar moment of inertia
3  Jp = integrate( integrate(f*r, (r, rL, rU)), (θ, 0, 2*sp.pi)).simplify()
4  display(Math(f" \\text{{Polar moment inertia formula of rings: }}{Jp}"))
```

Polar moment inertia formula of rings: $pi * (-r_L**4 + r_U**4)/2$

We obtained the formula for computing polar moment of inertia, J_p, of a ring region:

$$J_p = \frac{\pi \left(r_U^4 - r_L^4\right)}{2} \tag{5.19}$$

This is one of the most widely used formulas for problems in dynamics and mechanics involving disks and cylinders. If the disk or cylinder is solid (without hole), we simply set $r_L = 0$:

```
1  # Use the formula to compute some cases with specified parameters:
2  print(f' Polar moment of inertia of a disk = {Jp.subs({rL:0,rU:1})}')
3  print(f' Polar moment of inertia of a ring = {Jp.subs({rL:1,rU:2})}')
```

```
Polar moment of inertia of a disk = pi/2
Polar moment of inertia of a ring = 15*pi/2
```

5.4.4 *Example: Volume under a spherical dome*

Let us compute the volume under a spherical dome over a circular domain, as schematically shown in Fig. 5.4(a).

The function for the surface of the spheric dome is defined as

$$f(x,y) = \sqrt{R^2 - (x^2 + y^2)} + H \tag{5.20}$$

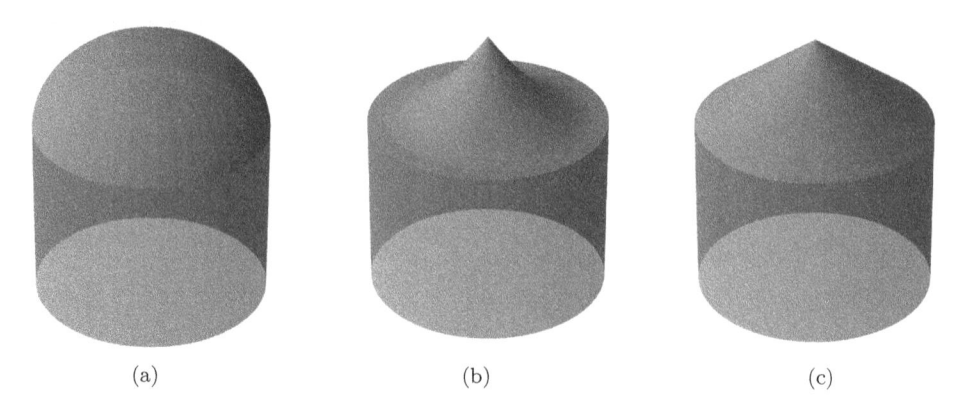

(a)   (b)   (c)

Figure 5.4. Schematical drawing of domes: (a) spherical dome on a cylinder; (b) parabolic cone on a cylinder; (c) cone on a cylinder.

where R is the radius of the dome's circular base and H is the height of the cylindrical section above the base. Equation (5.20) looks simple, but the integration cannot be done analytically in Cartesian coordinates. This is a perfect problem for polar coordinates because after the transformation, the integral in the polar coordinates becomes

$$
\begin{aligned}
I &= \int_{\theta_L}^{\theta_U} \int_{r_L}^{r_U} \left[\sqrt{R^2 - r^2} + H \right] r\,dr\,d\theta \\
&= 2\pi \int_0^R \left[\sqrt{R^2 - r^2} + H \right] \frac{1}{2}dr^2 \\
&= \pi \int_0^{R^2} \left[\sqrt{R^2 - \gamma} + H \right] d\gamma
\end{aligned}
\tag{5.21}
$$

In the last step, the square root eliminates the square of r within it. This is why many integrals involving terms like $x^2 + y^2$ become significantly easier to solve in polar coordinates. The integrand can be transformed into a simpler form because of the area scalar r. It can be absorbed into dr by converting it to dr^2. Here, we substitute $\gamma = r^2$ and $d\gamma = \frac{1}{2}dr^2$. This variable substitution technique (u-substitution) was discussed in Chapter 4. This "trick" is frequently employed to obtain analytical solutions for many integrals, as shown in various references (e.g., Ref. [3]).

The result from Eq. (5.21) is the analytic formula for computing the volume:

$$
V_{\text{spheric dome}} = \underbrace{\frac{2}{3}\pi R^3}_{\text{half-ball}} + \underbrace{\pi R^2 H}_{\text{cylinder}}
\tag{5.22}
$$

Let us use the following code to obtain the same formula:

```
1  θ = symbols('θ', real=True)
2  r, R, H = symbols('r, R, H', positive=True)
3
4  x = r*sp.cos(θ)
5  y = r*sp.sin(θ)
6  f = sp.sqrt(R**2-(x**2+y**2))+H    # function for spherical dome surface
7
8  V = integrate( integrate(f*r, (r, 0, R)), (θ, 0, 2*sp.pi))
9  display(Math(f" \\text{{Formula for volume under a spherical dome, \
10                       $V$ = }}{sp.latex(V)}"))
```

Formula for volume under a spherical dome, $V = \pi H R^2 + \dfrac{2\pi R^3}{3}$

We obtained the exact formula via integration in polar coordinates. It's important to note that SymPy performs the substitution automatically within its calculations. This formula consists of two parts: one for the cylinder under the cone and another for the half sphere (which is a well-known formula).

While it's possible to attempt the same problem using Cartesian coordinates for this dome, the analytical integration becomes quite challenging. Feel free to give it a try as an exercise to see the advantage of polar coordinates in this specific case.

5.4.5 *Example: Volume under a parabolic cone*

Let us compute the volume under a parabolic cone over a circular domain, as schematically shown in Fig. 5.4(b). The function of the surface of the cone is defined as

$$f(x, y) = h\left(1 - \frac{1}{R}\sqrt{x^2 + y^2}\right)^2 + H \tag{5.23}$$

where R is the radius of the cone base, h is the height of cone measured from its base, and H is the height of the cylinder beneath the cone base. The analytical antiderivative cannot be found for the same reasons mentioned in the previous example. We need to use the polar coordinate and integrate it over a circular domain. The code is given as follows:

```
1  r, R, h, H = symbols('r,R,h,H ',positive=True) # h: height of cone part
2                                                 # H: height of cylinder
3  x = r*sp.cos(θ); y = r*sp.sin(θ)
4  f = h*(1 - sqrt(x**2 + y**2)/R)**2 + H      # surface of parabolic cone
5
6  V = integrate( integrate(f*r, (r, 0, R)), (θ, 0, 2*sp.pi))
7  display(Math(f" \\text{{Formula for volume under a parabolic cone, \
8                       $V$ = }}{sp.latex(V)}"))
```

Formula for volume under a parabolic cone, $V = \pi H R^2 + \dfrac{\pi R^2 h}{6}$

The same u-substitution was done within SymPy. This formula consists of two parts: one for the cylinder under the cone base and one for the parabolic cone.

```
1  # compute some specific cases:
2  print(f' Volume under a cone (H=0) = {V.subs({R:1,h:1,H:0}).evalf(5)}')
3  print(f' Volume under a cone (H=1) = {V.subs({R:1,h:1,H:1}).evalf(5)}')
```

```
Volume under a cone (H=0) = 0.52360
Volume under a cone (H=1) = 3.6652
```

5.4.6 *Example: Volume under a cone dome*

Let us compute the volume under a cone dome over a circular domain, as schematically shown in Fig. 5.4(c). The function of the surface of the dome is defined as

$$f(x, y) = h\left(1 - \frac{1}{R}\sqrt{x^2 + y^2}\right) + H \tag{5.24}$$

where R is the radius of the cone base, h is the height of cone measured from the its base, and H is the height of the cylinder beneath the cone base. Similar to the previous two examples, the polar coordinate needs to be used, which allows the u-substitution. The code is given as follows:

```
1  f = h*(1 - sqrt(x**2 + y**2)/R) + H          # surface of cone dome
2  V = integrate( integrate(f*r, (r, 0, R)), (0, 0, 2*sp.pi))
3  display(Math(f" \\text{{Formula for volume under a cone dome, \
4                          $V$ = }}{sp.latex(V)}"))
```

Formula for volume under a cone dome, $V = \pi H R^2 + \dfrac{\pi R^2 h}{3}$

This formula consists of two parts: one for the cylinder under the cone base and another for the cone which is a well-known formula.

```
1  # compute some specific cases:
2  print(f' Volume under a cone (H=0) = {V.subs({R:1,h:1,H:0}).evalf(5)}')
3  print(f' Volume under a cone (H=1) = {V.subs({R:1,h:1,H:1}).evalf(5)}')
```

```
Volume under a cone (H=0) = 1.0472
Volume under a cone (H=1) = 4.1888
```

5.5 Integration in 2D natural coordinates

5.5.1 *Formulation*

For 2D problems with irregular domains (not a rectangle), we perform a coordinate transformation to enable integration over a regular domain, typically a square. A common type of mapping applies to quadrilateral domains with four vertices (Q4) in (x, y) and transforms them into a standard square in (ξ, η), as illustrated schematically in Fig. 3.15. This transformation is achieved using a bilinear transformation expressed as

$$x = \sum_{k=1}^{4} x_k N_k(\xi, \eta)$$

$$y = \sum_{k=1}^{4} y_k N_k(\xi, \eta) \tag{5.25}$$

where (x_k, y_k) is the coordinate of the kth node and $N_k(\xi, \eta)$ is the kth nodal bilinear shape function that can be generated using the code given in Chapter 3. As shown in Eq. (5.25), these shape functions are dimensionless.

The mapping is **bijective**, implying that there's a one-to-one correspondence between any point in the original domain and its transformed domain (as discussed in Section 3.10.1). This bijective property allows us to simplify all integral limits to a standard range of -1 to 1 since we're integrating over a standard square. Equation (5.11) becomes

$$I = \int_{D_{Q4}} f(x, y) dx dy$$

$$= \int_{-1}^{1} \int_{-1}^{1} f(\xi, \eta) \underbrace{\det(\mathbf{J})}_{\text{area scalar}} d\zeta d\eta \tag{5.26}$$

The double integral becomes nested two-layer 1D integrals. The Jacobian matrix, defined in Eq. (3.49), plays a crucial role here. Its determinant, $\det(\mathbf{J})$, acts as an area scalar that accounts for the stretching, compression, and skewing of the domain caused by the transformation. The differential area in the transformed coordinates is then expressed as $\det(\mathbf{J}) d\xi \, d\eta$.

5.5.2 *Elementwise computation*

Care is needed in computing the Jacobian matrix. When performing a mapping from, for example, polar to Cartesian or spherical to Cartesian (discussed later), the Jacobian matrix is only a function of the polar or spherical coordinates, respectively.

However, in the mapping from natural to physical coordinates, the Jacobian matrix becomes a function not only of ξ and η but also of x_k and y_k due to Eq. (5.25). This implies that the determinant of the Jacobian matrix relates ξ and η as well as x_k and y_k. As a result, the area scalar, $\det(\mathbf{J})$, needs to be computed for each Q4 domain if their shapes differ. This computation is performed **domainwise**.

The detailed process for a typical Q4 domain is provided in the following Python code snippets. First, create bilinear shape functions for the Q4 domain:

```
1  ξ, η =symbols('ξ, η')                        # define2D natural coordinate
2  nodexi = [-1, 1]; nodeyj = [-1, 1]
3  lxs,lys,Nij =gr.N_Lagrange_fC(nodexi,nodeyj)         # shape functions
4  print(f'Lagrange interpolators: {lxs}')
5  print(f'Lagrange interpolators: {lys}')
6
7  Nij = sp.Matrix(Nij)
8  Nij.T                       # [N1, N2, N3, N4] numbered counterclockwise
```

```
Lagrange interpolators: [1/2 - ξ/2, ξ/2 + 1/2]
Lagrange interpolators: [1/2 - η/2, η/2 + 1/2]
```

$$\left[\frac{(\eta-1)(\xi-1)}{4} \quad -\frac{(\eta-1)(\xi+1)}{4} \quad \frac{(\eta+1)(\xi+1)}{4} \quad -\frac{(\eta+1)(\xi-1)}{4} \right]$$

Since the node numbering of the Q4 domain is counterclockwise (CCW), as shown in Fig. 3.15, we create a Python function to directly form these four bilinear shape functions in that specific order:

```
1  def NQ4C(ξ, η):
2      '''
3          Shape function for Q4 elements, all sympy objects.
4          return a row vector for four nodal shape functions.
5          Nodes arrangement, Counterclockwise:  4    3
6                                                 1    2      '''
7      ξ, η =sp.symbols('ξ, η')
8      N4=sp.Matrix([[(1 - η)*(1 - ξ)], [(1 - η)*(ξ + 1)],
9                    [(η + 1)*(ξ + 1)], [(1 - ξ)*(η + 1)]])/4
10     return N4.T
```

We can now compute the expressions for x and y as functions of ξ and η:

```
 1  x1, x2, x3, x4 =symbols('x1, x2, x3, x4') # define symbolic coordinates
 2  y1, y2, y3, y4 =symbols('y1, y2, y3, y4')
 3  X = sp.Matrix([x1, x2, x3, x4])                # form vector of coordinates
 4  Y = sp.Matrix([y1, y2, y3, y4])
 5
 6  N4 = NQ4C(ξ, η)                    # Generate 4 bilinear shape functions
 7  x = (N4 @ X)[0]             # Expression for coordinate x and y, in ξ and η
 8  y = (N4 @ Y)[0]
 9  display(Math(f"{latex(x)}"))   # x
10  display(Math(f"{latex(y)}"))   # y
```

$$\frac{x_1 \left(1-\eta\right)\left(1-\xi\right)}{4} + \frac{x_2 \left(1-\eta\right)\left(\xi+1\right)}{4} + \frac{x_3 \left(\eta+1\right)\left(\xi+1\right)}{4}$$
$$+ \frac{x_4 \left(1-\xi\right)\left(\eta+1\right)}{4}$$

$$\frac{y_1 \left(1-\eta\right)\left(1-\xi\right)}{4} + \frac{y_2 \left(1-\eta\right)\left(\xi+1\right)}{4} + \frac{y_3 \left(\eta+1\right)\left(\xi+1\right)}{4}$$
$$+ \frac{y_4 \left(1-\xi\right)\left(\eta+1\right)}{4}$$

Next, compute the Jacobian matrix following Eq. (5.13):

```
 1  J = sp.Matrix([[x.diff(ξ), x.diff(η)],
 2                 [y.diff(ξ), y.diff(η)]])
 3  J
```

$$\begin{bmatrix} -\frac{x_1(1-\eta)}{4} + \frac{x_2(1-\eta)}{4} + \frac{x_3(\eta+1)}{4} - \frac{x_4(\eta+1)}{4} & -\frac{x_1(1-\xi)}{4} - \frac{x_2(\xi+1)}{4} + \frac{x_3(\xi+1)}{4} + \frac{x_4(1-\xi)}{4} \\ -\frac{y_1(1-\eta)}{4} + \frac{y_2(1-\eta)}{4} + \frac{y_3(\eta+1)}{4} - \frac{y_4(\eta+1)}{4} & -\frac{y_1(1-\xi)}{4} - \frac{y_2(\xi+1)}{4} + \frac{y_3(\xi+1)}{4} + \frac{y_4(1-\xi)}{4} \end{bmatrix}$$

This is the analytical formula for the Jacobian matrix. It contains functions of the natural coordinates ξ and η, which also depend on the nodal coordinates of the four vertices of the domain in the physical coordinates.

5.5.3 *Code for the Jacobian matrix of Q4 domains*

Let's write a SymPy function to compute the Jacobian matrix for a quadrilateral (Q4) domain for later use:

```
1  def jacobQ4C(x1, y1, x2, y2, x3, y3, x4, y4):
2      '''
3      Compute the Jacobian matrix of a Q4, quadrilateral with 4 nodal
4      coordinates. Nodes arrangement is Counterclockwise: 4   3
5                                                          1   2   '''
6      ξ, η =sp.symbols('ξ, η')
7      J = sp.Matrix([
8              [-x1*(1-η)/4 + x2*(1-η)/4 + x3*(η+1)/4 - x4*(η+1)/4,
9               -x1*(1-ξ)/4 - x2*(ξ+1)/4 + x3*(ξ+1)/4 + x4*(1-ξ)/4],
10             [-y1*(1-η)/4 + y2*(1-η)/4 + y3*(η+1)/4 - y4*(η+1)/4,
11              -y1*(1-ξ)/4 - y2*(ξ+1)/4 + y3*(ξ+1)/4 + y4*(1-ξ)/4]])
12     return J
```

```
1  # Compute the Jacobian matrix and its determinante:
2  J = jacobQ4C(x1, y1, x2, y2, x3, y3, x4, y4)      # form Jacobian matrix
3  detJ = sp.det(J).simplify()
4  #detJ                                              # The output is Lengthy
```

We are ready to derive the formula for computing the area of a Q4:

```
1  f = 1                             # unit function for area computation
2  fI = f*detJ
3  A_Q4 = integrate(integrate(fI,(ξ,-1,1)),(η,-1,1))
4  A_Q4
```

$$\frac{x_1 y_2}{2} - \frac{x_1 y_4}{2} - \frac{x_2 y_1}{2} + \frac{x_2 y_3}{2} - \frac{x_3 y_2}{2} + \frac{x_3 y_4}{2} + \frac{x_4 y_1}{2} - \frac{x_4 y_3}{2}$$

This formula can be rearranged as follows:

$$A_{Q4C} = \frac{1}{2}\Big[(x_1 y_2 + x_2 y_3 + x_3 y_4 + x_4 y_1)$$

$$- (y_1 x_2 + y_2 x_3 + y_3 x_4 + y_4 x_1)\Big] \tag{5.27}$$

We can observe a clear pattern in the bilinear shape functions: the terms within the first parenthesis involve a dot product of the vector $[x_1, x_2, x_3, x_4]^\top$ with the vector $[y_2, y_3, y_4, y_1]^\top$ (rolled one position backward). Similarly, the terms within the second parenthesis involve the dot product of vector $[y_1, y_2, y_3, y_4]^\top$ with vector $[x_2, x_3, x_4, x_1]^\top$ (also rolled one position backward). This computation can be efficiently expressed using the following formula:

$$A_{\text{polygon}} = \big(\mathbf{x}@\text{roll}(\mathbf{y}, -1) - \mathbf{y}@\text{roll}(\mathbf{x}, -1)\big)/2 \tag{5.28}$$

Here, roll($\mathbf{v}, n$) is a function commonly used in modern programming languages like Python. It performs a cyclic permutation of the elements in array $\mathbf{v}$ by n positions. For instance, for an array $\mathbf{v} = (1, 2, 3, 4)$, roll($\mathbf{v}, -1$) rolls the elements backward by 1, resulting in a new vector $\mathbf{v} = (2, 3, 4, 1)$. This precisely aligns with the operation needed in Eq. (5.27). The dot-product operation can often be optimized for parallel processing, leading to faster computation.

More importantly, Eq. (5.28) potentially provides a general formula applicable to various polygonal shapes with multiple edges, including triangles, quadrilaterals (like our current case), pentagons, and so on. We explore its application to other shapes in later examples.

This formula can also be simplified manually as follows:

$$A_{Q4C} = \frac{1}{2}\left[-(x_3 - x_1)(y_2 - y_4) + (x_2 - x_4)(y_3 - y_1) \right] \tag{5.29}$$

This is a particularly simple and efficient formula for calculating the area of a quadrilateral domain. It requires only the coordinates of the four nodes defining the Q4 domain and leverages coordinate differences of the diagonal nodes, leading to just two multiplications. Additionally, the formula exhibits a clear and consistent pattern.

The following code snippet implements dot-product formula in Eq. (5.28) using NumPy. We also provide a separate NumPy function (gr.area_Q4C()) based on Eq. (5.29) as an alternative:

```
1  def polygonA_xy(x, y, absOn=True):
2      '''
3      Compute the area of a polygon domain/element of any number of edges,
4      given its vertices coordinates in arrays x and y. Formulas for Q4:
5      area = (1/2) |(x1*y2+x2*y3+x3*y4+x4*y1)-(y1*x2+y2*x3+y3*x4+y4*x1)|
6                   cyclic permutation using np.roll()
7      if ABS == True, take the absolute value, otherwise the raw value.
8      The raw value provides additional information on node numbering.
9      '''
10     if absOn:
11         area=np.abs(x@np.roll(y,1)-y@np.roll(x,1))/2.
12     else:
13         area=(x@np.roll(y,-1)-y@np.roll(x,-1))/2. # roll backwards by 1
14
15     return area
```

In general, when using integration to compute area, the result is a signed area. This arises essentially because of the direction dependence of an integration, as discussed in Chapter 4. The sign depends on how the nodes are numbered and the shape of the Q4 element (convex or concave). If the absolute value is needed, it can be taken as done in many texts. However, the author chooses to leave the sign as it is. A negative value indicates that the node numbering might be incorrect or the shape of the polygon is concave. This information can be quite useful. While taking the absolute value is always an option, understanding why a negative sign might arise is often important. Improper numbering can lead to entirely wrong answers, not just a sign difference. The following examples reveal some possible scenarios.

5.5.4 *Example: Area of some quadrilaterals*

Let us use the formula derived to compute the area of three quadrilaterals using given coordinates. We study the effects of node numbering. First, we arrange the nodes for each of the Q4 elements in CCW manner:

$$4 \longleftarrow 3$$
$$\uparrow \qquad (5.30)$$
$$1 \longrightarrow 2$$

In the computation, we turn off the abs() operation so that we can see the signed area:

```
 1  # coordinates for Q4-1, a square [0, 1]×[0, 1] in x-y coordinate:
 2  # Nodes:          1  2  3  4
 3  x_CCW=np.array([0, 1, 1, 0])   # coordinates of simple cases for test
 4  y_CCW=np.array([0, 0, 1, 1])
 5  print(f"Area of Q4-1 = {polygonA_xy(x_CCW, y_CCW, absOn=False)}")
 6
 7  # coordinates for Q4-2: a quadrilateral in x-y coordinate:
 8  x_CCW = np.array([0., 2., 3.,  1.])
 9  y_CCW = np.array([0., 1., 2.5, 3.])
10  print(f"Area of Q4-2 = {polygonA_xy(x_CCW, y_CCW, absOn=False)}")
11
12  # coordinates for Q4-3, as square [-1, 1]×[-1, 1] in x-y coordinate:
13  x_CCW=np.array([-1, 1, 1,-1])
14  y_CCW=np.array([-1,-1, 1, 1])
15
16  x_CCW_=x_CCW.copy(); y_CCW_=y_CCW.copy()
17  print(f"Area of Q4-3 = {polygonA_xy(x_CCW, y_CCW, absOn=False)}")
```

```
Area of Q4-1 = 1.0
Area of Q4-2 = 4.25
Area of Q4-3 = 4.0
```

It is seen that the signed areas for all the three Q4 elements are positive because all the nodes are arranged CCW, following the same setting when deriving the formula. Let us swap the numbering of the nodes for Q4-3, making it clockwise:

$$2 \longrightarrow 3$$
$$\uparrow \qquad \downarrow \qquad (5.31)$$
$$1 \qquad 4$$

```
1  x_CW =x_CCW_.copy(); y_CW =y_CCW_.copy()  # make copies of the oringals
2
3  x_CW[1], x_CW[3] = x_CW[3], x_CW[1]  # swap the nodal coordinate values
4  y_CW[1], y_CW[3] = y_CW[3], y_CW[1]
5
6  print("  1  2  3  4")
7  print(x_CW); print(y_CW)
8  print(f"Area of Q4-3={polygonA_xy(x_CW, y_CW,absOn=False)}") # new area
```

```
   1  2  3  4
[-1 -1  1  1]
[-1  1  1 -1]
Area of Q4-3=-4.0
```

We obtained a negative value because of the clockwise numbering. Let us number it following an N-path:

$$
\begin{array}{cc}
2 & 4 \\
\uparrow \searrow \uparrow \\
1 & 3
\end{array}
\tag{5.32}
$$

```
1  x_N =x_CW.copy(); y_N =y_CW.copy()                   # make copies
2  x_N[2], x_N[3] = x_N[3], x_N[2]        # swap the nodal coordinate values
3  y_N[2], y_N[3] = y_N[3], y_N[2]
4
5  print("  1  2  3  4")
6  print(x_N); print(y_N)
7  print(f"Area of Q4-3 = {polygonA_xy(x_N, y_N, absOn=False)}")
```

```
   1  2  3  4
[-1 -1  1  1]
[-1  1 -1  1]
Area of Q4-3 = 0.0
```

This time, we got zero, which is expected. The area of the left-lower triangle is negative because it is clockwise numbered and that of the right-upper triangle is positive because of its CCW numbering. Their sum becomes zero, reflecting the overall area of the quadrilateral. This confirms that the formula correctly captures the concept of signed area.

These three examples effectively demonstrate the meaning of **signed area** in the context of integration. They also illustrate the potential dangers of simply using the absolute value function (**abs ()**) to make the result positive. A negative signed area isn't just a matter of sign; it can indicate incorrect node numbering or a concave shape, as discussed earlier. Understanding these concepts is crucial for accurate area calculations using integration.

5.5.5 *Example: Area of some triangles*

Note that Eq. (5.28) can be easily used to compute the area of a triangle with three nodes (T3), simply by providing only three nodes. We also study the effects of node numbering for T3 elements by turning off the abs() operation. First, we arrange the nodes in CCW manner:

$$3$$
$$\downarrow \nwarrow \qquad\qquad\qquad (5.33)$$
$$1 \longrightarrow 2$$

Here, we simply form a T3 using Q4-3:

```
1  print(f"T3 area, CCW={polygonA_xy(x_CCW[:-1], y_CCW[:-1],absOn=False)}")
```

T3 area, CCW=2.0

The area obtained is positive and is half the area of the Q4-3, as expected. Let us examine a T3 with clockwise nodal numbering:

$$3$$
$$\uparrow \searrow \qquad\qquad\qquad (5.34)$$
$$1 \longleftarrow 2$$

```
1  print(f"T3 area, CW = {polygonA_xy(x_CW[:-1], y_CW[:-1], absOn=False)}")
```

T3 area, CW = -2.0

We got a negative value because of the clockwise numbering.

5.5.6 *Example: Area of a pentagon*

While Eq. (5.28) was previously derived for quadrilaterals, the author's unpublished work demonstrates that it can be extended to general polygons with n edges, subject to CCW numbering. We acknowledge that this might not be a general formula applicable to all pentagons, and we explore the theoretical underpinnings of this extension in future discussions.

We have successfully tested this formula for triangles in the previous examples and obtained correct and expected results. In the meantime, let's proceed further using the same formula to compute the area of a pentagon consisting of five nodes. The code can also be exactly the same. All we need is to provide the coordinates for all these nodes arranged CCW:

$$5 \longleftarrow 4$$
$$\nwarrow$$
$$1 \longrightarrow 2 \longrightarrow 3$$

(5.35)

The code snippet for the pentagon is as follows:

```
1  # coordinates for a pentagon (P5), in x-y coordinates:
2  # Nodes:           1  2  3  4  5
3  x_CCW=np.array([-1, 1, 3, 1,-1])
4  y_CCW=np.array([-1,-1,-1, 1, 1])
5
6  x_CCW_=x_CCW.copy(); y_CCW_=y_CCW.copy()
7  print(f"Area of P5 = {polygonA_xy(x_CCW, y_CCW, absOn=False)}")
```

```
Area of P5 = 6.0
```

The area obtained for the pentagon (P5) is correct, which is the sum of the areas of the quadrilateral (Q4) and triangle (T3), as expected. Readers may try other polygons. The provided code can be a valuable tool in these cases, but careful consideration of the numbering scheme and potential limitations is necessary.

Always using CCW numbering remains a good practice for predictable results.

5.5.7 *Example: Volume of a function over a quadrilateral*

Using mapping, we can compute the volume under the surface defined by a function over a quadrilateral (Q4) domain. A schematic view for this problem is shown in Fig. 5.1. In this example, we consider a simple function $f(x, y) = \exp(x + y)$ and use the following code to get it done:

```
1  xcords = sp.Matrix([0, 2, 3, 1])            # Coordinates of the Q4
2  ycords = sp.Matrix([0, 1, 2, 3])
3
4  # Use the determinant formula of the Jacobian matrix obtained earlier:
5  detJ_ = detJ.subs(zip(X,xcords)).subs(zip(Y,ycords))
6  detJ_                            # use _ for variables after substitution
```

$$\frac{\eta}{4} - \frac{\xi}{2} + 1$$

```
1  x_ = x.subs(zip(X,xcords))      # use _ for variables after substitution
2  y_ = y.subs(zip(Y,ycords))
3  f = sp.exp(x_ + y_)                      # function defined over Q4
4
5  fI = f*detJ_
6  V_Q4 = integrate(integrate(fI,(ξ,-1,1)),(η,-1,1))
7  print(f" The volume found = {V_Q4.evalf(8)}")
8  display(Math(f" \\text{{Formula to compute the volume: \
9                          $V$ = }}{sp.latex(V_Q4)}"))
```

The volume found = 124.14205

$$\text{Formula to compute the volume: } V = -\frac{7e^4}{4} - \frac{e^3}{6} + \frac{5}{12} + \frac{3e^5}{2}$$

Since the given function is simple, we obtained the analytic solution. For more complicated functions, we may need to resort to numerical integrations.

5.5.8 *Example: Volume of a bilinear surface*

If the function can be approximated using a bilinear function, we can compute the volume of the function over a quadrilateral domain with ease using the mapping technique. This is because bilinear functions can be themselves defined in terms of natural coordinates (ξ and η) for a quadrilateral domain. The integration process becomes significantly more straightforward within the mapped domain. This approach leverages the convenient properties of the bilinear function within the transformed space.

A bilinear surface over a quadrilateral can be expressed as

$$f(x,y) = \sum_{k=1}^{4} f(x_k, x_k) N_k(\xi, \eta) = \sum_{k=1}^{4} f_k N_k(\xi, \eta) \tag{5.36}$$

We use the following code to get it done. First, consider a domain that is a simple unit square:

```
1  # A unit square domain: [0, 1]×[0, 1] in x-y coordinate, A=1:
2  xcords=sp.Matrix([0, 1, 1, 0])     # coordinates of unit square for test
3  ycords=sp.Matrix([0, 0, 1, 1])                         # counterclockwise
4
5  f1, f2, f3, f4 =symbols('f1, f2, f3, f4')
6  fn = sp.Matrix([f1, f2, f3, f4])            # function values at the nodes
7
8  N4 = NQ4C(ξ, η)                   # generate 4 bilinear shape functions
9  f_bl = (N4 @ fn)[0]                      # bilinear function in ξ and η
10 display(Math(f"{latex(f_bl)}"))
```

$$\frac{f_1 (1 - \eta) (1 - \xi)}{4} + \frac{f_2 t (1 - \eta) (\xi + 1)}{4} + \frac{f_3 (\eta + 1) (\xi + 1)}{4} +$$
$$\frac{f_4 (1 - \xi) (\eta + 1)}{4}$$

```
1  J = jacobQ4C(x1, y1, x2, y2, x3, y3, x4, y4)
2  detJ = sp.det(J).simplify()
3
4  detJ_ = detJ.subs(zip(X,xcords)).subs(zip(Y,ycords))
5
6  fI = f_bl*detJ_                # function to be integrated over the square
7  V_Q4 = integrate(integrate(fI,(ξ,-1,1)),(η,-1,1))
8  display(Math(f" \\text{{$V_{{Q4}}$ = }}{sp.latex(V_Q4)}"))
```

$$V_{Q4} = \frac{f_1}{4} + \frac{f_2}{4} + \frac{f_3}{4} + \frac{f_4}{4}$$

The result shows that the volume under this bilinear function over the unit square with $A = 1$ is the area-averaged function value at the four nodes. Each of the four nodal function values carries the same weight (contributes equally) because the shape is a regular square. This is intuitively reasonable, as a uniform area like a square implies equal weighting in the averaging process.

Let us now change the domain to a general quadrilateral (not a rectangle) to explore how the result might change:

```
1  xcords = sp.Matrix([0, 2, 3, 1])
2  ycords = sp.Matrix([0, 1, 2, 3])
3  detJ_ = detJ.subs(zip(X,xcords)).subs(zip(Y,ycords))
4
5  fI = f_bl*detJ_                # function to be integrated over the square
6  A_Q4 = integrate(integrate(detJ_,(ξ,-1,1)),(η,-1,1))
7  display(Math(f" \\text{{$A_{{Q4}}$ = }}{sp.latex(A_Q4)}"))
8  V_Q4 = integrate(integrate(fI,(ξ,-1,1)),(η,-1,1))
9  display(Math(f" \\text{{$V_{{Q4}}$ = }}{sp.latex(V_Q4)}"))
```

$$A_{Q4} = 4$$
$$V_{Q4} = \frac{13 f_1}{12} + \frac{3 f_2}{4} + \frac{11 f_3}{12} + \frac{5 f_4}{4}$$

This time, the Q4 shape is not regular, and its area is 4. Due to this irregularity, the four nodal function values now carry different weights in the volume calculation. However, the sum of these weights remains 4, which is precisely the area of the quadrilateral. This highlights a key principle of conservation.

This type of function approximation and integration is widely used in many computational methods, especially in the finite element method (FEM) at the element level. By breaking down complex domains into smaller elements (often quadrilaterals or triangles), we can perform the integrations efficiently on each element. The results from these element-level integrations are then assembled to obtain an approximate solution for the entire domain. This strategy allows us to tackle problems that would be intractable using analytical methods for complex geometries. For a deeper understanding of this concept in FEM and how the integral technique discussed here is actually used in FEM, one may refer to Ref. [6].

5.6 Integration in 3D simple domains

This section discusses the 3D integration over three simple domains: brick, spherical, and cylindrical domains.

5.6.1 *Integration over brick domains*

5.6.1.1 *Formulation*

By simple extension of the 2D case, the definite integral of a 3D function over a regular 3D domain (brick) can be expressed as follows. Consider a function $f(x, y, z)$ defined in 3D Cartesian coordinate space (x, y, z) over a regular brick domain, $D = [x_L, x_U] \times [y_L, y_U] \times [z_L, z_U]$, and it is continuous in $\mathcal{D}$. The definite integral can be expressed as

$$I = \int_{\mathcal{D}^n} f(\mathbf{x}) d\mathbf{x} = \int_{[x_L, x_U] \times [y_L, y_U] \times [z_L, z_U]} f(\mathbf{x}) d\mathbf{x}$$

$$= \int_{x_L}^{x_U} \left\{ \int_{y_L}^{y_U} \left[\int_{z_L}^{z_U} f(x, y, z) dz \right] dy \right\} dx \quad (5.37)$$

This means that we can carry out the integration in three stages, each a 1D integral:

1. **Integrate over z:** For a fixed $x \in [x_L, x_U]$ and $y \in [y_L, y_U]$, integrate $f(x, y, z)$ over interval $[z_L, z_U]$, leading to an intermediate antiderivative $F(x, y)$ that is a function of x and y.
2. **Integrate over y:** For a fixed $x \in [x_L, x_U]$, integrate $F(x, y)$ over interval $[y_L, y_U]$. This gives another intermediate antiderivative $F(x)$ that is a function of x only.
3. **Integrate over x:** Finally, integrating $F(x)$ over interval $[x_L, x_U]$ gives the final integral value I, which represents the signed hyper-volume under the surface defined by function $f(x, y, z)$.

Note that the sequence of the integration layers can change, depending on the convenience in carrying out the integration. The process effectively transforms the triple integral into three nested 1D integrations. Since the integral domain is regular, there is no special technical issue except that solving triple integrals can be a time-consuming task.

Let us look directly at some examples using Python code to evaluate these integrals.

5.6.1.2 *Example: 3D integral of a linear function*

In this example, we consider a simple linear function in 3D and compute its hypervolume under the plane surface defined by this function. The following code snippet reveals the detailed procedure for 3D definite integration:

```
1  x, y, z, a, b, c, d = symbols('x, y, z, a, b, c, d', real=True)
2
3  fxyz = a*x+b*y+c*z+d                # Liear function to be integrated
4  xL, xU = 0, 1.;  yL, yU = 0, 1.;  zL, zU = 0, 1.   # 3D integral domain
5  # xL: x lower bound, xU: x upper bound ...
6  V_domain = (xU-xL)*(yU-yL)*(zU-zL)
7
8  Fxy = integrate(fxyz, (z, zL,zU)) # along z over [zL,zU], x and y fixed
9  display(Math(f" ∫_0^1({sp.latex(fxyz)})dz = {sp.latex(Fxy)}"))
10
11 Fx  = integrate(Fxy,  (y, yL,yU))        # along y over [yL,yU], x fixed
12 display(Math(f" ∫_0^1({sp.latex(Fxy)})dy = {sp.latex(Fx)}"))
13
14 V_f = integrate(Fx,   (x, xL,xU))               # along x over [xL, xU]
15 display(Math(f" ∫_0^1({sp.latex(Fx)})dy - {sp.latex(V_f)}"))
16                                 # Hypervolume under f(x,y,z)
17 print(f' Integrated result, I = {V_f.subs({a:.0, b:.0, c:0., d:1.})}')
```

$$\int_0^1 (ax + by + cz + d)dz = 1.0ax + 1.0by + 0.5c + 1.0d$$

$$\int_0^1 (1.0ax + 1.0by + 0.5c + 1.0d)dy = 1.0ax + 0.5b + 0.5c + 1.0d$$

$$\int_0^1 (1.0ax + 0.5b + 0.5c + 1.0d)dy = 0.5a + 0.5b + 0.5c + 1.0d$$

```
Integrated result, I = 1.00000000000000
```

The volume of the domain was obtained using the unit function. Each integral reduces one independent variable for the integrand, reading to a final fixed number.

For a linear integrand, we set the following:

```
1  dict_ = {a:3., b:2., c:1.5, d:1.}
2  xyz_center = {x:(xU-xL)/2, y:(yU-yL)/2, z:(zU-zL)/2}
3  print(f' For linear f,   I = {V_f.subs(dict_)}')
4  print(f' For linear f_center*V_domain = {fxyz.subs(dict_|xyz_center)}')
```

```
For linear f,   I = 4.25000000000000
For linear f_center*V_domain = 4.25000000000000
```

We obtained the correct solution by two means: triplet integration and the formula for volume under a planar hyper-surface.

When the function is linear within the integral domain, the volume obtained via integration is the same as that obtained using the function value at the centroid of the domain multiplied by the volume of the domain. This is true for all linear functions, as discussed earlier.

5.6.1.3 *Example: Hyper-volume under a 3D parabolic function*

Let us compute the hyper-volume under a parabolic hyper-surface over a cube. The function for the surface is defined as

$$f(x, y, z) = R^2 - (x^2 + y^2 + z^2) \tag{5.38}$$

where R is the radius of the paraboloid.

```
1  R = symbols('R ', positive=True)
2
3  fxyz = R**2 - (x**2 + y**2 + z**2)               # surface of paraboloid
4  xL, xU = -1., 1.;   yL, yU = -1., 1.;   zL, zU = -1., 1.   # for 3D domain
5
6  fxy = integrate(fxyz, (z, zL,zU))        # integral along z over [zL,zU]
7  fx  = integrate(fxy,  (y, yL,yU))                # along y over [yL,yU]
8  V_f = integrate(fx,   (x, xL,xU))                # along x over [xL,xU]
9  display(Math(f" \\text{{Hyper-volume under fxyz = }}{sp.latex(V_f)}"))
```

Hyper-volume under fxyz $= 8.0R^2 - 8.0$

This is the formula to compute the hyper-volume over a standard cube. For a given R, we obtain the following:

```
1  print(f' Hyper-volume of a given 3D cone, I = {V_f.subs({R:2})}')
```

```
Hyper-volume of a given 3D cone, I = 24.0000000000000
```

5.6.2 *Integration over spheric domains*

5.6.2.1 *Formulation*

Triplet integrals can be done in spheric coordinates, which relate to the Cartesian coordinates as

$$x = r \sin \theta \, \cos \varphi$$

$$y = r \sin \theta \, \sin \varphi$$

$$z = r \cos \theta \tag{5.39}$$

where r, θ, and φ are, respectively, the radial, polar angle, and azimuthal angle coordinates. We used the same coordinates in Chapter 4 when dealing with the line integrals. The Jacobian matrix and its determinant can be obtained using the following code:

```
1  θ, φ = symbols('θ, φ', real=True)      # polar angle and azimuthal angle
2  r, R = symbols('r, R', positive=True)        # radial coordinate, Radius
3
4  x = r*sp.sin(θ)*sp.cos(φ)
5  y = r*sp.sin(θ)*sp.sin(φ)
6  z = r*sp.cos(θ)
7
8  J = sp.Matrix([[x.diff(r), x.diff(θ), x.diff(φ)],
9                 [y.diff(r), y.diff(θ), y.diff(φ)],
10                [z.diff(r), z.diff(θ), z.diff(φ)]])
11 J
```

$$\begin{bmatrix} \sin(\theta)\cos(\varphi) & r\cos(\theta)\cos(\varphi) & -r\sin(\theta)\sin(\varphi) \\ \sin(\theta)\sin(\varphi) & r\sin(\varphi)\cos(\theta) & r\sin(\theta)\cos(\varphi) \\ \cos(\theta) & -r\sin(\theta) & 0 \end{bmatrix}$$

```
1  # Determinant of the Jacobian matrix:
2  detJ = sp.det(J).simplify()
3  detJ                                         # the volume scalar
```

$$r^2 \sin(\theta)$$

With the volume scalar obtained, the integral in the spheric coordinates can be expressed as

$$I = \int_{r_L}^{r_U} r^2 \left\{ \int_{\theta_L}^{\theta_U} \sin \theta \left[\int_{\varphi_L}^{\varphi_U} f(r, \theta, \varphi) d\varphi \right] d\theta \right\} dr \tag{5.40}$$

Depending on the integrand $f(r, \theta, \varphi)$, the sequence of the integral may change for convenience and effectiveness in carrying out the integration.

Equation (5.40) is nested three standard 1D definite integrals over a regular domain similar to Eq. (5.37).

5.6.2.2 *Example: Volume, surface area of a ball*

Using the spherical coordinates, we can easily compute the volume or partial volume, the surface or partial surface of a ball. The codes are given as follows:

```
1  # Compute the volume of a ball with radius of R via integration:
2  f = 1                          # unit function for volume integration
3  detJ = r**2*sin(θ)                        # volume scalar
4  fI = f*detJ
5  V=integrate(integrate(integrate(fI,(r,0,R)),(θ,0,sp.pi)),(φ,0,2*sp.pi))
6  display(Math(f" \\text{{Formula for ball volume = }}{sp.latex(V)}"))
```

Formula for ball volume $= \dfrac{4\pi R^3}{3}$

```
1  # Compute partial volume of a ball with radius of R:
2  θL = 0; θU = sp.pi/3                     # ↓ partial polar angle
3  V=integrate(integrate(integrate(fI,(r,0,R)), (θ,θL,θU)), (φ,0,2*sp.pi))
4
5  display(Math(f" \\text{{Formula for partial (θ: 0~}}{sp.latex(θU)}) \
6                          \\text{{ volume = }}{sp.latex(V)}"))
```

Formula for partial $\left(\theta : \dfrac{\pi}{3}\right)$ volume $= \dfrac{\pi R^3}{3}$

```
1  # Compute the surface area of a ball with radius of R via integration:
2  f = 1                          # unit function for area integration
3  fI = f*detJ.subs(r,R)
4  A = integrate(integrate(fI, (θ,0,sp.pi)),(φ,0,2*sp.pi))
5  display(Math(f" \\text{{Formula for surface area of a ball = }}\
6                          {sp.latex(A)}"))
```

Formula for surface area of a ball $= 4\pi R^2$

```
1  # Compute the partial surface area of a ball with radius of R:
2  fI = f*detJ.subs(r,R)
3  θL = 0; θU = sp.pi/2                        # partial polar angle
4  A_partial = integrate(integrate(fI, (θ, θL, θU)),(φ, 0,2*sp.pi))
5  display(Math(f" \\text{{Formula for partial (θ: 0~}}{sp.latex(θU)}) \
6          \\text{{ surface area of a ball = }}{sp.latex(A_partial)}"))
```

Formula for partial $\left(\theta : 0 \sim \dfrac{\pi}{2}\right)$ surface area of a ball $= 2\pi R^2$

5.6.3 *Integration over cylindrical domains*

Cylindric coordinates are useful in dealing with systems that are axial symmetric. The cylindric coordinate system with z-axis being the axial symmetric axis is shown in Fig. 5.5.

The cylindrical coordinates fit well for axial symmetrical domains such as a cylinder or a cone, as shown in Fig. 5.6.

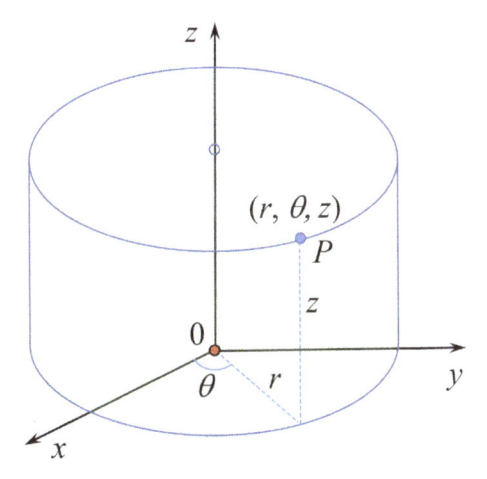

Figure 5.5. Cylindrical coordinates with radial distance r, polar angle θ, and vertical axis z.

Figure 5.6. A cylinder with a cone inside. Both the cylinder and the cone are all axial symmetric.

5.6.3.1 *Formulation*

Three-dimensional integrals can be done in cylindric coordinates, which relate to the Cartesian coordinates as

$$x = r\cos\theta; \quad y = r\sin\theta; \quad z = z \tag{5.41}$$

where r and θ are, respectively, the radial and polar angle coordinates. Axis z is not changed. The Jacobian matrix and its determinant can be obtained using the following code:

```
1  z, θ = symbols('z, θ', real=True)
2  r, R = symbols('r, R', positive=True)
3
4  x = r*sp.cos(θ)                         # same as the polar coordinates
5  y = r*sp.sin(θ)
6  z = z                                   # no change in z-axis
7
8  J_cyl = sp.Matrix([[x.diff(r), x.diff(θ), x.diff(z)],
9                     [y.diff(r), y.diff(θ), y.diff(z)],
10                    [z.diff(r), z.diff(θ), z.diff(z)]])
11 #gr.printM(J_cyl, "The Jacobian matrix for cylindrical coordinate:")
```

We found the Jacobian matrix for cylindrical coordinate as

$$\begin{bmatrix} \cos(\theta) & -r\sin(\theta) & 0 \\ \sin(\theta) & r\cos(\theta) & 0 \\ 0 & 0 & 1 \end{bmatrix} \tag{5.42}$$

The determinant of the Jacobian matrix is computed using the following:

```
1  detJ_cyl = sp.det(J_cyl).simplify()     # Determinant of Jacobian matrix
2  detJ_cyl                                 # the volume scalar
```

r

The volume scalar is found simply as r. The integral in the cylindrical coordinates can be expressed as

$$I = \int_{r_L}^{r_U} r \left\{ \int_{\theta_L}^{\theta_U} \left[\int_{z_L}^{z_U} f(r,\theta,z)dz \right] d\theta \right\} dr \tag{5.43}$$

The sequence of the integration may change depending on convenience. As seen, the integration over cylindrical coordinates is largely the same as that over polar coordinates except one more dimension along z-axis.

5.6.3.2 *Example: Volume and partial volume of a cylinder*

Using cylindrical coordinates, we can efficiently compute the volume or partial volume of a cylinder as well as its surface area or partial surface area. Consider a cylinder with radius R and height H, as shown in Fig. 5.6 (without the cone inside). The codes for computing the volume and partial volume of a cylinder are given in the following:

```
1  # Compute the volume of a cylinder with radius of R via integration:
2  z, θ = symbols('z, θ', real=True)
3  R, H, Θ = symbols('R, H, Θ', positive=True)    # [0, θ]: interval for θ
4  f = 1                                          # function for volume integration
5  fI = f*detJ_cyl
6  V_cyl = integrate(integrate(integrate(fI,(r,0,R)),(θ,0,Θ)),(z,0,H))
7  display(Math(f" \\text{{Formula for computing volume of a cylinder = }}\
8                                                  {sp.latex(V_cyl)}"))
```

Formula for computing volume of a cylinder $= \dfrac{HR^2\Theta}{2}$

```
1  # Set dimensional parameters:
2  print(f' Volume of a full cylinder = {V_cyl.subs({R:1,H:2,θ:2*sp.pi})}')
3  print(f' Volume of a half cylinder = {V_cyl.subs({R:1,H:2,θ: sp.pi})}')
```

```
Volume of a full cylinder = 2*pi
Volume of a half cylinder = pi
```

```
1  # Compute the surface area of a cylinder with radius of R via integral:
2  f = 1                                          # unit function
3  fI = (f*detJ_cyl).subs(r,R)
4  A_cyl = integrate(integrate(fI, (θ,0,Θ)),(z,0,H))
5  print(f' Formula for computing surface area of a cylinder = {A_cyl}')
```

```
Formula for computing surface area of a cylinder = H*R*θ
```

```
1  # Set dimensional parameters:
2  print(f' Surface area, full cylinder={A_cyl.subs({R:1,H:2,θ:2*sp.pi})}')
3  print(f' Surface area, half cylinder={A_cyl.subs({R:1,H:2,θ: sp.pi})}')
```

```
Surface area, full cylinder=4*pi
Surface area, half cylinder=2*pi
```

5.6.3.3 *Example: Volume and partial volume of a cone*

Consider a cone schematically shown in Fig. 5.6 (inside the cylinder).

The function for its surface is given by

$$r = R\left(1 - \frac{z}{H}\right) \tag{5.44}$$

where H is the height of the cone and R is the base radius of the cone. We compute the volume and partial volume of the cone, using the following code:

```
1  # Compute the volume of a cone with radius of R and height H:
2  z, θ = symbols('z, θ', real=True)
3  R, H, θ = symbols('R, H, θ', positive=True)    # [0, θ]: interval for θ
4
5  f = 1                              # function for volume integration
6  fI = f*detJ_cyl
7  frz = integrate(fI,(θ, 0, θ))                  # integrate along θ
8
9  # r needs to staying being a variable, because it relates to z:
10 fz  = integrate(frz, (r, 0, R*(1-z/H)))
11 display(Math(f" \\text{{{$f(z)$ = }} {sp.latex(fz)}"))
```

$$f(z) = \frac{R^2 \Theta \left(1 - \frac{z}{H}\right)^2}{2}$$

Here, we used Eq. (5.44) for the upper limit of r. The formula obtained is a function of z. Integrating it along z gives the following:

```
1  V_cone = integrate(fz, (z, 0, H))
2  display(Math(f" \\text{{Formula for computing volume of a cone = }}\
3                                          {sp.latex(V_cone)}"))
```

$$\text{Formula for computing volume of a cone} = \frac{HR^2 \Theta}{6}$$

```
1  # substitute specific varlue of θ to the formula:
2  display(Math(f" \\text{{Formula for the volume of the full cone = }}\
3                              {sp.latex(V_cone.subs({θ:2*pi}))}"))
```

$$\text{Formula for the volume of the full cone} = \frac{\pi HR^2}{3}$$

This is the equation we obtained earlier. For this example, one can also integrate over θ and z first, use Eq. (5.44) to convert z to a function of r, and finally integrate over r. The result will be the same. Readers may give it a try.

5.6.3.4 *Example: Polar moment of inertia of a cone and cylinder*

Consider first a solid cone schematically shown in Fig. 5.6 (inside the cylinder). The function for computing the polar moment of inertia with respect to the vertical z-axis is $f(r) = r^2$. The integral for computing the polar moment of inertia becomes

$$I = \int_{z_L}^{z_U} \left\{ \int_{r_L}^{r_U} r \left[\int_0^{2\pi} r^2 d\theta \right] dr \right\} dz \qquad (5.45)$$

Assume that the height of the cone is H. We write the following code to do the task:

```
1  # Compute the polar moment of inertia of a cone with height H:
2  z, θ = symbols('z, θ', real=True)
3  H = symbols('H ', positive=True)
4  f = r**2                           # function for polar moment of inertia
5  fI = f*detJ_cyl
6  frz = integrate(fI,(θ, 0, 2*sp.pi))              # integrate along θ
7  display(Math(f" \\text{{{$f(r,z)$ = }} {sp.latex(frz)}"))
8
9  # r needs to staying being a variable, because it relates to z:
10 fz  = integrate(frz, (r, 0, R*(1-z/H)))  # use (r, 0, R) for a cylinder
11 display(Math(f" \\text{{{$f(z)$ = }} {sp.latex(fz)}"))
```

$$f(r, z) = 2\pi r^3$$
$$f(z) = \frac{\pi R^4 \left(1 - \frac{z}{H}\right)^4}{2}$$

Here, we also used Eq. (5.44) for the upper limit of r. The formula obtained is a function of z. Integrating it along z gives the following:

```
1  Jz = integrate(fz, (z, 0, H))
2  display(Math(f" \\text{{{Formula for polar moment of inertia of a \
3                            cone = }}{sp.latex(Jz)}"))
```

Formula for polar moment of inertia of a cone $= \dfrac{\pi H R^4}{10}$

Considering now a cylinder with inner radius of R_L and outer radius of R_U, we integrate frz (obtained after the integration along θ) directly along z:

```
1  # r needs to staying being a variable, because it relates to z:
2  r, RL, RU = symbols('r,R_L,R_U', positive=True) # R interval: [R_L,R_U]
3
4  fz = integrate(frz, (r, RL, RU))                 # frz: obtained earlier
5  Jz = integrate(fz, (z, 0, H))
6  display(Math(f" \\text{{{Formula for polar moment of inertia of a \
7                            cylinder = }}{sp.latex(Jz)}"))
```

$$\text{Formula for polar moment of inertia of a cylinder} = H\left(-\frac{\pi R_L^4}{2} + \frac{\pi R_U^4}{2}\right)$$

This is essentially the same formula obtained earlier using the polar coordinates. The only difference is that the height of the cylinder is now taken into account.

```
1  # substitute specific varlues for these parameters:
2
3  print(f' Jz, cylinder = {Jz.subs({RL:0.5, RU:1, H:1})}')
```

```
Jz, cylinder = 0.46875*pi
```

5.7 Integration over irregular 3D domain in natural coordinates

5.7.1 *Coordinate transformation in 3D*

Consider a hexahedral domain with eight points/nodes (H8) at its vertices. Since an H8 domain is in general not regular, we need to perform a mapping between the Cartesian coordinates (x, y, z) and the natural coordinates (ξ, η, ζ) defined over a standard cube. It occupies $[-1, 1] \times [-1, 1] \times [-1, 1]$, as shown in Fig. 5.7.

It is a 1D extension from the 2D case. All the mapping-related formulas obtained previously apply. The most important formula is the Jacobian matrix, whose general expression is given in Eq. (5.7). The determinant of the Jacobian matrix accounts for the distortion introduced when transforming the physical domain (H8) to the standard cube in natural coordinates. This allows us to perform numerical integration over the standard cube using the

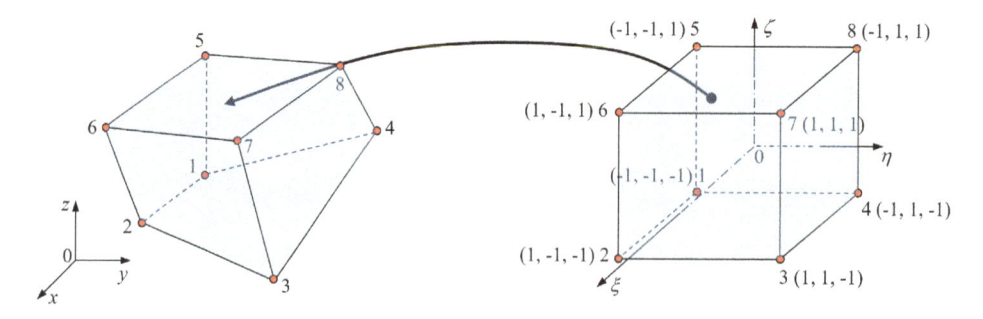

Figure 5.7. Coordinate mapping in 3D. Left: a general hexahedron with eight nodes (H8) in the physical x-, y-, z- coordinates; right: the standard cube in the natural ξ-, η-, ζ- coordinates.

Jacobian determinant. In this section, we leverage trilinear mapping created using the Lagrangian nodal shape functions. These shape functions define the variation of the mapped coordinates within the H8 domain. The numerical sampling and integration are then performed along straight lines parallel to ξ, η, ζ, respectively.

The formula for the mapping can be written as

$$x = \sum_{k=1}^{8} x_k N_k(\xi, \eta, \zeta)$$

$$y = \sum_{k=1}^{8} y_k N_k(\xi, \eta, \zeta)$$

$$z = \sum_{k=1}^{8} z_k N_k(\xi, \eta, \zeta) \tag{5.46}$$

where (x_k, y_k, z_k) are the coordinates of the kth node and $N_k(\xi, \eta, \zeta)$ is the kth nodal shape function, which is trilinear for an H8 domain.

Using Eqs. (5.7) and (5.46), the Jacobian matrix is computed using

$$\mathbf{J} = \begin{bmatrix} \sum_{k=1}^{8} x_k \frac{\partial N_k(\xi,\eta,\zeta)}{\partial \xi} & \sum_{k=1}^{8} x_k \frac{\partial N_k(\xi,\eta,\zeta)}{\partial \eta} & \sum_{k=1}^{8} x_k \frac{\partial N_k(\xi,\eta,\zeta)}{\partial \zeta} \\ \sum_{k=1}^{8} y_k \frac{\partial N_k(\xi,\eta,\zeta)}{\partial \xi} & \sum_{k=1}^{8} y_k \frac{\partial N_k(\xi,\eta,\zeta)}{\partial \eta} & \sum_{k=1}^{8} y_k \frac{\partial N_k(\xi,\eta,\zeta)}{\partial \zeta} \\ \sum_{k=1}^{8} z_k \frac{\partial N_k(\xi,\eta,\zeta)}{\partial \xi} & \sum_{k=1}^{8} z_k \frac{\partial N_k(\xi,\eta,\zeta)}{\partial \eta} & \sum_{k=1}^{8} z_k \frac{\partial N_k(\xi,\eta,\zeta)}{\partial \zeta} \end{bmatrix} \tag{5.47}$$

The Jacobian matrix is a function of both the natural coordinates (ξ, η, ζ) and the coordinates (x_k, y_k, z_k) of the eight vertices (nodes) of the H8 domain. These nodal coordinates define the specific location and "shape" of the H8 domain in the physical domain. The Jacobian matrix can be computed using the derivatives of the nodal shape functions for a given H8 domain, which are relatively straightforward to calculate, as shown in Chapter 3.

5.7.2 *Integration in natural coordinates*

With coordinate mapping, 3D triplet integrals can now be done over a standard cube. All we need is to add the volume scalar in the integral. The formula for the mapped integration is

$$I = \int_{\mathcal{D}} f(x, y, z)dxdydz$$

$$= \int_{\xi_L}^{\xi_U} \int_{\eta_L}^{\eta_U} \int_{\zeta_L}^{\zeta_U} f(\xi, \eta, \zeta)\det(\mathbf{J})d\zeta d\eta d\zeta$$

$$= \int_{-1}^{1} \int_{-1}^{1} \int_{-1}^{1} f(\xi, \eta, \zeta) \underbrace{\det(\mathbf{J})}_{\text{volume scalar}} d\zeta d\eta d\zeta \qquad (5.48)$$

The integral is now transformed into a nested three-layer 1D integral over a standard cube thanks to the Jacobian matrix. This offers significant advantages for numerical computation. The integral limits are fixed from -1 to 1 in all directions, making the integration process convenient. Additionally, sampling points along all three directions can be determined in a fixed manner.

Since the determinant of the Jacobian matrix depends on the specific shape of the H8 domain, which is defined by its nodal coordinates (refer to Eq. (5.46)), we need to calculate $\det(\mathbf{J})$ for each individual H8 domain that has different shape.

5.7.3 *Trilinear shape functions*

Similar to 2D cases, we first compute the nodal shape functions. For an H8 domain in 3D, we use trilinear shape functions that can be created using the Lagrange interpolators, as we did in Chapter 3. Naturally, the formulas to be obtained will be quite long:

```
1  ξ, η, ζ =symbols('ξ, η, ζ')                    # define 3D natrual coordinates
2
3  nodexi = [-1, 1]; nodeyj = [-1, 1]; nodeyk = [-1, 1]       # Compute Nk
4  lxs, lys, lzs, N8 = gr.N_Lagrange3D_f(nodexi, nodeyj, nodeyk)
5
6  print(f'Lagrange interpolators: {lxs}')
7  print(f'Lagrange interpolators: {lys}')
8  print(f'Lagrange interpolators: {lzs}')
9  #print(f'Formed shape functions N:{N8}')
10 N8[:3]                                          # display 3 out of 8
```

```
Lagrange interpolators: [1/2 - ξ/2, ξ/2 + 1/2]
Lagrange interpolators: [1/2 - η/2, η/2 + 1/2]
Lagrange interpolators: [1/2 - ζ/2, ζ/2 + 1/2]
```

$$\left[-\frac{(\zeta - 1)(\eta - 1)(\xi - 1)}{8}, \ \frac{(\zeta - 1)(\eta - 1)(\xi + 1)}{8}, \ -\frac{(\zeta - 1)(\eta + 1)(\xi + 1)}{8} \right]$$

These trilinear nodal shape functions are all linear in only three directions. We can now compute the expressions for x, y, and z as functions of ξ,
η, and ζ and the coordinates (x_k, y_k, z_k):

```
1  # define symbolic coordinate variables:
2  x0, x1, x2, x3, x4, x5, x6, x7 =sp.symbols('x0,x1,x2,x3,x4,x5,x6,x7')
3  y0, y1, y2, y3, y4, y5, y6, y7 =sp.symbols('y0,y1,y2,y3,y4,y5,y6,y7')
4  z0, z1, z2, z3, z4, z5, z6, z7 =sp.symbols('z0,z1,z2,z3,z4,z5,z6,z7')
5
6  nodeX = sp.Matrix([x0,x1,x2,x3,x4,x5,x6,x7])     # vector of coordinates
7  nodeY = sp.Matrix([y0,y1,y2,y3,y4,y5,y6,y7])
8  nodeZ = sp.Matrix([z0,z1,z2,z3,z4,z5,z6,z7])
9
10 N8M = sp.Matrix(N8).T
11 xξηζ = (N8M @ nodeX)[0]                          # expression of x in terms of ξ,η,ζ
12 yξηζ = (N8M @ nodeY)[0]
13 zξηζ = (N8M @ nodeZ)[0]
14 #x.simplify()                                    # display one case (quite lengthy)
```

5.7.4 *Computation of the Jacobian matrix in 3D*

Next, compute the Jacobian matrix:

```
1  H8J = sp.Matrix([[xξηζ.diff(ξ), xξηζ.diff(η), xξηζ.diff(ζ)],
2                   [yξηζ.diff(ξ), yξηζ.diff(η), yξηζ.diff(ζ)],
3                   [zξηζ.diff(ξ), zξηζ.diff(η), zξηζ.diff(ζ)]])
4  #H8J[:,0]          # 1st column of the Jacobian matrix, still out off page
```

Before computing the determinant of the Jacobian matrix, we substitute
the coordinate values to the Jacobian matrix expression. This reduces drastically the length of the expression for the obtained determinant. Otherwise,
the length would be about 20 pages, although SymPy is able to derive the
expression for us:

```
1  # Compute the Jacobian:
2  H8detJ = sp.det(H8J).simplify()
3  #!!! Donot display, executed in 1m 36.4s, finished 17:04:01 2024-07-12
```

The H8detJ (det($\mathbf{J}$) for the H8 element) is a polynomial of the natural coordinates, making it relatively straightforward to integrate during the
following integration process. As we learned previously, polynomials can be
integrated with ease. We are now ready to derive the formula for computing
the volume of an H8 element.

5.7.5 *Formula for volume of a hexahedron*

We can now perform 3D definite integration over the standard cube in the natural coordinates to obtain the volume of an H8. The integration is a nested three-layer 1D integral:

```
 1  f = 1                              # unit function for volume computation
 2  fI = f*H8detJ                            # multiply the volume scalar
 3
 4  #          three-layers nested 1D integrals:
 5  V_H8xyz = integrate(integrate(integrate(fI,(ξ,-1,1)),(η,-1,1)),(ζ,-1,1))
 6
 7  # Examine these expression:
 8  print(len(V_H8xyz.as_ordered_terms()),'terms')# terms in the expression
 9  print(str(V_H8xyz)[:70]+"...")         # display first part of expression
10  sp.sympify(str(V_H8xyz)[:110])            # print part of it latex-like
11  #V_H8xyz    # took 2m 21s, expression is ~ 1-page long; display 1st line
```

```
144 terms
-x0*y1*z2/12 - x0*y1*z3/12 + x0*y1*z4/12 + x0*y1*z5/12 + x0*y2*z1/12 -...
```

$$-\frac{x_0 y_1 z_2}{12} - \frac{x_0 y_1 z_3}{12} + \frac{x_0 y_1 z_4}{12} + \frac{x_0 y_1 z_5}{12} + \frac{x_0 y_2 z_1}{12} - \frac{x_0 y_2 z_3}{12} + \frac{x_0 y_3 z_1}{12} + \frac{x_0 y_3 z_2}{12}$$

We have successfully obtained the formula for computing the volume of a general hexahedron with eight nodes (H8). As expected, the formula is a function of the physical coordinates of the eight nodes and is quite lengthy with 144 terms. Let's call this formula the long formula for later reference.

Our aim now is to simplify this formula. This takes a number of steps.

To start, we leverage the fact that the volume of an H8 domain should not depend on its absolute location within the coordinate system. Thus, we can perform a coordinate translation. We relocate the origin of the coordinate system to coincide with node-0 of the H8. In other words, we set the coordinates of node-0 to (0, 0, 0). Consequently, the coordinates of all other nodes are re-expressed relative to node-0 using its coordinates (x_0, y_0, z_0) as the reference point. The formulas to update the coordinates of these nodes are

$$x_i := x_i - x_0; \quad y_i := y_i - y_0; \quad z_i := z_i - z_0; \quad i = 0, 1, 2, \ldots, 7 \quad (5.49)$$

This translation eliminates the dependence of the formula on the absolute location of the H8 element, leading to a simpler expression without affecting the value of the volume of H8.

```
1  V_H8_0 = V_H8xyz.subs({nodeX[0]:0, nodeY[0]:0, nodeZ[0]:0}).simplify()
2  print(len(V_H8_0.as_ordered_terms()),'terms') # terms in the expression
3  sp.sympify(str(V_H8_0)[:110])               # print part of it latex-like
4  #print(V_H8_0)                   # still long about one page; display 1st line
```

90 terms

$$-\frac{x_1 y_2 z_3}{12} + \frac{x_1 y_2 z_5}{12} + \frac{x_1 y_2 z_6}{12} + \frac{x_1 y_3 z_2}{12} - \frac{x_1 y_4 z_5}{12} - \frac{x_1 y_5 z_2}{12} + \frac{x_1 y_5 z_4}{12} - \frac{x_1 y_5 z_6}{12}$$

We have successfully reduced the formula size from 144 to 90 terms. However, there's room for further simplification.

5.7.6 *A vMv dot-product formula for H8 volume*

We observe some interesting properties in the formula's structure that might allow for a more compact representation. First, all terms share a coefficient of 12. Second, each term is trilinear in the 3D nodal coordinates. By factoring out the common coefficient (12) and rearranging the terms based on their trilinear nature, we might be able to obtain a significantly simpler expression:

$$V_{H8} = \frac{1}{12} \mathbf{x}_{nd}^{\top} \mathbf{Z}_{nd} \mathbf{y}_{nd} \tag{5.50}$$

where $\mathbf{x}_{nd}$: vector containing nodal x-coordinates (all x_i values), $\mathbf{y}_{nd}$: vector containing nodal y-coordinates (all y_i values) and $\mathbf{Z}_{nd}$: matrix containing nodal z-coordinates (arranged in some way).

Equation (5.50) is a **vector–matrix–vector** (vMv) dot product. It is a form of matrix-modulated dot product, which drastically reduces the operation counts and can be executed efficiently in most modern computers via parallelization.

To find such an expression, in particular the $\mathbf{Z}_{nd}$ matrix, we need to perform two stages of factorization using SymPy. The following codes may require a good understanding of SymPy. Readers may just take a look at the procedure and the outcome for now if there is difficulty to follow the coding.

First, we factor out x_i:

```
1  V_H8_12 = V_H8_0*12                    # remove 1/12 for simpler expressions
2  yz=[V_H8_12.factor(nodeX[i]).args[0].coeff(nodeX[i])
3                          for i in range(1,8)]
4
5  print(len(yz[0].as_ordered_terms()),'terms')  # terms in the expression
6  sp.sympify(str(yz[0]))                      # print part of it latex-like
7  # The results are 7 expressions, half a page long;  display the 1st one
```

10 terms

$$-y_2 z_3 + y_2 z_5 + y_2 z_6 + y_3 z_2 - y_4 z_5 - y_5 z_2 + y_5 z_4 - y_5 z_6 - y_6 z_2 + y_6 z_5$$

This is the expression of coefficient x_1. We have seven of these. Next, factor out y_i for each of these:

```
1  y_Z = [yzi.factor(nodeY) for yzi in yz]   # all the terms in expression
2  print(len(y_Z[0].as_ordered_terms()),'terms, the first term is:')
3  sp.sympify(str(y_Z[0]))                    # print part of it latex-like
4  # The results 5 expressions; display the 1st expression
```

5 terms, the first term is:

$$-y_2 \left(z_3 - z_5 - z_6 \right) + y_3 z_2 - y_4 z_5 - y_5 \left(z_2 - z_4 + z_6 \right) - y_6 \left(z_2 - z_5 \right)$$

Now, we are ready to form the matrix $\mathbf{Z}$ by collecting these coefficients of y_i:

```
1  Z = sp.zeros(len(nodeZ), len(nodeZ))
2  for i in range (1, 8):
3      Z[i,1:]=sp.Matrix([y_Z[i-1].coeff(nodeY[j]) for j in range(1, 8)]).T
4
5  #gr.printM(Z[1:,1:],"The Z matrix containing z-coordinates") # over page
```

We obtain matrix $\mathbf{Z}$ and it is given in Eq. (5.51). It is intentionally made smaller to fit in the page for easy examination:

$$
\begin{bmatrix}
0 & -z_3+z_5+z_6 & z_2 & -z_5 & -z_2+z_4-z_6 & -z_2+z_5 & 0 \\
z_3-z_5-z_6 & 0 & -z_1+z_6+z_7 & 0 & z_1-z_6 & z_1-z_3+z_5-z_7 & -z_3+z_6 \\
-z_2 & z_1-z_6-z_7 & 0 & z_7 & 0 & z_2-z_7 & z_2-z_4+z_6 \\
z_5 & 0 & -z_7 & 0 & -z_1+z_6+z_7 & -z_5+z_7 & z_3-z_5-z_6 \\
z_2-z_4+z_6 & -z_1+z_6 & 0 & z_1-z_6-z_7 & 0 & -z_1-z_2+z_4+z_7 & z_4-z_6 \\
z_2-z_5 & -z_1+z_3-z_5+z_7 & -z_2+z_7 & z_5-z_7 & z_1+z_2-z_4-z_7 & 0 & -z_2-z_3+z_4+z_5 \\
0 & z_3-z_6 & -z_2+z_4-z_6 & -z_3+z_5+z_6 & -z_4+z_6 & z_2+z_3-z_4-z_5 & 0
\end{bmatrix}
$$

$$(5.51)$$

It contains only the z-coordinates of the H8. We can make the following observations:

1. The matrix is antisymmetric with zero diagonal terms. This suggests a connection to the cross product, as discussed in Ref. [5].
2. The volume of an H8 can now be computed very efficiently using Eq. (5.50). It involves a matrix–vector product and a dot product, both of which can be parallelized effectively.
3. This form of triple formula always produces a scalar (volume).
4. On the author's laptop, computing the volume via coordinate mapping (involving computation of the determinant of the Jacobian and 3D integration) takes about 4 minutes. Using volume_H8() that codes Eq. (5.50), it takes about 2 seconds. This is about 100 times faster.

Equation (5.50) is an explicit analytical formula for computing the volume of a general hexahedron with eight nodes. It is obtained for the first time by the author.

Let us write a function in SymPy for computing the volume of a general hexahedron, using its eight nodal coordinates (H8) with node-0 removed.

5.7.7 *Python code for H8 volume using vMv formula*

```
 1  def volume_H8(nodeX, nodeY, nodeZ):
 2      '''
 3      Compute the volume of a general hexahedron, using its eight nodal
 4      coordinates of an H8, with node-0 removed.
 5      It uses the vetor-matrix-vector (vMv) dot-product formula.
 6      '''
 7      z1, z2, z3, z4, z5, z6, z7 =  nodeZ                # node-0 removed
 8
 9      r=z3-z5; s=z1+z2; t=z2-z4; u=z2-z7; k=z5-z7    # to simplify matrix
10      v=z2-z5; w=z1-z6; p=z3-z6; q=z1-z7; g=z4-z6; h=z4+z7
11
12      Z = sp.Matrix([[0,    -r+z6,     z2,   -z5,-t-z6,    -v,     0],
13                     [r-z6,     0, -w+z7,     0,     w,   q-r,    -p],
14                     [-z2,  w-z7,     0,    z7,     0,     u,  t+z6],
15                     [ z5,     0,   -z7,     0,-q+z6,    -k, r-z6],
16                     [t+z6,   -w,     0,  q-z6,    0,  -s+h,     g],
17                     [v,    -q+r,    -u,     k,   s-h,     0,  -t-r],
18                     [0,       p,  -t-z6, -r+z6,   -g,   t+r,     0]])
19
20      V = (sp.Matrix(nodeX).T@Z@sp.Matrix(nodeY))[0]
21
22      return V/12, Z
```

5.7.8 *Equivalent vMv formulas*

Equation (5.50) can be written in the following equivalent forms:

$$V_{H8} = \frac{1}{12}\mathbf{x}_{nd}^\top \mathbf{Z}_{nd}\,\mathbf{y}_{nd} = -\frac{1}{12}\mathbf{y}_{nd}^\top \mathbf{Z}_{nd}\,\mathbf{x}_{nd}$$

$$= -\frac{1}{12}\mathbf{x}_{nd}^\top \mathbf{Y}_{nd}\,\mathbf{z}_{nd} = \frac{1}{12}\mathbf{z}_{nd}^\top \mathbf{Y}_{nd}\,\mathbf{x}_{nd}$$

$$= -\frac{1}{12}\mathbf{z}_{nd}^\top \mathbf{X}_{nd}\,\mathbf{y}_{nd} = \frac{1}{12}\mathbf{y}_{nd}^\top \mathbf{X}_{nd}\,\mathbf{z}_{nd} \qquad (5.52)$$

where

- matrix $\mathbf{X}_{nd}$ has the same configuration of matrix $\mathbf{Z}_{nd}$, but all the entries z_i in $\mathbf{Z}_{nd}$ are replaced, respectively, by x_i;
- matrix $\mathbf{Y}_{nd}$ has the same configuration of matrix $\mathbf{Z}_{nd}$, but all the entries z_i in $\mathbf{Z}_{nd}$ are replaced, respectively, by y_i.

The rule for these equivalent forms is as follows: any single swap between any two of x_i, y_i, and z_i flips the sign. Therefore, twice swap results in no sign change. The root reason for this rule is because of the determinant of $\mathbf{J}$ whose sign change follows the same rule. This is also because of the cross product involved in the process, whose result changes the sign for any order swap of the vectors.

5.7.9 *Examples: vMv dot product for the volume of a cube*

Using our volume_H8() function, we compute first the $\mathbf{Z}_{nd}$ matrix for a unit cube treated as an H8 element and then calculate the volume of the unit cube. Since the actual volume of a unit cube is 1, this computation serves as a validation test for volume_H8(). The code snippet is as follows:

```
1  # Nodal coordinates for a unit cube in the physical domain (xk, yk, zk):
2  Cube_nodes=[[0, 0, 0], [1, 0, 0], [1, 1, 0], [0, 1, 0],         # 8 nodes
3             [0, 0, 1], [1, 0, 1], [1, 1, 1], [0, 1, 1]]
4
5  Cube_nodes_np = np.array(Cube_nodes)
6  Cube7nodes = Cube_nodes_np[1:,:]-Cube_nodes_np[0,:]    # remove node-0
7
8  # the xZy form:              x                Z              y
9  V, Z = volume_H8(Cube7nodes[:, 0], Cube7nodes[:, 1], Cube7nodes[:, 2])
10 print(f"The volume of the cube, xZy = {V}")
```

```
11  display(Math(f" \\text{{Z matrix for the cube = }}{latex(Z)}"))
12  #gr.printM(Z, "Z matrix for the cube:")
13  # the zXy form:           z              X              y
14  V, X = volume_H8(Cube7nodes[:, 1], Cube7nodes[:, 0], Cube7nodes[:, 2])
15  print(f"The volume of the cube, zXy = {V}")    # notice the sign change!
16  #display(Math(f" \\text{{X matrix for the cube = }}{latex(X)}"))
```

The volume of the cube, xZy = 1

$$Z \text{ matrix for the cube} = \begin{bmatrix} 0 & 2 & 0 & -1 & 0 & 1 & 0 \\ -2 & 0 & 2 & 0 & -1 & 0 & 1 \\ 0 & -2 & 0 & 1 & 0 & -1 & 0 \\ 1 & 0 & -1 & 0 & 2 & 0 & -2 \\ 0 & 1 & 0 & -2 & 0 & 2 & 0 \\ -1 & 0 & 1 & 0 & -2 & 0 & 2 \\ 0 & -1 & 0 & 2 & 0 & -2 & 0 \end{bmatrix}$$

The volume of the cube, zXy = -1

The $\mathbf{Z}_{nd}$ (or $\mathbf{Y}_{nd}$ or $\mathbf{X}_{nd}$) matrix is antisymmetric with zero diagonals and has a clear pattern.

We can also use volume_H8() to produce an explicit computation-count reduced formula. The output will still be quite long. The code snippet to get this done is given as follows. Readers may give it a try by uncommenting the following two code lines:

```
1  #V, Z = volume_H8(nodeX[1:], nodeY[1:], nodeZ[1:])
2  #gr.printM(V, "Simplified formula for computing the volume of H8:")
```

Finally, let us check whether the simplified formula is the same as the original long formula:

```
1  print(f"Is the same? {(V-V_H8_0).simplify()==0}")
```

Is the same? True

5.7.10 *Volume of a parallelepiped*

If the opposite surfaces of an H8 are parallel, the H8 becomes a parallelepiped, as shown in Fig. 5.8.

In this case, the coordinates at nodes, 3, 6, 7, and 8 are related to those at nodes 2, 4, and 5, as shown also in Fig. 5.8. We need only three vectors that originated at zero. We thus have the following substitutions:

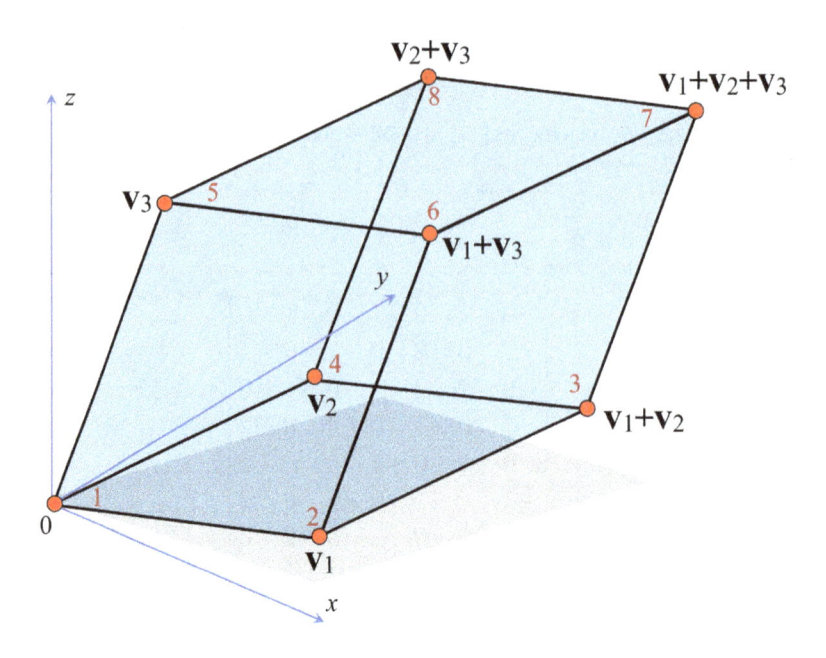

Figure 5.8. A parallelepiped in a local x-, y-, z- coordinates. A special case of H8.

```
 1  dict_ ={nodeX[2]:nodeX[1]+nodeX[3], nodeX[5]:nodeX[1]+nodeX[4],
 2         nodeX[6]:nodeX[1]+nodeX[3]+nodeX[4], nodeX[7]:nodeX[3]+nodeX[4],
 3
 4         nodeY[2]:nodeY[1]+nodeY[3], nodeY[5]:nodeY[1]+nodeY[4],
 5         nodeY[6]:nodeY[1]+nodeY[3]+nodeY[4], nodeY[7]:nodeY[3]+nodeY[4],
 6
 7         nodeZ[2]:nodeZ[1]+nodeZ[3], nodeZ[5]:nodeZ[1]+nodeZ[4],
 8         nodeZ[6]:nodeZ[1]+nodeZ[3]+nodeZ[4], nodeZ[7]:nodeZ[3]+nodeZ[4]}
 9
10  V_P8 = V_H8xyz.subs(dict_).subs({nodeX[0]:0, nodeY[0]:0, nodeZ[0]:0})\
11                              .simplify()
12  V_P8                               # the volume of the parallelpiped
```

$$x_1 y_3 z_4 - x_1 y_4 z_3 - x_3 y_1 z_4 + x_3 y_4 z_1 + x_4 y_1 z_3 - x_4 y_3 z_1$$

The volume of the parallelepiped can be computed by simply using the **triplet product** formula [5] that uses only the three vectors $\mathbf{v}_1$, $\mathbf{v}_2$, and $\mathbf{v}_3$ shown in Fig. 5.8:

$$V_{P8} = \det([\mathbf{v}_1, \mathbf{v}_2, \mathbf{v}_3]) \tag{5.53}$$

Let us confirm this using the following code:

```
1  #                        v1           v2           v3
2  triplet_product = sp.Matrix([[nodeX[1], nodeX[3], nodeX[4]],
3                               [nodeY[1], nodeY[3], nodeY[4]],
4                               [nodeZ[1], nodeZ[3], nodeZ[4]]])
5
6  V_parallelepiped = sp.det(triplet_product)
7  print(f"Is the volume of P8 same as that given by the triplet_product?"
8          f" {V_P8==V_parallelepiped}")
9  V_parallelepiped
```

```
Is the volume of P8 same as that given by the triplet_product? True
```

$$x_1 y_3 z_4 - x_1 y_4 z_3 - x_3 y_1 z_4 + x_3 y_4 z_1 + x_4 y_1 z_3 - x_4 y_3 z_1$$

5.7.11 *Formulas for volume of a tetrahedron*

Base on the simple geometry, the volume of a tetrahedron with four nodes (T4) is just one-sixth of the volume of the parallelepiped:

$$V_{T4} = \frac{1}{6} \det\left([\mathbf{v}_3, \mathbf{v}_2, \mathbf{v}_1]\right) \quad \text{(order-sensitive)}$$

$$= \frac{1}{6}\left(x_1 y_3 z_4 - x_1 y_4 z_3 - x_3 y_1 z_4 + x_3 y_4 z_1 + x_4 y_1 z_3 - x_4 y_3 z_1\right) \quad (5.54)$$

Equation (5.54) can also be written in the following vMv form:

$$V_{T4} = \frac{1}{6} \mathbf{x}_{nd}^\top \mathbf{Z}_{nd}\, \mathbf{y}_{nd}$$

$$= \frac{1}{6} \begin{bmatrix} x_1 & x_3 & x_4 \end{bmatrix} \begin{bmatrix} 0 & z_4 & -z_3 \\ -z_4 & 0 & z_1 \\ z_3 & -z_1 & 0 \end{bmatrix} \begin{bmatrix} y_1 \\ y_3 \\ y_4 \end{bmatrix} \quad (5.55)$$

We obtained two formulas for computing the volume of a tetrahedron (T4):

- **Equation (5.54):** This is a well-established formula that finds application in various areas.
- **Equation (5.55):** This is a vMv dot-product formula derived by the author. The derivation procedure is similar to the one used for the H8 volume formula. The code snippets for deriving this formula is given as follows.

Note that the equivalence given in Eq. (5.52) applies to Eq. (5.55) for the same reason mentioned earlier.

```
1  # Factor out xi from the volume formula for P8:
2  yz=[V_P8.factor(nodeX[i]).args[0].coeff(nodeX[i]) for i in [1,3,4]]
3  yz
```

$$[y_3 z_4 - y_4 z_3, \ -y_1 z_4 + y_4 z_1, \ y_1 z_3 - y_3 z_1]$$

```
1  # Create the matrix Z:
2  Z_P8 = sp.zeros(3,3)
3  for i in range(3):
4      Z_P8[i,:]= sp.Matrix([yz[i].coeff(nodeY[j]) for j in [1, 3, 4]]).T
5
6  gr.printM(Z_P8, "The Z matrix containing z-coordinates:")
```

The Z matrix containing z-coordinates:

$$\begin{bmatrix} 0 & z_4 & -z_3 \\ -z_4 & 0 & z_1 \\ z_3 & -z_1 & 0 \end{bmatrix}$$

Let us verify whether Eq. (5.54) is the same as Eq. (5.55):

```
1  T4_nodeX = sp.Matrix([x1, x3, x4])
2  T4_nodeY = sp.Matrix([y1, y3, y4])
3
4  V_T4_triplet = (T4_nodeX.T@Z_P8@T4_nodeY)[0].simplify()
5  print(f"Are the same? {(V_parallelepiped-V_T4_triplet).simplify()==0}")
6  V_T4_triplet
```

Are the same? True

$$-y_1 (x_3 z_4 - x_4 z_3) + y_3 (x_1 z_4 - x_4 z_1) - y_4 (x_1 z_3 - x_3 z_1)$$

5.7.12 *Python code for T4 volume using vMv formula*

```
1  def volume_T4(nodeX, nodeY, nodeZ):
2      '''
3      Compute the volume of a tetrahedron, using the three nodal
4      coordinates of an T4 (with node-0 removed).
5      It uses the vetor-matrix-vector (vMv) dot-product formula.
6      Nodes arrangment: counterclockwise on the base triangle, and the
7      thumb points to the tip node.
8      '''
9      #z1, z2, z3
```

```
10      z1,  z3, z4 = nodeZ                    # nodeZ has z1, z2, z3: z3·(z1×z2)
11
12      Z = sp.Matrix([[   0,   z4,  -z3],
13                     [-z4,    0,   z1],
14                     [ z3,  -z1,    0]])
15
16      V = (sp.Matrix(nodeX).T@Z@sp.Matrix(nodeY))[0]
17
18      return V/6, Z
```

Let's write a function in NumPy to compute the volume of a tetrahedron (T4) using its nodal coordinates:

```
1  def T4_V3(v0, v3, v2, v1, absOn=True):
2      """
3      Computes the volume of a tetrahedron with 4 nodes (T4): v0,v1,v2,v3,
4      all in numpy array (3,) each containing 3 coordinates (x, y, z).
5      Nodes arrangment: counterclockwise on the base triangle, and the
6      thumb points to the tip node.
7      returns: the volume of T4, float.
8      """
9      # Form the moment matrix using 4 nodes:
10     vs = np.vstack([v3-v0, v2-v0, v1-v0])# see above on node arrangment
11     det = np.linalg.det(vs)
12     V = -np.linalg.det(vs)/6.0
13     if absOn: V = abs(V)
14
15     return V                                       # Volume of T4
```

5.7.13 *Example: Volume of a tetrahedron*

We use the above code to compute the volume of the tetrahedron formed using the three vectors v_1, v_2, and v_3 of a parallelepiped (P8) shown in Fig. 5.8 but assume the P8 is a unit cube so that we know the correct answer:

```
1  nodes = [np.array([0, 0, 0]), np.array([ 1, 0, 0.1]),
2           np.array([0,.9, 0]), np.array([0.1, 0,  .8])]
3  T4_formedBy_v123=T4_V3(nodes[0],nodes[3],nodes[2],nodes[1],absOn=False)
4  T4_formedBy_v123
```

0.1185

The volume can also be computed using the vMv formula coded in volume_T4():

```
1  nodeX = [xi[0] for xi in nodes[1:]]
2  nodeY = [yi[1] for yi in nodes[1:]]
3  nodeZ = [zi[2] for zi in nodes[1:]]
4  T4_formedBy_v123_, _ = volume_T4(nodeX, nodeY, nodeZ)
5  T4_formedBy_v123_
```

0.1185

We obtained the same result using two different algorithms.

5.7.14 *Case study: Hyper-volume of a function over a hexahedron*

Consider an H8 domain with eight points/nodes: (x_1, y_1, z_1), (x_2, y_2, z_2), ..., (x_8, y_8, z_8). For a particular case, we assume the following:

$$(x_1, y_1, z_1) = (-1, -1, -1); (x_2, y_2, z_2) = (1, -1, -1.2)$$

$$(x_3, y_3, z_3) = (0.8, 0.8, -1.); \quad (x_4, y_4, z_4) = (-1.5, 1.2, -1.5)$$

$$(x_5, y_5, z_5) = (-1, -1, 1.2); (x_6, y_6, z_6) = (1, -1, 1.3)$$

$$(x_7, y_7, z_7) = (1, 1.2, 1); (x_8, y_8, z_8) = (-1, 1.5, 1.1) \tag{5.56}$$

Clearly, the shape of this H8 is not regular. Our tasks are to compute

1. the volume of the irregular H8;
2. the hyper-volume of under a function over the H8. The function is given as

$$f(x, y, z) = \cos(x + y + z) \tag{5.57}$$

Since the H8 domain is not regular, coordinate mapping is needed so that the integration can be done over the standard cube of $[-1, 1] \times [-1, 1] \times [-1, 1]$ in the natural coordinates (ξ, η, ζ).

5.7.14.1 *Python code to plot hexahedrons*

We first use the following code to plot this hexahedron:

```
1  # Eight nodal coordinates for the H8 in physical domain (xk, yk, zk):
2
3  H8nodes = [[-1.,-1.,-1.], [1.,-1.,-1.2], [0.8,0.8,-1.], [-1.5,1.2,-1.5],
4            [-1.,-1.,1.2], [1.,-1., 1.3], [1., 1.2, 1.], [-1.0,1.5, 1.1]]
5
6  np_8nodes = np.array(H8nodes)  # use NumPy array for vertices of the H8
```

```python
1  # To plot the H8 with 8 nodes and 12 edges:
2  # written based on a suggestion from ChatGPT.
3
4  from mpl_toolkits.mplot3d import Axes3D
5  from mpl_toolkits.mplot3d.art3d import Poly3DCollection
6
7  plt.figure(); plt.ioff()
8  fig = plt.figure(figsize=(8, 6))
9  ax = fig.add_subplot(111, projection='3d')
10
11 # Get the x, y, and z coordinates from np_8nodes given:
12 # np_8nodes0 = np_8nodes - np_8nodes[0,:]      # remove node-0; optional
13 x = np_8nodes[:,0]; y = np_8nodes[:,1]; z = np_8nodes[:,2]
14
15 # edge-nodes connectivity (there are 12 edges, and 8 nodes):
16 edges = [(0, 1), (1, 2), (2, 3), (3, 0), (4, 5), (5, 6),      # edge 1-6
17          (6, 7), (7, 4), (0, 4), (1, 5), (2, 6), (3, 7)]      # edge 7-12
18
19 # face-nodes connectivity: nodes to form the 6 faces (ccw):
20 faces = np.array([[0, 1, 2, 3],          # Bottom
21                   [4, 5, 6, 7],          # Top
22                   [0, 3, 7, 4],          # Left
23                   [1, 5, 6, 2],          # Right
24                   [0, 4, 5, 1],          # Front
25                   [2, 3, 7, 6]])         # Back
26
27 H8 = Poly3DCollection(np_8nodes[faces], facecolors='skyblue', alpha=0.3)
28 ax.add_collection3d(H8)
29
30 for edge in edges:                       # plot the edges with nodes
31     ax.plot(xs=[x[edge[0]], x[edge[1]]], ys=[y[edge[0]], y[edge[1]]],\
32             zs=[z[edge[0]], z[edge[1]]], marker='o',color='b',zorder=4)
33
34 numbering = True                         # number the nodes
35 if numbering:
36     for i in range(len(np_8nodes)):
37         ax.text(x[i]+0.05, y[i]+0.05, z[i]+0.05, str(i), fontsize=8)
38
39 ax.set_xlabel('x', labelpad=-8); ax.set_ylabel('y', labelpad=-8)
40 ax.set_zlabel('z', labelpad=-8)
41 ax.tick_params(axis='x',pad=-3); ax.tick_params(axis='y',pad=-3)
42 ax.tick_params(axis='z',pad=-2)
43 plt.savefig('imagesDI/H8_element.png', dpi=500)
44 #plt.show()
```

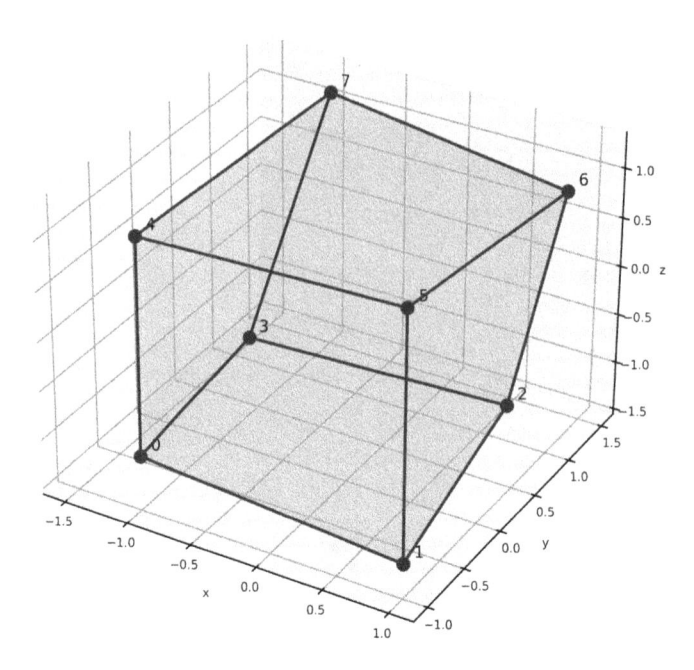

Figure 5.9. An eight-noded hexahedral domain (H8) in x-, y-, z-coordinates.

The 3D domain with eight nodes are plotted in Fig. 5.9 in the 3D coordinate system.

5.7.14.2 *Computation of the volume of an H8*

We use Eq. (5.50), which is coded as volume_H8():

```
1  np_8nodes = np.array(H8nodes)
2  H7nodes = np_8nodes[1:,:]-np_8nodes[0,:]    # remove node-0 (tanslation)
3  #                      x              z              y
4  V, Z = volume_H8(H7nodes[:, 0], H7nodes[:, 1], H7nodes[:, 2])
5  print(f"The volume of the H8 using vMv formula = {V}")
6  gr.printM(Z, "Z matrix for the H8:")
```

The volume of the H8 using vMv formula = 10.4373333333333
Z matrix for the H8:

$$
\begin{bmatrix}
0 & 4.8 & 0 & -2.3 & 0.2 & 2.3 & 0 \\
-4.8 & 0 & 4.3 & 0 & -2.2 & 0.5 & 2.5 \\
0 & -4.3 & 0 & 2.1 & 0 & -2.1 & -0.2 \\
2.3 & 0 & -2.1 & 0 & 4.3 & -0.2 & -4.8 \\
-0.2 & 2.2 & 0 & -4.3 & 0 & 4.5 & 0.2 \\
-2.3 & -0.5 & 2.1 & 0.2 & -4.5 & 0 & 5.0 \\
0 & -2.5 & 0.2 & 4.8 & -0.2 & -5.0 & 0
\end{bmatrix}
$$

To compare the results of the vMv form (Eq. 5.50), let's also compute the H8 volume using the long formula derived earlier by computing the Jacobian matrix determinant and integration. Since the long formula uses nodal coordinate variables, we need to create a dictionary to pair these variables with the specific coordinate values provided. This dictionary creation requires a few lines of code:

```
1   # Form coordinate variable groups:
2   xyzi = [[xi, yi, zi] for xi, yi, zi in zip(nodeX, nodeY, nodeZ)]
3   xyzi = [item for sublist in xyzi for item in sublist]       # unpack it
4
5   unpacked_H8nodes = [item for sublist in H8nodes for item in sublist]
6
7   dictH8 = {}
8   for key, value in zip(xyzi, unpacked_H8nodes):       # pairing the values
9       dictH8[key] = value
10
11  #del dict                 # Remove the earlier defined dict, if conflict
12  print(dict(list(dictH8.items())[:8]))     # print out first 8 data pairs
13  #dictH8                    # full dictionary with paired data, quite long
```

```
{x0: -1.0, y0: -1.0, z0: -1.0, x1: 1.0, y1: -1.0, z1: -1.2, x2: 0.8, y2: 0.8}
```

Now, we can use the long formula to obtain the results by substitution:

```
1   print(f"Volume of the H8, by int(detJ) = {V_H8xyz.subs(dictH8)}")
```

```
Volume of the H8, by int(detJ) = 10.4373333333333
```

The result is the same as what we obtained using volume_H8() which codes the vMv formula.

5.7.14.3 *Hyper-volume under a function over an H8 domain*

Using coordinate mapping, we can now compute the hyper-volume under some simple enough function over a hexahedron via definite integration analytically. The formula is Eq. (5.48), in which $\det(\mathbf{J})$ must be evaluated. This is done in the following steps. SymPy needs to be used, or the work can be laborious:

```
1   # Define the function to be integrated:
2   x, y, z = symbols('x, y, z', real=True)     # independent variables in 3D
3   f = x + y**2 + z**3                          # function to be integrated in x,y,z
4
5   fξηζ = f.subs({x:xξηζ, y:yξηζ, z:zξηζ}).subs(dictH8)       # mapping
6   #fξηζ                      # function expression in terms of ξ,η,ζ; half a page long
```

We can first compute the volume via integration using the unit function:

```
1  fI = 1*H8detJ.subs(dictH8)   # unit func. compute volume via integration
2  V_H8=integrate(integrate(integrate(fI,(ξ,-1,1)),(η,-1,1)),(ζ,-1,1))
3  print(f"The volume the H8 via integration = {V_H8}")
```

```
The volume the H8 via integration = 10.4373333333333
```

The result obtained is the same as that obtained earlier. Let us now compute the hyper-volume under the function over the H8:

```
1  fI = fξηζ*H8detJ.subs(dictH8)        # combined function to be integrated
2  hypperV = integrate(integrate(integrate(fI,(ξ,-1,1)),(η,-1,1)),(ζ,-1,1))
3  print(f"The hyper-volume under the function over H8 = {hypperV}")
4  # executed in 35.1s, finished 18:17:19 2024-06-29
```

```
The hyper-volume under the function over H8 = 2.97692788444445
```

5.7.15 *On 3D integration using natural coordinates*

As demonstrated, analytically integrating over an irregular 3D domain poses significant challenges. The Jacobian matrix is 3×3, with each entry spanning about two lines. Although the determinant can also be computed analytically, the resulting formula exceeds 20 pages in length. Additionally, we have derived a formula for calculating the volume of a hexahedral element (H8), which alone spans nearly two pages.

In practical applications, the domain must be divided into a large number of hexahedral elements. Given the complexity of these calculations for even a single element, performing analytical integration for millions of elements is not feasible. Therefore, numerical methods are often the most practical approach for performing integrations, as employed in the FEM [6].

5.8 Surface integral of a scalar function

5.8.1 *Formulation*

A surface integral is an extension of the 1D line integral discussed in Chapter 4. In line integration, we introduce a single parameter and determine the differential length ds. For surface integrals, we introduce two parameters and determine the differential area dS on the surface to be integrated over.

Consider a smooth surface $\mathcal{S}$, as shown in Fig. 5.10.

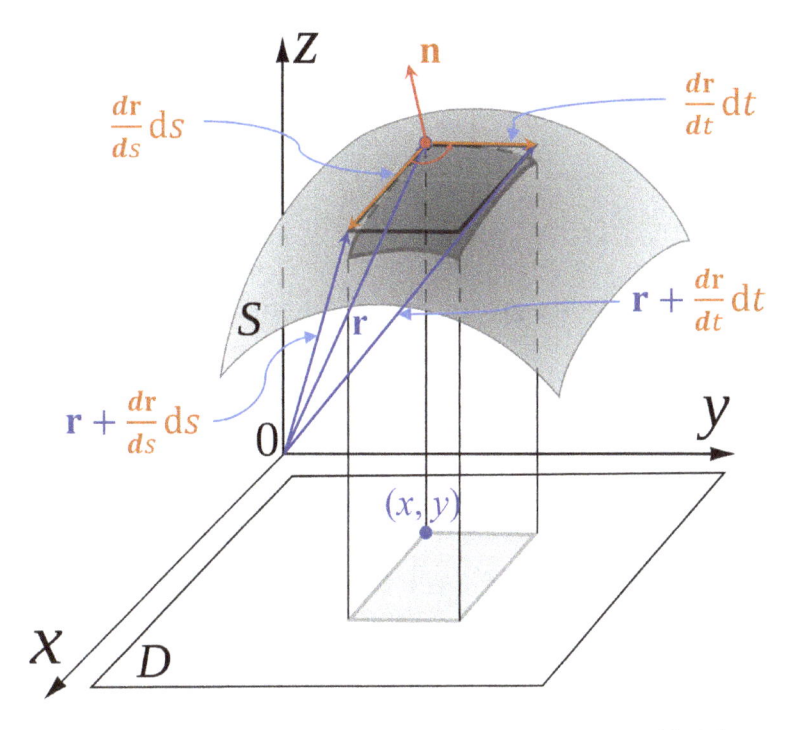

Figure 5.10. Schematic illustration of a differential parallelogram dS and vectors that form dS on surface $\mathcal{S}$.
Source: Images modified based on the one from en.wikipedia Wikimedia Commons by Cronholm144 under the CC BY-SA 3.0 license.

We first parameterize $\mathcal{S}$ with 2D curvilinear coordinates (t, s) on $\mathcal{S}$. The position vector can be written generally as

$$\mathbf{r}(x, y, z) = \mathbf{r}\big(x(s,t), y(s,t), z(s,t)\big) = \mathbf{r}(s,t) \tag{5.58}$$

The surface integral for a given integrand $f(x, y)$ can be written as

$$\iint_{\mathcal{S}} f(x,y)\,\mathrm{d}S = \iint_{\mathcal{D}_p} f(\mathbf{r}(s,t)) \underbrace{\left\| \frac{\partial \mathbf{r}}{\partial s} \times \frac{\partial \mathbf{r}}{\partial t} \right\| \mathrm{d}s\,\mathrm{d}t}_{dS} \tag{5.59}$$

where $\mathcal{D}_p$ is the domain with (s,t) coordinates corresponding to the surface $\mathcal{S}$. Here, we use the cross product of the two vectors $\frac{\mathrm{d}\mathbf{r}}{\mathrm{d}s}\mathrm{d}s$ and $\frac{\mathrm{d}\mathbf{r}}{\mathrm{d}t}\mathrm{d}t$, which gives the area of the differential parallelogram, denoted as dS, as shown in Fig. 5.10. In this case, $\left| \frac{\partial \mathbf{r}}{\partial s} \times \frac{\partial \mathbf{r}}{\partial t} \right|$ is the area scalar.

Using Eq. (5.58), we have

$$
\frac{\partial \mathbf{r}}{\partial s} = \begin{bmatrix} \frac{\partial x}{\partial s} \\ \frac{\partial y}{\partial s} \\ \frac{\partial z}{\partial s} \end{bmatrix} ; \quad
\frac{\partial \mathbf{r}}{\partial t} = \begin{bmatrix} \frac{\partial x}{\partial t} \\ \frac{\partial y}{\partial t} \\ \frac{\partial z}{\partial t} \end{bmatrix}
\tag{5.60}
$$

The normal vector $\mathbf{n}$ is given by the cross product [5]:

$$
\mathbf{n} = \frac{\partial \mathbf{r}}{\partial s} \times \frac{\partial \mathbf{r}}{\partial t} = \begin{vmatrix} \mathbf{i} & \mathbf{j} & \mathbf{k} \\ \frac{\partial x}{\partial s} & \frac{\partial y}{\partial s} & \frac{\partial z}{\partial s} \\ \frac{\partial x}{\partial t} & \frac{\partial y}{\partial t} & \frac{\partial z}{\partial t} \end{vmatrix} = \begin{bmatrix} \frac{\partial y}{\partial s}\frac{\partial z}{\partial t} - \frac{\partial z}{\partial s}\frac{\partial y}{\partial t} \\ \frac{\partial z}{\partial s}\frac{\partial x}{\partial t} - \frac{\partial x}{\partial s}\frac{\partial z}{\partial t} \\ \frac{\partial x}{\partial s}\frac{\partial y}{\partial t} - \frac{\partial y}{\partial s}\frac{\partial x}{\partial t} \end{bmatrix}
\tag{5.61}
$$

The cross product can be worked out using the following:

```
1  xs, ys, zs, xt, yt, zt = symbols('x_s,y_s,z_s,x_t,y_t,z_t') # x_s=dx/ds
2  r_s = Matrix([xs, ys, zs])
3  r_t = Matrix([xt, yt, zt])
4  gr.cross2vectors(r_s, r_t, sympy=True).T
```

$$
\begin{bmatrix} y_s z_t - y_t z_s & -x_s z_t + x_t z_s & x_s y_t - x_t y_s \end{bmatrix}
$$

The area scalar $\left\| \frac{\partial \mathbf{r}}{\partial s} \times \frac{\partial \mathbf{r}}{\partial t} \right\|$ is computed using the following:

```
1  (gr.cross2vectors(r_s, r_t, sympy=True)).norm()
```

$$
\sqrt{|x_s y_t - x_t y_s|^2 + |x_s z_t - x_t z_s|^2 + |y_s z_t - y_t z_s|^2}
$$

The differential parallelogram area becomes

$$
dS = \|\mathbf{n}\| ds\, dt
$$

$$
= \sqrt{\left(\frac{\partial y}{\partial s}\frac{\partial z}{\partial t} - \frac{\partial z}{\partial s}\frac{\partial y}{\partial t}\right)^2 + \left(\frac{\partial z}{\partial s}\frac{\partial x}{\partial t} - \frac{\partial x}{\partial s}\frac{\partial z}{\partial t}\right)^2 + \left(\frac{\partial x}{\partial s}\frac{\partial y}{\partial t} - \frac{\partial y}{\partial s}\frac{\partial x}{\partial t}\right)^2}\, ds\, dt
$$

$$
\tag{5.62}
$$

If we are only interested in computing the area of the surface, we simply set $f(x,y) = 1$.

When the surface function is given as $z(x, y)$, we simply let $t = x$ and $s = y$, then Eq. (5.59) becomes

$$A = \iint_S dS = \iint_{D_p} \left\| \frac{\partial \mathbf{r}}{\partial x} \times \frac{\partial \mathbf{r}}{\partial y} \right\| dx\, dy \tag{5.63}$$

In this simpler parameterization, we have

$$\mathbf{r} = \begin{bmatrix} x \\ y \\ z \end{bmatrix}; \quad \frac{\partial \mathbf{r}}{\partial x} = \begin{bmatrix} 1 \\ 0 \\ \frac{\partial z}{\partial x} \end{bmatrix}; \quad \frac{\partial \mathbf{r}}{\partial y} = \begin{bmatrix} 0 \\ 1 \\ \frac{\partial z}{\partial y} \end{bmatrix} \tag{5.64}$$

Their cross product produces the normal vector $\mathbf{n}$ [5], which is given by (without $dxdy$)

$$\mathbf{n} = \frac{\partial \mathbf{r}}{\partial x} \times \frac{\partial \mathbf{r}}{\partial y} = \begin{vmatrix} \mathbf{i} & \mathbf{j} & \mathbf{k} \\ 1 & 0 & \frac{\partial z}{\partial x} \\ 0 & 1 & \frac{\partial z}{\partial y} \end{vmatrix} = \begin{bmatrix} -\frac{\partial z}{\partial x} \\ -\frac{\partial z}{\partial y} \\ 1 \end{bmatrix} \tag{5.65}$$

Calculating the norm of $\mathbf{n}$, Eq. (5.63) becomes

$$A = \iint_S dS = \iint_{D_p} \sqrt{1 + \left(\frac{\partial z}{\partial x}\right)^2 + \left(\frac{\partial z}{\partial y}\right)^2}\, dx\, dy \tag{5.66}$$

All these can be done using the following code:

```
1  zx, zy - symbols('z_x, z_y ')              # z_x = dz/dx
2  r_x = Matrix([1, 0, zx])
3  r_y = Matrix([0, 1, zy])
4  (gr.cross2vectors(r_x, r_y, sympy=True)).norm()
```

$$\sqrt{|z_x|^2 + |z_y|^2 + 1}$$

It is advised to read this section in comparison with Section 4.14 on line integrals. The concepts that lead to the formations are very similar.

Let us look at an example.

5.8.2 *Example: Integral of a function over a cone surface*

Assume that the curved surface is given as

$$z(x, y) = \sqrt{x^2 + y^2} \tag{5.67}$$

A scalar integrand function $f(x, y)$ is defined as

$$f(x, y) = \sin(x + y) \tag{5.68}$$

1. Compute the area of this surface in region of $\mathcal{D}_p = [0, 1] \times [0, 1]$.
2. Compute the integral of $f(x, y)$ over the surface $z(x, y)$ over the region of $\mathcal{D}_p = [0, 1] \times [0, 1]$.

Solution: We use the following code to get this done:

```
1  x, y = symbols('x, y')
2  f = sin(x+y)
3  z = sp.sqrt(x**2+y**2)
4  dS = sp.sqrt(1+z.diff(x)**2 + z.diff(y)**2)
5
6  # Answer to question 1:
7  A = integrate(integrate(dS, (y, 0, 1)), (x, 0, 1)).evalf()
8  print(f" Answer to question 1 = {A}")
9
10 # Answer to question 2:
11 I_f = integrate(integrate(f*dS, (y, 0, 1)), (x, 0, 1)).evalf()
12 print(f" Answer to question 2 = {I_f}")
```

```
Answer to question 1 = 1.41421356237310
Answer to question 2 = 1.09409860486971
```

5.8.3 *Example: Integral of a function over a spherical surface*

Assume that the curved surface is given as

$$x^2 + y^2 + z^2 = R^2 \tag{5.69}$$

where R is the radius of the sphere. The scalar integrand $f(x, y, z)$ is defined as

$$f(x, y, z) = x^2 + 2y + 8z \tag{5.70}$$

1. Compute the area of the surface of the sphere.
2. Compute the integral of $f(x, y, z)$ over the sphere surface.

Solution: Let us use a spherical coordinate system (R, θ, φ), where θ is the polar angle and φ is the azimuthal angle, as we did in Chapter 4 (see Fig. 4.26):

$$x = R\sin(\theta)\,\cos(\varphi)$$

$$y = R\sin(\theta)\,\sin(\varphi)$$

$$z = R\cos(\theta) \tag{5.71}$$

We use the following code to complete these tasks:

```
1  x, y, z, θ, φ = symbols('x, y, z, θ, φ', real=True)
2  R = symbols('R', positive=True)
3
4  f = x**2 + 2*y  + 8*z                              # the integrand
5
6  x_ = R*sin(θ)*cos(φ)    # x_, y_, and z_ are x,y,z on the sphere surface
7  y_ = R*sin(θ)*sin(φ)
8  z_ = R*cos(θ)
9
10 r_θ = Matrix([x_.diff(θ), y_.diff(θ), z_.diff(θ)])          # Jacobian
11 r_φ = Matrix([x_.diff(φ), y_.diff(φ), z_.diff(φ)])
12
13 n = gr.cross2vectors(r_θ, r_φ, sympy=True)         # normal vector on S
14 dS = n.norm()                                       # area scalar
15
16 # Answer to question 1:
17 A = integrate(integrate(dS, (φ, 0, sp.pi)), (θ, 0, 2*sp.pi))
18 display(Math(f" \\text{{Answer to question 1, A = }}{sp.latex(A)}"))
19
20 # Answer to question 2:
21 dic = {x:x_, y:y_, z:z_}
22 I_f = integrate(integrate(f.subs(dic)*dS, (φ,0,sp.pi)), (θ,0,2*sp.pi))
23 display(Math(f" \\text{{Answer to question 2, I_f = }}{sp.latex(I_f)}"))
```

Answer to question 1, $A = 4\pi R^2$

Answer to question 2, $I\_f = \dfrac{4\pi R^4}{3}$

5.9 Indefinite integrations in high dimension

We discussed indefinite integration in 1D in detail in Chapter 4. A similar concept applies to n-dimensional (nD) functions to obtain antiderivative functions. However, discussions on this topic are rare in the open literature. This may be because, when integrating an nD integrand, the integral constants become functions of the other coordinates not being integrated. This creates difficulties in performing multiple indefinite integrals. This section presents the following:

1. a complete definition of indefinite integration for nD functions;
2. a procedure to find the antiderivative function and determine these integral functions under a given set of conditions.

5.9.1 *Definition*

Consider an integrand function $f(\mathbf{x})$ defined in an nD Cartesian coordinate system with independent variables: $\mathbf{x} = (x_1, x_2, \ldots, x_n)$. The indefinite integral is defined as

$$\underbrace{\int \cdots \int f(x_1, x_2, \ldots, x_n)\, dx_1 \cdots dx_n}_{\int \cdots \int f(\mathbf{x}) d\mathbf{x}}$$

$$= \underbrace{F(x_1, x_2, \ldots, x_n)}_{F(\mathbf{x})} + C_1(\mathbf{x}\backslash x_1) + C_2(\mathbf{x}\backslash x_2) + \cdots + C_n(\mathbf{x}\backslash x_n) \quad (5.72)$$

where $\mathbf{x}\backslash x_i$ is the difference of the independent variable set $\mathbf{x}$ and x_i, which gives $x_1, x_2, \ldots, x_{i-1}, x_{i+1}, \ldots, x_n$. $C_i(\mathbf{x}\backslash x_i)$ is the ith **integral function** resulting from the indefinite integration along x_i-axis. Equation (5.72) can be simply expressed as

$$\int \cdots \int f(\mathbf{x}) d\mathbf{x} = \underbrace{F(\mathbf{x}) + \sum_i^n C_i(\mathbf{x}\backslash x_i)}_{\text{general antiderivative}} \quad (5.73)$$

Function $F(\mathbf{x})$ is the base antiderivative function that is obtained without considering the integral constant functions. Function $F(\mathbf{x}) + \sum_i^n C_i(\mathbf{x}\backslash x_i)$ is the general antiderivative function.

It is easy to prove that our definition obeys the standard antiderivative definition, implying that

$$\frac{d^n \left[F(\mathbf{x}) + \sum_i^n C_i(\mathbf{x}\backslash x_i) \right]}{dx_1, \ldots, dx_n} = \underbrace{\frac{d^n F(\mathbf{x})}{dx_1, \ldots, dx_n}}_{f(\mathbf{x})} + \sum_i^n \underbrace{\frac{d^n C_i(\mathbf{x}\backslash x_i)}{dx_1, \ldots, dx_n}}_{=0} = f(\mathbf{x}) \quad (5.74)$$

Note that in this definition, we assume the general antiderivative function satisfies Clairaut's theorem. This means that the function is continuously differentiable to the nth order, ensuring that all mixed partial derivatives are equal and the sequence of performing cross partial differentials is immaterial.

Let us take a look at a 3D case.

5.9.2 *Example: Indefinite integral in 3D*

Let us use SymPy to create some functions in 3D and demonstrate our definition given in Eq. (5.72). We first define necessary independent variables in 3D and integral functions:

```
1  x, y, z = symbols('x, y, z', real=True)    # independent variables in 3D
2
3  Cx = Function('C_x' )(y, z)      # integral functions, constant w.r.t x
4  Cy = Function('C_y' )(z, x)               #  constant w.r.t y
5  Cz = Function('C_z' )(x, y)               #  constant w.r.t z
6  Cx                                        # display an example
```

$$C_x(y, z)$$

Now, define an integrand function and then perform three layers of integration without integration intervals to obtain the base antiderivation function:

```
1  f = x*y*z        # define an integrand function, any integrable function
2
3  #    three layers of indefinite integration:
4  F = integrate(integrate(integrate(f,x),y),z) # integrals without limits
5  display(Math(f" \\text{{Base antiderivative $F(x,y,z)$ = }}\
6                  ∫∫∫({latex(f)})dxdydz = {latex(F)}"))
```

$$\text{Base antiderivative } F(x,y,z) = \int\int\int (xyz)dxdydz = \frac{x^2y^2z^2}{8}$$

Next, add in three arbitrary integral functions to form the general antiderivative:

```
1  F_general = F + Cx + Cy + Cz
2  display(Math(f" \\text{{General antiderivative $F_G(x,y,z)$ = }}\
3                  {latex(F_general)}"))
```

$$\text{General antiderivative } F_G(x,y,z) = \frac{x^2y^2z^2}{8} + C_x(y,z) + C_y(z,x) + C_z(x,y)$$

Finally, perform three-stage differentiations:

```
1  d3F = F_general.diff(y).diff(z).diff(x)
2  display(Math(f" \\frac{{∂^3}}{{∂x∂y∂z}}\\big({latex(F_general)}\\big) = \
3                  {latex(d3F)}"))
```

$$\frac{\partial^3}{\partial x \partial y \partial z}\left(\frac{x^2y^2z^2}{8} + C_x(y,z) + C_y(z,x) + C_z(x,y)\right) = xyz$$

We have successfully obtained the integrand function.

5.9.3 *Case study: Determination of the integral functions*

In our discussion on antiderivatives for 1D functions, we presented a procedure to determine the integral constants in the general antiderivative functions. For nD functions, we need to determine these integral functions, which can be challenging.

First, we need additional conditions about the general antiderivative nD function. Such conditions can help determine the forms of these integration functions, which may still include some integral constants. We then require boundary conditions to determine these constants.

The best way to describe this process is through an example problem.

5.9.3.1 *Setting of the problem*

The following is a simple example to demonstrate the process, considering a 2D problem. The integrand is simply a linear function given as

$$f(x, y) = ax + by \tag{5.75}$$

where a and b are given constants. We need to find the antiderivative function for the following indefinite integral:

$$F(x, y) = \int \int f(x, y) dx dy \tag{5.76}$$

We assume that $F(x, y)$ must satisfy the following condition:

$$\frac{d^2 F(x, y)}{dx^2} = \frac{d^2 F(x, y)}{dy^2} \tag{5.77}$$

The boundary conditions for $F(x, y)$ are assumed to be

$$\frac{dF(x, y)}{dx} = 0 \quad \text{at} \quad x = 0, y = 0$$

$$\frac{dF(x, y)}{dy} = 0 \quad \text{at} \quad x = 0, y = 0$$

$$F(x, y) = 0 \quad \text{at} \quad x = 0, y = 0$$

$$F(x, y) = 0 \quad \text{at} \quad x = 1, y = 1 \tag{5.78}$$

5.9.3.2 *Python code to solve the problem*

We write the following code in SymPy to find the general form of $F(x, y)$:

```
1  # Definte symbolic vriables and functions:
2  x, y = symbols('x', 'y', real=True)              # indepedent variables
3  a, b = symbols('a', 'b', real=True)              # given constants
4  C1, C2, C3, C = symbols('C1', C2, C3, C',real=True) # integral constants
5
6  Cx  = Function('C_x' )(y)    # integral function of y, constant w.r.t x
7  Cy  = Function('C_y' )(x)    # integral function of x, constant w.r.t y
8  FCy = Function('F_Cy')(x)    # antiderivative of Cy, constant w.r.t x
9  Cx
```

$$C_x(y)$$

Integrating once

```
1  # Define an integrand function:
2  f = a*x+b*y                                      # given integrand
3
4  # Integrate along y-axis:
5  Fx_base = integrate(f, y)                        # the base anti-direvative
6  Fx = Fx_base + Cy        # add in an integral function of x, const. wrt x
7  Fx
```

$$axy + \frac{by^2}{2} + C_y(x)$$

Here, we added a function $C_y(x)$ to the base antiderivative after the first layer of integration along y-axis. This is because $C_y(x)$ is an integral constant to y, which is an unknown function of x. The justification is

$$\frac{\mathrm{d}\left(axy + \frac{by^2}{2} + C_y(x)\right)}{\mathrm{d}y} = ax + by \tag{5.79}$$

which gives back the original integrand.

Integrating twice

Next, we perform integration along the x-axis. We integrate Fx_base and $C_y(x)$ separately. We also need to add in an unknown function $C_x(y)$, which is an integration constant with respect to x. The code is as follows:

```
1  F_base = integrate(Fx_base, x)                   # base anti-direvative
2  F_base + integrate(Cy, x) + Cx   # integral function of y, const. wrt x
```

$$\frac{ax^2y}{2} + \frac{bxy^2}{2} + C_x(y) + \int C_y(x)\,dx$$

The last term is the antiderivative of $C_y(x)$, which is also a function of x. Let's denote it as $F_{Cy}(x)$. Therefore, the general form of the antiderivative of Eq. (5.76) becomes

$$F_{\text{general}}(x, y) = \frac{ax^2 y}{2} + \frac{bxy^2}{2} + F_{Cy}(x) + C_x(y) \tag{5.80}$$

where both $F_{Cy}(x)$ and $C_x(y)$ are arbitrary unknown functions. The justification is

$$\frac{\mathrm{d}^2 \left(\frac{ax^2 y}{2} + \frac{bxy^2}{2} + F_{Cy}(x) + C_x(y) \right)}{\mathrm{d}y\mathrm{d}x} = ax + by \tag{5.81}$$

which gives back the original integrand.

5.9.3.3 *Determination of the integral functions*

To determine the unknown integral functions, we first use condition in Eq. (5.77):

```
1  F_general = F_base + FCy + Cx    # form the general antiderivative func.
2  dFdx2 = F_general.diff(x, 2)          # compute the 2nd derivative wrt x
3  dFdx2.args
```

$$\left(ay, \ \frac{d^2}{dx^2} F_{Cy}(x) \right)$$

```
1  dFdy2 = F_general.diff(y, 2)          # compute the 2nd derivative wrt y
2  dFdy2.args
```

$$\left(bx, \ \frac{d^2}{dy^2} C_x(y) \right)$$

Equation (5.77) gives

$$ay + \frac{d^2 F_{Cy}(x)}{dx^2} = bx + \frac{d^2 C_x(y)}{dy^2} \tag{5.82}$$

It can be rearranged to

$$\underbrace{\frac{d^2 F_{Cy}(x)}{dx^2} - bx}_{\text{function of only } x} = \underbrace{\frac{d^2 C_x(y)}{dy^2} - ay}_{\text{function of only } y} \tag{5.83}$$

The left side is a function of only x, and the right side is a function of only y. For this equation to hold, both must equal a common constant. Let us denote it as C. We thus obtain two separated equations:

$$\frac{d^2 F_{Cy}(x)}{dx^2} - bx = C$$

$$\frac{d^2 C_x(y)}{dy^2} - ay = C \tag{5.84}$$

Each of these two equations are now 1D. The first equation in Eq. (5.84) can be solved for $F_{Cy}(x)$ by twice indefinite integration along the x-axis using the following code:

```
1  eq_left = C + dFdy2.args[0]
2  sln_FCy = integrate(integrate(eq_left, x)+C2, x)+C1
3  sln_FCy
```

$$\frac{Cx^2}{2} + C_1 + C_2 x + \frac{bx^3}{6}$$

The solution carries two additional integral constants.

Similarly, the second equation in Eq. (5.84) can be solved for $C_x(y)$ by twice indefinite integration along the y-axis using the following:

```
1  eq_right = C + dFdx2.args[0]
2  sln_Cx = integrate(integrate(eq_right, y)+C2, y)+C1
3  sln_Cx
```

$$\frac{Cy^2}{2} + C_1 + C_2 y + \frac{ay^3}{6}$$

It carries two additional integral constants.

5.9.3.4 *Determination of the integral constants*

We put these solutions together and locate these terms:

```
1  Cxy = sln_FCy + sln_Cx
2  Cxy.args                                    # Locate the terms
```

$$\left(2C_1, \ C_2 x, \ C_2 y, \ \frac{Cx^2}{2}, \ \frac{Cy^2}{2}, \ \frac{ay^3}{6}, \ \frac{bx^3}{6} \right)$$

We can now form the general form of the antiderivative function using Eq. (5.76), which gives the following:

```
1  F_general = F_base + sum(Cxy.args[3:]) + C1*x + C2*y + C3
2  F_general
```

$$\frac{Cx^2}{2} + \frac{Cy^2}{2} + C_1x + C_2y + C_3 + \frac{ax^2y}{2} + \frac{ay^3}{6} + \frac{bx^3}{6} + \frac{bxy^2}{2}$$

Here, we have regrouped and renamed the integral constants, leading to a neater form with a total of four constants C_1, C_2, C_3, and C to be determined.

Next, using the boundary conditions for $F(x, y)$ given in Eq. (5.78), we can establish a set of four algebraic equations and solve for these four constants. The code is as follows:

```
1  eq1 =  F_general.subs({x:0,y:0})
2  eq2 = (F_general.diff(x)).subs({x:0,y:0})
3  eq3 = (F_general.diff(y)).subs({x:0,y:0})
4  eq4 =  F_general.subs({x:1,y:1})
5
6  sln_C = sp.solve([eq1, eq2, eq3, eq4], [C1, C2, C3, C])
7  sln_C
```

$$\left\{ C : -\frac{2a}{3} - \frac{2b}{3}, \ C_1 : 0, \ C_2 : 0, \ C_3 : 0 \right\}$$

5.9.3.5 *Final solution of the antiderivative*

Finally, substituting these constants found back to the general form of the antiderivative, we obtain the following:

```
1  F_final = F_general.subs(sln_C)
2  F_final
```

$$\frac{ax^2y}{2} + \frac{ay^3}{6} + \frac{bx^3}{6} + \frac{bxy^2}{2} + \frac{x^2\left(-\frac{2a}{3} - \frac{2b}{3}\right)}{2} + \frac{y^2\left(-\frac{2a}{3} - \frac{2b}{3}\right)}{2}$$

This is the final antiderivative function. Let us check whether it satisfies Eq. (5.76):

```
1  F_final.diff(x).diff(y)
```

$$ax + by$$

This completes our task.

This example demonstrates how a 2D antiderivative function can be found. It is a good exercise for the integration of nD functions and involves various techniques and utilities of integral functions and constants. These techniques are widely used in solving mechanics problems. Real-life applications can be found in Chapter 3 of a widely used classic textbook on mechanics [6].

5.10 Remarks

We summarize this chapter with the following remarks:

1. For regular integral domains in the Cartesian coordinate system, integration techniques for 1D can be extended to nD functions. This involves n nested 1D integrals, as discussed in Chapter 4.
2. For irregular domains, we apply coordinate transformation to convert the integral domain to a regular one, allowing the integration to be done as n nested 1D integrals. The volume scalar is the determinant of the Jacobian matrix, representing the differential relations between the two coordinate systems.
3. For circular domains, we use polar coordinates, where the area scalar is simply r. The same scalar applies to cylindrical coordinates.
4. For spherical domains, we use spherical coordinates, where the volume scalar is $r^2 \sin \theta$.
5. For an arbitrary hexahedral domain (H8), we use natural coordinates and nodal shape functions for mapping. The volume scalar is a function of the natural coordinates and the nodal coordinate values of the eight nodes of the hexahedral domain. It can be computed using the $v\boldsymbol{M}v$ formula developed by the author.
6. For domains with general complicated geometry, we subdivide the domain into Q4 or H8 elements, perform the integral for each element, and then sum them up. For better domain adaptation, we also use triangular (2D) or tetrahedral (3D) elements. In these cases, numerical techniques are often used, which is a practical and intensive subject for future discussions.
7. We also defined antiderivatives for nD integrand functions. An antiderivative consists of a base antiderivative function and a set of integral constant functions, each of which is a constant with respect to one coordinate. A 2D example is presented to demonstrate how such an antiderivative function can be found and determined using given conditions.

The applications of derivatives and integrations of functions are vast. The numerous examples given in this book are far from exhaustive.

References

[1] Y.C. Fan, *Advanced Mathematics*, 1979.

[2] L.X. Yang and B.Y. Bi, *Analytic Mathematics Practices, by Boris Demidovich*, 2005.

[3] Gilbert Strang, *Calculus*, 1991. Available online (http://ocw.mit.edu/OcwWeb/resources/RES-18-001Spring-2005/ResourceHome/index.htm).

[4] G.R. Liu, *Numbers and Functions: Theory, Formulation, and Python Codes*, World Scientific, 2024.

[5] G.R. Liu, *Mechanics of Materials: Formulations and Solutions with Python*, World Scientific, 2024.

[6] G.R. Liu and Siu Sin Quek, *The Finite Element Method: A Practical Course*, Butterworth-Heinemann, 2013.

Index

local approximation approach, 93
local change, 10
local extremum, 84
local features, 213
local slope, 14
locality behavior, 11
locality of the derivative, 10, 22
locally indeterminate, 159
logarithmic function, 80
loss function, 100, 146

M

Machin's formula, 95
mapping, 315
maximum point, 197
mean value theorem (MVT), 40, 42, 214
milder delta function, 235
minimization of functions, 97
minimizer, 97
minimum point, 197
minkowski inequality, 269
ML model, 146
monomial function, 17
multi-dimensional (nD) functions, 299
multi-dimensions, 123–124
mutually inverse, 54

N

N-path, 321
natural coordinates, 315, 336–337
nD Taylor series expansion, 159
negative definite, 160
Newton iteration, 64, 66, 83
Newton's method, 120
nodal shape functions, 337
non-differentiable point, 27
non-integrable, 208, 257
number of jumping points, 205
number of minima, 145

O

odd function, 37
odd integrand, 227

one-to-one correspondence, 283
oscillatory values, 266

P

p-convergence, 63
parabolic cone, 313
parameterization, 357
parameterized curves, 293
partial derivative, 127–128, 171
partial derivative approximation, 185
partial fraction decomposition, 249
partitions of unity, 164
physical coordinates, 316
piecewise continuous, 227, 242
piecewise differentiation, 29
piecewise linear approximations, 62
piecewise linearization, 60
planar curve, 283
polar moment of inertia, 311
polar to Cartesian, 315
polygonal shapes, 319
polynomial derivatives, 106
polynomial function, 105, 190
polynomial in high dimensions, 189
polynomial space, 183
position vector parametrized, 282
positive definite, 160
power functions, 261
power integrand, 260
power rule, 18, 31
presentation of a function, 41
product rule, 30
product rule of differentiation, 18
properties of integrals, 266

Q

quadratic approximation, 62
quadratic function, 55
quadratic polynomial, 102
quadratic shape functions, 170–171

R

Ramanujan's second approximation, 293
rate of change, 10

*9 7 8 9 8 1 9 8 0 1 0 0 8 *